Oren Day Pomeroy

The Diagnosis and Treatment of Diseases of the Ear

Oren Day Pomeroy

The Diagnosis and Treatment of Diseases of the Ear

ISBN/EAN: 9783337140403

Printed in Europe, USA, Canada, Australia, Japan

Cover: Foto ©berggeist007 / pixelio.de

More available books at **www.hansebooks.com**

THE

DIAGNOSIS AND TREATMENT

OF

DISEASES OF THE EAR

BY

OREN D. POMEROY, M.D.

SURGEON TO THE MANHATTAN EYE AND EAR HOSPITAL ; OPHTHALMIC AND AURAL SURGEON
TO THE N. Y. INFANT ASYLUM ; CONSULTING SURGEON TO THE PATERSON EYE
AND EAR INFIRMARY ; MEMBER OF THE NATIONAL OPHTHALMOLOGICAL
AND OTOLOGICAL SOCIETIES, ETC.

With one hundred illustrations

NEW YORK
BERMINGHAM & CO.
1883

PREFACE.

One of the first things which the reader will notice in the perusal of this book is the absence of Anatomy and Physiology. I have occasionally mentioned an anatomical or physiological point, where it has seemed essential to the understanding of the subject. Where it is desired to study the anatomy and physiology of the ear thoroughly, the books devoted to such subjects treat the matter more exhaustively than ordinary text-books on the diseases of the ear can possibly do.

As my time for the preparation of this book has been limited, I have done what I could to produce an acceptable work for the use of the Practitioner, and it is hoped that some of the younger Aural Surgeons may find something in it to aid them in their practice. From the nature of things, a text-book cannot be exhaustive; neither would it be desired, if it were possible.

The original illustrations in this book are from pen-drawings by Dr. J. O. Tansley, of New York, whose initials may be found appended to the cuts.

I can only hope for the indulgent kindness of my friends in overlooking the defects, which may readily be found in this book.

Oren D. Pomeroy.

Lexington Avenue, near 38th St.
New York, April, 1883.

CONTENTS.

INSTRUMENTS FOR THE EXAMINATION OF THE
THROAT AND NARES, WITH SUGGESTIONS AS TO

DISEASES OF THE MIDDLE EAR.

MASTOID AFFECTIONS.

UNCLASSIFIED DISEASES.

DISEASES OF THE EAR MOSTLY OR WHOLLY CONFINED TO THE LABYRINTH OR ACOUSTIC NERVE.

INSTRUMENTS FOR AIDING THE HEARING.

DISEASES OF THE EAR.

INSTRUMENTS USED IN THE EXAMINATION OF THE EAR.

THE first step in the study of diseases of the ear is its proper examination. The auricle and the outer portion of

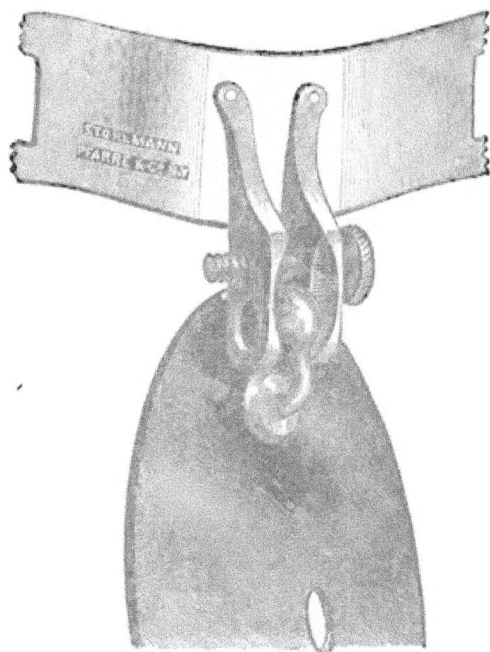

FIG. 1.
Pomeroy's forehead mirror.

the meatus externus require no apparatus for their examination; the deeper parts of the meatus and the membrana tympani, however, require special appliances for

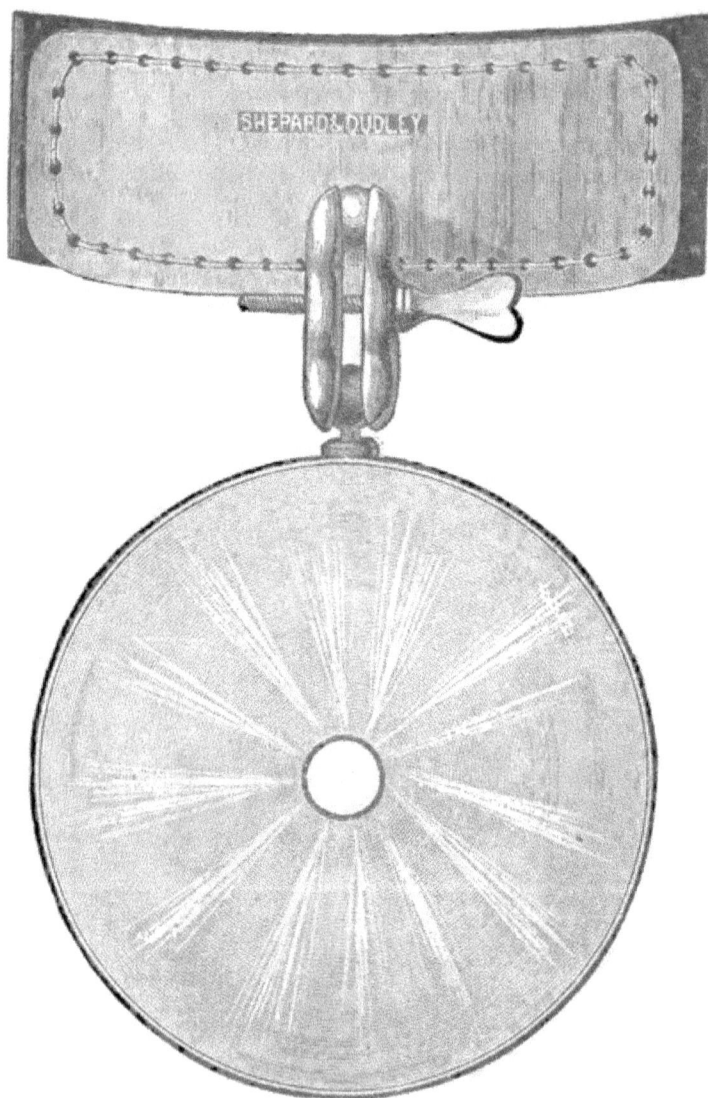

Fig. 2.
Dr. Smallwood's forehead mirror.

examination, and some directions as to the best methods. In the first place a concave mirror of about seven

inches focus and three to three and a half inches in diameter is needed. The one here figured, and known as Pomeroy's (Fig. 1.), I am in the habit of using, although there are many others which fully answer the purpose. Fig. 2, known as Dr. Smallwood's mirror, with double ball-and-socket joints, is a very admirable instrument. In

FIG 3.
Wilde's speculum.

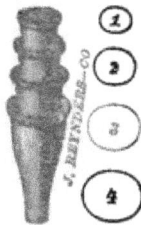

FIG. 5.
Gruber's speculum.

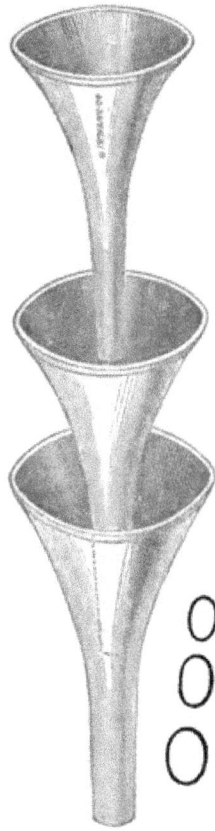

FIG. 4.
Toynbee's speculum.

all the mirrors which fairly meet the indications, an appliance for fastening them to the forehead should be used. This requires a ball attached to its back, to which a clamp is applied, so as to produce a universal joint. The author's mirror is attached to the head by an inelastic band of silk. A handle may be attached to the mirror

when it is not found necessary to use it against the forehead; but this is hardly necessary, it being quite easy to
hold the mirror by the forehead piece, and the handle
may be dispensed with.   The next requirement is a speculum.   Three kinds may be recommended: Wilde's, Toynbee's, and Gruber's (Figs. 3, 4, and 5).   As the meatus
externally is oval, Gruber has thought fit to make his speculum oval to more readily adjust itself to the shape of the
canal.   I am not in the habit of using Gruber's speculum,
for the reason that it *is* oval; as I insert the speculum with
a twirling movement, this shape would be faulty.   Again, the
meatus being externally cartilaginous, is sufficiently mobile
to adjust itself to a circular speculum without difficulty.
The Toynbee speculum is funnel-shaped: that is, the outer
surface represents a concave line extending from tip to rim,
and is very easily inserted, and may be moved from side to
side, and be easily directed to any part of the canal or
membrane we may desire to inspect.   It is, however, obviously a poor speculum to operate with.   The Gruber
speculum bulges at the outer portion, which presents a roomy
space to operate in.   The Wilde speculum manifestly com-
bines the characteristics to some extent, of both the previous
varieties.   All the forms of bivalve specula I am inclined to
discard, except in a few instances when operating.   Kramer's
(Fig. 6) or Speir's (Fig. 7) are as good as any.   Miliken's
Self-retaining Speculum (Fig. 8), will sometimes be found
useful.   For purposes of magnifying, a variety of otoscopes
have been devised, such as Hinton's (Fig. 9), or Brunston's
(Fig. 10), or Hassenstein's (Fig. 11), or Simrock's (Fig. 12).
I am not, however, in favor of these instruments ordinarily.   The Hinton and Brunston instrument sare useful for
demonstrating the appearances of the membrana, but this
mode of studying the ear is at best a poor one; the student
should overcome the difficulties of making an examination
without these aids.   Too much assistance is evidently un-
profitable.   The forehead mirror being in position, I place
a $+ \frac{1}{3}$ or $+ \frac{1}{4}$ lens against the opening of the speculum,
somewhat obliquely, so as not to be annoyed by the
reflections, and then by a to-and-fro movement focus the
parts perfectly.   Moreover, changing the position of the
glass in this manner aids in determining the perspective.
Another method is to place a clip behind the mirror, and
insert a lens which will sufficiently magnify.   A $+ \frac{1}{12}$ or
$+ \frac{1}{10}$ will sufficiently enlarge the image.   It is also con-

FIG. 6.
Kramer's bivalve speculum.

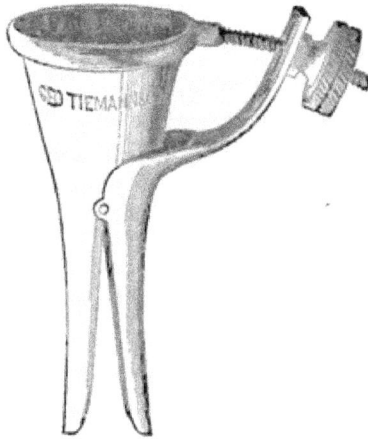

FIG. 7.
Speir's bivalve speculum.

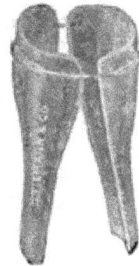

FIG. 8.
Miliken's self retaining speculum.

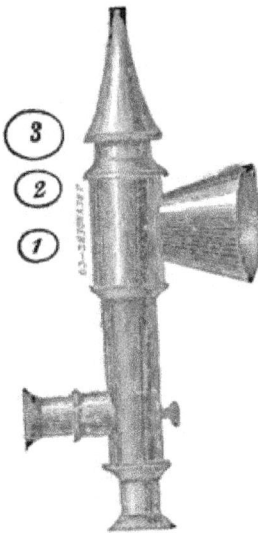

FIG. 9.
Hinton's speculum.

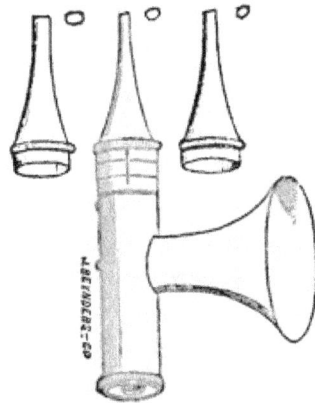

FIG. 10.
Brunston's speculum.

venient if the surgeon happens to be far-sighted or pres-
byopic.

THE KIND OF ILLUMINATION TO BE SELECTED is a very important matter. I am firmly of the opinion that daylight
when at its best is the most satisfactory means of illumination. I would define this condition to be a cloudy day with
very thin white clouds; a room admitting light from one window, or, what would be better, from a single pane of glass.
The argument then turns on the frequency with which we
obtain this condition. The direct sunlight is manifestly
bad. The light from the blue sky is of faint illuminating
power. The light reflected from a white house is good,

FIG. 11.
Hassenstein's speculum.

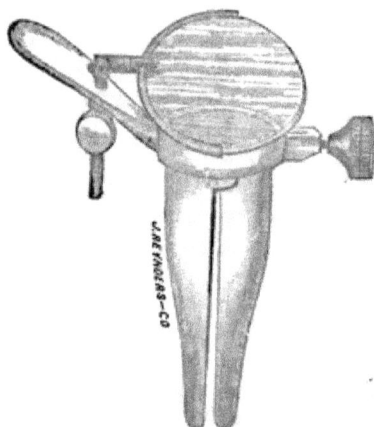

FIG. 12
Simrock's speculum.

but it is bad from a house of any other color. On the other
hand, some form of gas, coal oil, or candle-light is always at
our service, and is adequate to a proper examination of the
ear. The only objection to this kind of illumination is the
reddish-yellow color of the light, which gives to the membrana tympani a slightly changed appearance; that is, a reddened membrane will look a litte too red by gas-light.
Even this could be obviated by placing a blue chimney
upon the lamp. The advantage of the artificial light is,
that it is always of the same quality, and if the normal

membrane is studied by gas-light, there is no difficulty in detecting any deviation resulting from disease.

HINTS AS TO THE PROPER MODE OF EXAMINING THE EAR.

If daylight only is used, the mirror requires a focal length of about ten or twelve inches. It is not difficult, to understand how a lamp placed near the ear, and giving off divergent rays, would require a mirror of a shorter focus; hence the recommendation of a seven-inch mirror in the first instance. In ordinary examination of the ear, the patient may be placed opposite a window; if daylight be used, the rays of light should be caught on the mirror, and thrown focused upon the membrana.

If, however, the patient is placed at some distance from the window, more light will fall upon the mirror, as there will be added to this light, that which has been reflected from the side of the room corresponding to the window; moreover, being in the darker part of the room, some benefit will result from the contrast between the somewhat darkened part where the patient is sitting and the quite bright reflex from the mirror. Again, the much diminished size of the reflected image of the window from the distance, will appear to sharpen and brighten it. Unless operating, I would generally recommend to hold the mirror in the hand, and not place it on the forehead. As the meatus externus has practically two curves—that is, the inferior wall has a convexity looking upward, and the anterior wall a convexity looking backward—it is evident that if the canal were straightened a better view of the drum membrane could be obtained. This is possible, as the outer portion of the canal is cartilaginous and somewhat movable. It is accomplished in this manner: Grasp the auricle by its upper part, between the first two fingers; then draw it upward, backward, and outward, when the membrana is well exposed to view. The outward movement of the auricle is indicated to be made when the canal is somewhat collapsed, as in certain inflammatory affections. In elderly persons also it may be somewhat collapsed from flaccidity of its soft parts.

The speculum may now be introduced, first moistened in water, by a gently screwing motion, being careful not to reach beyond the middle of the canal, for when it impinges against the osseous portion a feeling of discomfort or even pain is experienced. When pushed far enough inward, it is held in position by the thumb and forefinger if it be the right ear, and the thumb and third finger if the left. The inner portion of the canal is examined in detail by inclining the speculum first to one side, then to the other, then up and down. A Toynbee speculum will obviously do better for this purpose.

THE EUSTACHIAN CATHETER AND THE METHOD OF ITS USE.

Ordinarily, three sizes of the catheter may be selected, with the occasional addition of a fourth smaller size. The largest-sized catheter which will readily enter the nostril should be chosen. Catheters may be made of pure silver, which are very readily bent to any desired shape. German silver is also used, as well as other and less expensive metals, which are afterwards plated with silver or nickel. One advantage of the metallic catheter is, that from its greater weight it more readily falls into position when introduced. The hard rubber catheter is inexpensive, light, does not become corroded or soiled when remedial agents are sent through it. It is readily bent to any shape after immersing in hot water, or exposing it for a few seconds over the flame of a lamp or gas-jet. However heated, the catheter, when bent, must be held in this form until cool, in order that the new shape may be permanently maintained. If it be immersed in cold water this may be accomplished immediately. I usually prefer catheters of this material for the reasons above stated. It is not necessary to have a bulbous tip to the catheter, as has been recommended by some. The curvature of the instrument must be sufficient to fill the meatus when introduced into the Eustachian tube, and may be sharp in some instances and gradual in others. This is determined by the readiness with which it is introduced; sometimes entering the Eustachian tube better with a short curve, and in other instances with a more gradual one. No two patients are likely to require a catheter of exactly the same curve. A ring or screw at the larger end of the catheter indicates the direction which the nozzle takes when it enters the Eustachian tube. (The addition of the screw is a device of Dr. Roosa.) The patient may be seated in a chair having a

high back to lean his head against when having the catheter introduced, or this may be dispensed with. The head
is thrown back, the tip of the nostril pushed slightly upward, so as to expose the inferior meatus, and the catheter,
previously dipped into water, is passed with the ring or
guide downward along the floor of the inferior meatus
until it reaches the posterior wall of the pharynx. It is
then withdrawn, in some instances a little more and in others
a little less than half an inch, then turned so that the
ring points towards the ear corresponding to the tube we
wish to enter; then, by a variety of manipulations, seek to
adjust the catheter to the mouth of the tube. To this end
the catheter may be moved upward, downward, or toward
either side. When it is fixed in the Eustachian tube the
sensation imparted to the hand of the operator gives
assurance as to this fact. This final position, the result
of a quarter revolution of the instrument, is not readily
attained, for the bend of the catheter needs to be so great
as to make it a little difficult to accomplish this requisite
turn. If the surgeon is not particular about fitting the
catheter well into the tube, but is only desirous to come
near to its aperture without passing within,—which often
does nearly as well,—a bend so moderate that the instrument can be turned in the nostril with perfect ease may be
used.

When the catheter is in position, the patient can talk,
whistle, etc., without danger of dislodging it. There are
many cases where, from malformation of the nares, it is
impossible to introduce the catheter by this method. It
will be proper in such exceptional cases to pass the beak
of the catheter in, pointing upward or sometimes toward
one side or the other—that is, in any direction in which we
may succeed in introducing it. Sometimes it will be advisable not to pass the instrument further than the fossa of
the Eustachian tube, and enter it at once—in other words,
use any manipulation whatever that succeeds in introducing
the catheter. Occasionally a very small catheter may be
introduced where a moderate-sized one would fail to pass.
Where one nostril will not allow a catheter to be introduced, the other may. Taking advantage of this circumstance, catheters of extra length of curve have been
constructed, so as to enter the tube through the opposite
nostril. This does not allow the beak of the catheter to
enter the mouth of the tube, but to impinge upon its an-

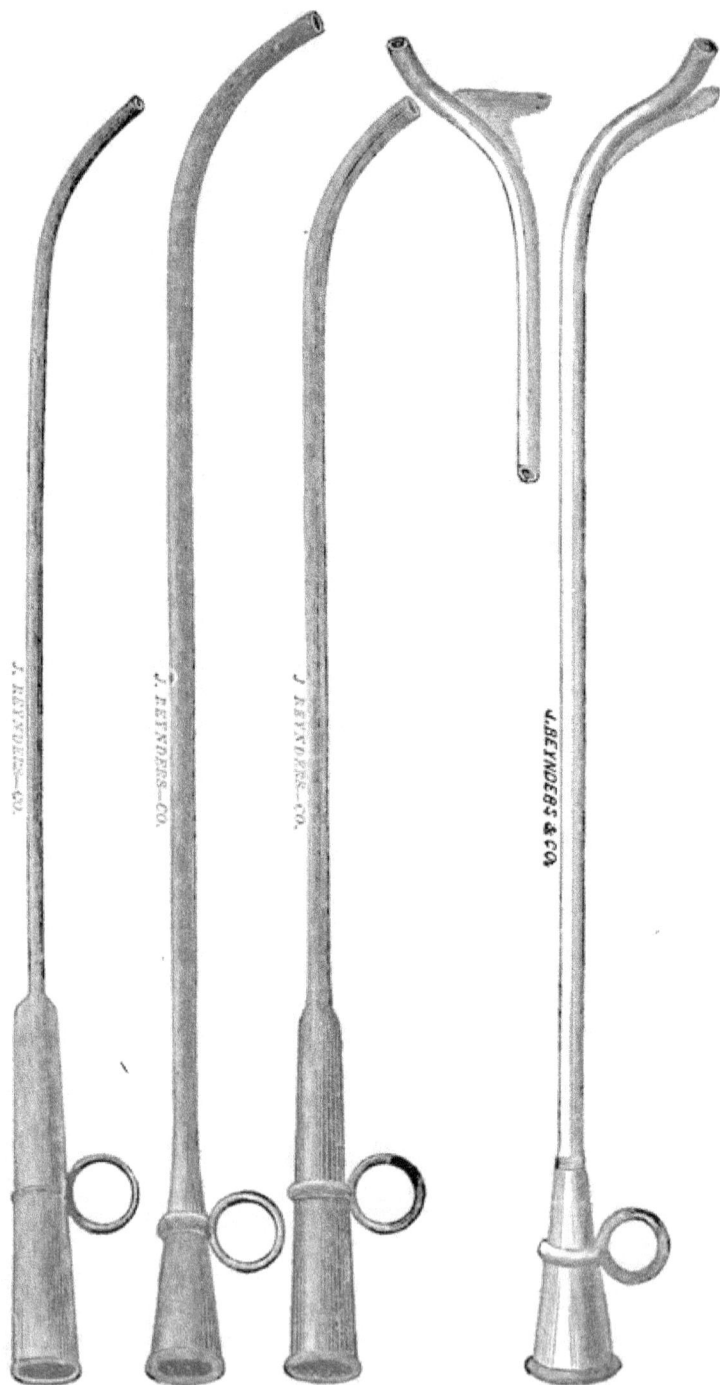

FIG. 13.

At the right may be seen Noyes's double curved Eustachian catheter, and to the left the ordinary Eustachian catheter.

terior wall. It is true that the tympanum may be inflated by this form of the catheter, but not as well as when introduced in the ordinary manner. To correct this faulty adaptation of the catheter to the mouth of the tube, Dr. Noyes of New York has suggested, in the *Tr. Am. Otol. Society*, 1870, a catheter with a double curve of extra length. The cut FIG. 13 will exhibit the manner in which the second curve is made to point exactly in the direction of the axis of the Eustachian tube, and enables it to be inserted properly. No particular directions are necessary for introducing this form of the catheter except to keep the beak as near to the nasal septum and floor of the inferior meatus as possible, until the neighborhood of the Eustachian tube is reached, when the same rules given for the ordinary introduction of the catheter will suffice. I might perhaps state in this connection that a very satisfactory way out of this difficulty in entering the nostrils is found in the use of the faucial Eustachian catheter. The Eustachian catheter may be used without a holder. For some purposes, however, as, for instance, in injecting steam into the middle ear or in any other prolonged use of the catheter, a holder will certainly be convenient if not indispensable. My own catheter holder (see Index) will answer this purpose. A description is hardly necessary: a pivot in the forehead piece allows the holder to swing first to one nostril, then to the other, besides allowing the clamp which holds the catheter to slide in a vertical direction. Before fastening the catheter, this part of the instrument is moved above its level, and when applied is slipped down, the blades of the clamp being placed on either side of the catheter. These are made to grasp the catheter by means of a small thumb-screw. When inflating the ear by means of the catheter, the surgeon's mouth may be applied to a rubber tube attached to the catheter by means of a tip which readily adjusts itself to the end of the catheter without pushing or in the least degree changing its position. Better than this, however, is the Politzer air-bag, with the tip made adjustable to the catheter. The bag may be used with an aperture, rather than an expensive valve, liable to get out of order. During an attempt at inflation the patient may be directed to swallow, or resort to the various manœuvres recommended under the head of Politzer's inflation. Great care needs to be taken not to lacerate the mucous lining of the pharynx by the catheter, for on attempting to

inflate subsequently, a very annoying and possibly dangerous emphysema may result. I have reason to believe that
I once lacerated the mucous membrane in attempting to
introduce a catheter, and subsequently on inflating the ear
I produced emphysema of the cellular tissue surrounding
the posterior pharyngeal space, especially that of the velum
pendulum palati. The patient could not breathe well for
some minutes, and he was fearful that suffocation might
result. I made several minute punctures into the uvula,
which was acting as a valve to obstruct inspiration, with
almost immediate relief. I previously, however, had
directed him to breathe through the nose, so that the
velum might fall into a more natural position, and cause
less obstruction to the respiration. The two cases of death
from the introduction of the catheter, published many
years since, have done much to cause timidity in catheterization. I do not draw any important conclusion from
these cases, however, nor do I believe that life is jeopar-
dized by the use of the Eustachian catheter.

THE FAUCIAL EUSTACHIAN CATHETER, AND THE METHOD OF ITS USE.

In 1872 I published an account in the *Transactions of
the American Otological Society* of an instrument for inflating the middle ear, and making applications to the mouth
of the Eustachian tube by way of the throat. I had previously been using the Eustachian catheter, selecting instruments with a moderate curvature so as to be more
readily introduced. This was attached to the air-bag by
means of a rubber tube about twelve inches in length. Subsequently I made a right-angled bend in the catheter, passed
it behind the velum, and succeeded in readily inflating the
tympani by this means, the rubber tube and bag being used
as in the first instance. I found that with a good-sized
catheter, having a large calibre, I could force air enough
behind the velum to inflate both ears simultaneously, and
often excessively. Afterwards I used a catheter with a
smaller aperture so as to meet that difficulty. From that
instrument to the present one was but a step. This instrument (Fig. 14 reduced nearly one half) is made of hard rubber, seven and a half inches in length; breadth at its larger
extremity one fourth of an inch, gradually tapering to its

beak, which is a little more than one eighth of an inch in thickness. At the larger extremity is a tip for the adjustment of the rubber tube; beyond this, at the distance of about an inch and a half, is a *guide* placed perpendicularly to the shaft of the instrument, and pointing in an opposite direction to the bent extremity of the tube. This will indicate the direction of the beak of the instrument when placed behind the velum, and consequently out of sight. The bent portion of the tube is one inch and three sixteenths in length, and forms an angle of about 75° with the shaft of the instrument. Within a line from the extremity of the tube is an aperture the size of a No. 1 Bowman probe for the passage of air and fluids for injecting the tympani.

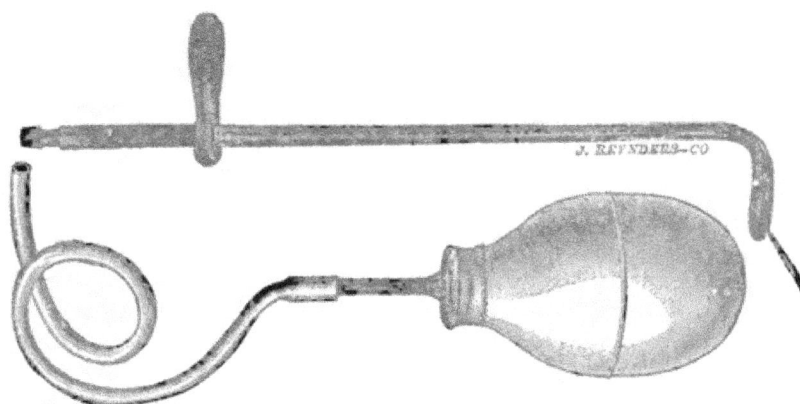

Fig. 14.—Pomeroy's Faucial Eustachian Catheter.

This is placed at an angle of about 65° to the bent portion of the tube. It will be seen that when the catheter is properly introduced, the aperture will point in the direction of the axis of the Eustachian tube. The instrument is not properly a *catheter*, for the beak does not enter the Eustachian tube, but simply comes in contact by its small aperture with the faucial opening of the tube.

It is introduced in the following manner: The instrument, with its perforated rubber bag and rubber tube, is caught by its larger extremity (but not by the guide), held lightly in the left hand, while the right hand holds the rubber bag, with the thumb on the aperture. No tongue depressor is ordinarily necessary. The instrument is passed in with the bent extremity pointing downward, as it must pass over the

tongue, which may rise up considerably, and present a
convex surface. No effort should ordinarily be used to depress the tongue, being careful, however, not to excite it to
resistance by rough or slow manipulation in the introduction of the instrument. In view of the fact that the patient
sometimes inclines to withdraw his head when the catheter
is about to be introduced, it is well to catch the lower jaw
with one of the fingers of the right hand—which may still
hold the bag—thereby fixing the patient's head. If the patient's head is placed against a solid support, he is likely to
cause annoyance to the operator by a sudden lateral movement to escape the instrument, which would be more embarrassing than if the head were left free. The catheter is then
turned quickly upward, and insinuated behind the velum.
During this manœuvre the patient may be requested to pronounce the letter a, prolonging the sound. If the velum is
thrown backward against the posterior wall of the pharynx,
the patient may be directed to breathe through the nostrils,
when it is likely to fall into a natural position, giving a sufficient space for the passage of the instrument behind it. If
the right tube is to be inflated, turn the beak of the instrument to the right of the throat so that it shall stand at an
angle of about 40°; draw the instrument gently forward and
inflate. If the velum is thrown backward in a spasm, the
catheter will require to be drawn forward rather forcibly,
but this condition of things is, if possible, to be avoided.
In order to inflate the left ear, the catheter may be introduced as described for the right, except that the curve of
the instrument must point to the left, 90° from the position
it previously occupied. It is a good manœuvre to pass the
instrument in at the right side and turn it to the left, when
the latter is to be inflated. If the velum resists this excursion, which it frequently does, then the patient may be
directed to breathe through the nose; and if that fails to
place the velum in position, the difficulty is overcome by
suddenly moving the handle of the catheter up and down
during the effort to turn it to the opposite side. By this
procedure the end of the instrument is, so to speak, un-
hooked from the spasmodic velum, and allowed to move
freely.

If it is needful only to inflate, this may be done for both
ears, without removing the catheter from the throat. This
may be accomplished in four or five seconds if the surgeon
is required to work rapidly and has a reasonable amount

of dexterity. For a number of years some of the members of the surgical staff at the Manhattan Eye and Ear Hospital have turned over patients to me where they had failed to inflate the tympani, and I have so far not failed to inflate in using this instrument. Of course no real stricture could have existed in any of these cases.

I can inflate the tympanic cavities more rapidly with this instrument than by Politzer's method or any of its modifications. The delay with the last-named methods arises from the fact that the patients will not always swallow or phonate or respire so as to make the operation successful, until considerable teaching has been expended on them, which of course consumes time. If the patient swallows exactly when an attempt is made to inflate, the operation is quickly done, but he does not often do this.

When it is desired to make an application to the mouths of the Eustachian tubes, my practice is to do it as follows: A dropper is used, by means of which a fractional part of a drop, or several drops, as may be required, is deposited on any non-absorbent surface. With the air-bag compressed, and the thumb on the aperture, the small extremity of the catheter at the opening is laid on the drop of fluid previously deposited, and the air-bag is then allowed to fill; the fluid is drawn within the catheter, and it is ready for introduction. If rapid work is desired, drops may be deposited in several places, to be ready for use. In applying the remedial agent, it is well not to force it in too strongly, as it may enter the tympanum, when the object is only to inject the Eustachian tube for a short distance, or even to apply it to the mouth of the tube.

I have never had any serious accident from injecting the tympanum by this method, but some years since Dr. Weir, of New York, injected the tympanum by means of this instrument, and excited an acute otitis media. If the fluid is sent with some little force a coarse spray is produced, which medicates a considerable portion of the side of the pharynx. If the application needs to be made with considerable exactness, I am in the habit of using only a small quantity of the agent. While testing the instrument, some years since, I was in the habit of employing a very small quantity of a saturated solution of silver nitrate, and I usually succeeded in making a minute white spot of cauterized mucous membrane in the fossa of the Eustachian tube. Ordinarily I use one or two drops of a forty to eighty grain solution of the nitrate. I have never receded from the ground I took years since, that in many cases of chronic catarrh of the tube no benefit resulted until a strong solution was used. I have within a week had a similar experience. It has been urged that this instrument is exceedingly difficult to use properly. Personally, I am unable to judge; I have used it so much that any difficulties in the way of its employment may have unconsciously disappeared. I cannot say that I have succeeded in teaching pupils to become as expert with it as I could have wished. I can, however, assert that I constantly use it, and in preference to other methods. In my hands this instrument is not as unpleasant to the patient as the Eustachian catheter, and I have latterly used that instrument much less frequently than formerly. In the case of children, the bend of the instrument may be about three fourths of an inch in length. As the instrument is made of hard rubber, it may be bent in any manner desired by heating it in boiling water, or placing it over the flame of a lamp, being careful not to burn the rubber.

THE USE OF THE TUNING-FORK IN DIAGNOSIS.

The tuning-fork is mainly used in differentiating between middle ear and labyrinth disease. Many varieties of the instrument are used, and the number of facts connected with the subject is so great that it would be foreign to the purpose of this book to go into detail. The subject will, however, be sufficiently developed, it is hoped, to be made available in ordinary practice. The kind of fork generally used is the one known as Politzer's, and is a "middle C" of 512 vibrations per second. It is of large size, being 8 inches in length, the prongs $\frac{1}{8}$ of an inch in width and $\frac{7}{16}$ of an inch in thickness. It gives a powerful resonance, which is quite necessary in many cases of obtunded sensibility of the nerve. Clamps may be used, which prevent the harsh metallic sound of the over-tone; this is a fourth above the ground tone of the instrument. They also add greatly to the power of the undulations. By successively moving the clamps from the extremity of the instrument to the opposite end, the pitch becomes about twelve tones higher; any intermediate tones are produced by fixing the clamps in the proper position. This is of great advantage, and makes a number of forks of different pitch less necessary. It will often be found that some of the fibres of Corti are destroyed, or at least are not active, when the fork vibrating in unison with such fibres will not be heard; hence the desirability of tuning-forks of different pitch. Many aural surgeons, however, of large practice use the tuning-fork without clamps. Dr. Blake, of Boston, has devised a hammer, one face of which is tipped with rubber. It is attached to the base of the fork by an elastic wire handle, which moves up and down through its point of attachment (Fig. 15). By drawing the hammer away a certain distance each time a blow is given, great uniformity of resonance is obtained. It is possible, however, to secure a sufficiently uniform blow by striking the fork on the knee while the leg is flexed upon the thigh, or even extending the palm of the hand and striking upon its fleshy part. It

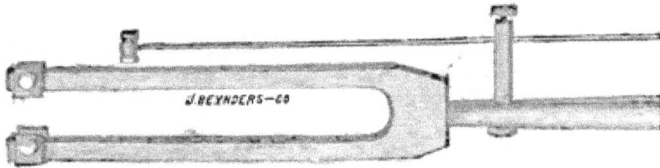

Fig. 15.—Blake's Tuning Fork.

is unnecessary to enlarge here on the desirability of great simplicity in instrumentation. The fork may be applied by its base to the front teeth, the forehead, temples, vertex, mastoid processes, and also held with the forked part near the ear. In the normal ear, the tuning-fork, when placed on the central incisors, is heard equally well in both ears ; the same is true if placed on the vertex or on the centre of the forehead. If placed on the mastoid process, it is heard better in the ear of the same side. If the patient has his tympani inflated with air the tuning-fork is not as well heard. Urbantschitsch states in his work on the ear that in some elderly people with normal hearing, the bone conduction is defective [Politzer]. It is heard longer when placed near the meatus than by bone conduction; that is, when placed on the teeth, etc. This is best tested by holding the fork on the teeth until no longer heard, when it will be distinctly audible placed near the meatus. This is explained by the fact that the most natural hearing is through the air in consequence of the vibratory mechanism of the tympanum. E. H. Weber has discovered that if the meatus is stopped by the finger, or covered by the hand even, that the tuning-fork is heard better, as well as longer, in that ear. If the finger is pressed too far into the ear, however, it spoils the test; this latter observation has been made by Dr. J. A. Andrews, and has been verified by myself. Many explanations are given for this improved hearing when the meatus is closed, but the one given by Politzer and Mach seems the most satisfactory, namely, by stopping the ear the sound waves are hindered from their passage outward to the open air, and also are reflected inward, so that the nerve receives an augmented impression. This test requires to be made with the greatest care to prevent deception. If the tuning-fork is placed on the vertex, or in the centre of the forehead, or upon the central incisor teeth, and the patient, with closed eyes, states by a fall of the hand the precise moment at which it is not heard in either ear, and the hands fall simultaneously, we shall be quite sure that bone conduction is equally good in each ear. This test may be made with both ears closed. Naturally this would prove that bone conduction was the same in each, and it would be left for the voice or watch to test whether the hearing was perfect.

Where there is some defect in the hearing, dependent on disease located in the middle or external ear, the tuning-fork placed by its base upon the teeth, or on any of the central portions of the skull, will be heard better in the diseased ear. If placed in the air, near the ear, it may not be heard as well (aërial conduction) as in the previous position (bone conduction); but if the disease in the middle and external ear is not excessive, it probably will be heard somewhat better by aërial than by bone conduction.

It is a better plan for the patient to signify the comparative length of time he hears the tuning-fork in each ear, it being more exact than to ask if he hears it better in one ear than in the other. It has been found in doubtful cases that if the fork is laid a little to one side of the median line, it assists to confirm a diagnosis. For instance, the patient thinks he hears the fork best in the right ear, and it is then moved to the left side a little; if he still hears it as well in the right ear, or even hears it equally well in both ears, there is then no question of his hearing it better in the right. If the suspected ear is closed, and there is little increase in the length of time the tuning-fork is heard, evidence grows stronger of middle ear disease. When both ears are simultaneously closed, and the nerves are active, there ought to be very little difference in the length of time the fork is heard, unless we accept a condition hereafter to be explained as intermittent bone conduction.

As to the mode in which the tuning-fork vibrations through the bones reach the nerve. There is no question but that sonorous impressions may be carried direct from the skull to the labyrinth. This seems to be proven by a case reported by Lucae in the *Arch. f. Ohrenh.*, xvi., p. 88, and alluded to by Knapp in the *Tr. Am. Otol. Soc.*, 1880, p. 408, in which a case of congenital absence of the external and middle ear, but with a normal labyrinth, had good bone conduction on that side, but considerably better on the other side, which was normal.

A much more important portion of the vibrations, however, reach the nerve by means of the apparatus of the middle ear. A better way to state the idea, perhaps, would be to say that the membrana and ossicula are much more readily thrown into vibrations than the immovable petrous bone surrounding the nerve. Other evidences of the agency of the middle ear mechanism in carrying undulations to the labyrinth will appear in the part of this article devoted to intermittent bone conduction.

When the deafness depends on disease of the parts beyond the external and middle ear, the tuning-fork is heard badly or not at all in the deaf ear, when placed at any of the central positions we have before indicated. If it is a case in which there is little difference in the hearing of either ear, the manœuvre previously referred to—moving the fork beyond the middle line—will aid us. If there is any hearing in the suspected ear it will be increased by stopping the meatus, although the increase may be very slight indeed. This is on the principle that stopping the ear adds to the number of vibrations falling upon the nerve, and consequently increases the hearing; if there is no hearing, there can, of course, be no increase by closing the meatus. If there is hearing by bone conduction, then there will be better hearing when the fork is placed near the meatus, as in middle-ear disease it may be heard worse under the same conditions. The explanation of this has already been given, and may here be repeated in this statement: if the middle ear is normal, the tuning-fork is best heard near the meatus; but if it be diseased, it may be best

heard by bone conduction. If both ears are stopped, the
tuning-fork will probably be heard better in the good
ear if it be a case of nervous deafness. Then, in general
terms, the tuning-fork will be heard by bone conduction
better in the bad ear in middle-ear disease, and worse in
the bad ear in labyrinth disease. In many cases of middle-
ear trouble, combined with labyrinth disease, the diagnosis
will be very difficult indeed, and all the rules laid down
must be applied, which will require great ingenuity and
judgment on the part of the surgeon. The subject is still
further complicated by recent developments pointing to
what has been called intermittent bone conduction. This
has been studied by Bürkner in the *A. f. O.*, xiv., 96, and by
Dr. J. A. Andrews in the *New York Medical Journal and Ob-
stetric Review* for February, 1882, and by some others. The
main points seem to be as follows:

1. A patient with catarrhal otitis, obstruction of the tube
and collapsed membrana, with defective aërial conduction,
may have also bad bone conduction, but on inflating the
tympanum and restoring the position of the membrana to
the normal, or at least a bettered one, which probably improves the aërial conduction, also improves the bone conduction, which may indeed be better in this ear than the
normal fellow, in accordance with the rule. This condition
may return to the former state if the tube becomes again
closed with collapse of the membrana.

2. Pressure upon the round or oval windows, from fluid
in the tympanum, masses of inspissated secretion, as dried
flakes of mucous, pus, blood-clots, etc., interfere with both
bone and aërial conduction, but on the removal of which
restoration of both forms of conduction may result.

3. Excessive hyperæmia of the tympanum may interfere
with both bone and aërial conduction, which may return
on the subsidence of the congestion.

4. Anything whatever that interferes with the free vibration of the membrana tympani, the membrane of the round
window, the ossicula, including the impacted base of the
stapes in the oval window, interferes with bone conduction,
which may be improved by the removal of the hindrance to
free vibration of these parts.

In connection with this subject of intermittent bone conduction, it may be interesting to discuss briefly the condition of the organ of hearing, where its function has been
interrupted by some disturbance situated in the tympa-

num—as to whether the lowered hearing is due to pressure
on the labyrinth fluids, and in its turn on the ultimate nervous apparatus, or to the non-vibratility of the ossicles
and membranes, so that a diminished impulse is conveyed
to the acoustic nerve. I am aware that I have held the
opinion that pressure on the nerve was the principal cause
of deafness (*Tr. Am. Otol. Soc.*, 1880). Pressure of the finger on the eye will cause temporary diminution of sight,
which is restored directly on removal of the pressure,
although it is not instantaneous. Whether this would be
the case if the pressure were continued for days or weeks I
am unable to say; probably not. In the case of the ear the
hearing is restored instantaneously on inflating the tympanum, even though this condition may have been present for
weeks together. The observations in intermittent bone
conduction seem to point conclusively to the fact that absence of free vibratility of the ossicles, the membrana and

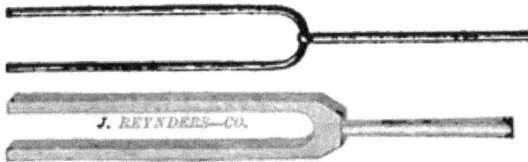

Fig. 16.—The lower represents the appearance of the ordinary clinical tuning-fork,
the upper, the round tuning-fork, which produces a very low tone. Both are much
diminished in size.

the membrane of the round window is the main cause of
defective hearing. To test the bone conduction with utter
thoroughness would seem to require tuning-forks representing every tone which is audible to the normal ear.

THE DIFFERENT MODES OF TESTING THE HEARING.

An exact standard of hearing is scarcely possible. The
very young hear better than the adult or aged; those of
high intelligence, especially where the ear has been trained,
hear better than others: musicians naturally belong to this
class. A patient will hear the tick of a watch further in a
small room than a large one; if there is a stone or wood
floor, and an absence of drapery at the windows, it enables
a person to hear at a greater distance than otherwise

Drafts of air through a room interfere with audition. When a person is in good health and spirits, with absence of any considerable fatigue or mental distress, the hearing is better than it otherwise would be. *The watch, the voice, the whisper, and the snapping together of the thumb-nail and that of one of the fingers,* are, with the exception of the tuning-fork, ordinarily sufficient to test the hearing. It is convenient to use the following abbreviations: R., right ear; L., left ear; H. D., hearing distance; w., watch; v., voice; wh., whisper; n., finger-nails; c., contact; p., pressed against the auricle, etc.

The distance a normal ear hears with any one of these tests may be placed in the denominator of a fraction, and the distance any given ear may hear is placed in the numerator. If the hearing is perfect the numerator equals the denominator of the fraction, and is equal to unity, which is expressive of perfect hearing. To illustrate : H. D. R., w., $\frac{20}{40}$ (assuming the watch to be heard forty inches), or hearing of the right ear, is by the watch only one half the normal. If the patient hears the watch pressed on the auricle, then place the letter p in the numerator of the fraction; or, if on contact, then the letter c. is placed there. Dr. Prout, of Brooklyn, and Dr. Roosa, of New York, have done much to perfect this method, which is analogous to the tests used for vision. The test of the hearing by the watch is very uncertain. My own watch may be heard about five feet. A medical student once heard it at a distance of twenty-three feet. More recently, a medical man came to me for treatment, saying that his right ear troubled him; he heard his own voice too distinctly on that side, and he thought the hearing was not quite as acute as it had been. I found he heard my watch at twenty-five feet. I then inflated his middle ear, and it sprung up to thirty-five feet, the highest degree of hearing I have ever met with. The test was made with the patient blindfolded, and the watch was alternately held in the air and then behind my back, and he readily detected the tick when held in the air, but failed to do so immediately when placed behind my back. The doctor remarked that in auscultation he often heard sounds that the most expert men with normal hearing failed to detect. We never expect our hospital patients to hear quite as far as others. *The voice* may be heard at least fifty feet in a closed room. The words should be spoken distinctly, in a middle tone, and not too rapidly.

There will sometimes be found a very great discrepancy between the voice hearing and that of the watch; naturally we place a higher value on the improvement to voice hearing. This test is a valuable one for children or malingerers, as it cannot be proven whether the watch is heard or not; but a pertinent answer to a question settles all doubts about the hearing. *The whisper* is a useful test, especially in a small room, where the voice is too well heard. *The snapping of the thumb and finger nails together* is useful as a test where the watch is not heard   I have seen thef ormula written, H. D. w., $\frac{0}{48}$, which expresses only a negative quantity. If the watch is heard in contact with the auricle, the finger-nails may be heard at about three feet.

It may be pertinent to observe here that the hearing for the voice at the near point is several times in excess of actual needs; the frequent failure of people to hear the voice in ordinary conversation being a matter of carelessness, inattention, or even dulness of comprehension. Many people lose considerable hearing without being aware of it. The above remarks only refer to aërial vibrations; the tuning-fork tests of hearing will appear elsewhere.

THE APPEARANCES OF THE NORMAL MEMBRANA TYMPANI.

In order to appreciate the appearances of the membrana properly, it is well to remember some of its characteristics. Its vertical and horizontal diameters are about the same— 8 to 9 mm., or 4 lines; its longest diameter passes obliquely from before backward and upward, and is about 1 mm. longer than the other diameters. *The obliquity of the membrane* is indicated by the following statements: the upper wall of the meatus is about $3\frac{1}{2}$ lines nearer the external opening than the lower, and the posterior wall $2\frac{1}{2}$ lines nearer the outer orifice than the anterior. The direction of the membrane then will be downward and inward, and forward and inward. It forms an angle of about 35° with the axis of the canal. In very young children this obliquity is much greater, so much so as to become nearly horizontal. Opposed to this proposition, however, stands the following statement from Politzer, in his " Lehrbuch der Ohrenheilkunde," B. i. S. 22:

" The inclination of the drum membrane in children has hitherto been described as nearly horizontal. By means of numerous measurements Dr. J. Pollak has shown this view to be erroneous, there being no perceptible difference in the inclination of this membrane in infancy and adult life."

The color of the membrana is of a pearly gray, translucent, and in very young children quite transparent. In the aged, however, it is often very opaque. Its color varies with that of the inner wall of the tympanum, which is capable of reflecting light through the membrane and adding to it its own peculiar tint. In the very young the membrane is sufficiently transparent to see the long shank of the incus passing downward and backward, parallel with the malleus handle, and terminating at about two thirds its distance; very rarely the lower portion of one ramus of the anvil may be seen. Other peculiarities of color may be observed, but will be mentioned elsewhere. *The light reflex* is a very important landmark indeed, and depends on the presence of two factors: 1st, a normal state of polish of the dermoid (outer) layer of the membrane; 2d, on placing the point from which the light reflex appears upon such a position of the membrane as to reflect the light back to the observer's eye. Manifestly, every part of a normal membrane reflects light, but the obliquity of most of the reflecting surface prevents the rays returning to the observer's eye. Hence it follows that if a membrane retains its lustre, and any part of its surface lies at right angles to the axis of the canal, a light reflex must needs be observed. The naturally oblique position of the membrana would reflect the light downward and forward, so as to impinge on the antero-inferior wall instead of passing out of the canal to reach the eye. The cause of a change in the central position of the membrana, so that a reflex appears at the umbo, is the traction of the malleus handle. If any membrane be stretched upon a ring placed obliquely at the end of a tube resembling the meatus externus, and a stick be fastened to the membrane so that one extremity reaches to the centre while the opposite end is at the periphery, and this is drawn toward the far end of the tube, the membrane will be found to be dimpled or umbilicated at its centre precisely as at the umbo of the membrana tympani. Or, take a pocket handkerchief, fasten it tightly to a circular hoop;

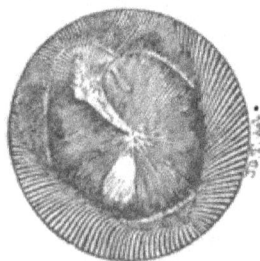

Fig. 17.
Normal membrana tympani.

place it very obliquely opposite a long tube, then catch it at the centre by the fingers and draw it sharply inward: the upper part will be more inclined than before, and the lower part will be so drawn inward that a small portion will present a surface at right angles to the axis of the tube, which, if a polished surface, would reflect a cone of light. The *shape* of the light spot will be somewhat triangular, with its apex towards the end of the malleus handle and its base extending nearly to the periphery of the membrane (its diameter is a little more than a line). A very frequent *position* of the light spot is nearly half-way between the centre of the membrane and its periphery, but somewhat nearer the centre (Fig. 17). The light spot will vary in *shape* with the regularity of curvature of the membrane, and may be a somewhat rounded spot, a few small points, or even several linear reflections; this reflex forms an obtuse angle with the malleus handle. From what has been said it will readily be anticipated that a light spot may be found wherever the membrane falls into a position at right angles to the axis of the canal. If there be a cup-shaped *depression* in the membrane, the bottom of it must needs be at some point at right angles to the axis of the canal and reflect a light spot to the eye. By the same law, any *elevation* from bulging of the membrane would give a reflex from its most projecting portion. The short process of the malleus, for this reason, sometimes gives a minute reflex. *The malleus handle* passes downward and backward at an inclination of about 45° or 50°, and terminates at a point near the centre of the membrane, dividing it into two halves, of which the anterior is somewhat larger than the posterior half. The *short process* is placed on the upper extremity of the malleus handle, and very near the periphery of the membrane. The neck of the malleus completes the area of membrane, while the head of the hammer projects far into the tympanum at its upper and somewhat anterior portion. The handle passes between the outer (radiate) fibres of the fibrous or middle layer and the circular (inner) fibres of the same layer. It terminates by a spatula-like extremity, which often appears as such on inspection. The short process seems, according to Tröltsch, to be covered only by the dermoid or external layer of the membrane. The membrane is maintained in its concave position at the umbo by the traction exerted on the malleus handle by means of the tendon of the tensor tympani; the latter emerges from an aperture in the anterior pyramid on the inner wall of the tympanum, passes across the cavity and is inserted into the malleus near the short process. By the settling of the membrane upon the knuckle-like projection of the short process, it is thrown into two slight folds—an anterior and posterior. The anterior fold commences at the short process, often with a rounded prominence resembling the short process itself, and passes downward, forming an acute angle with the manubrium, and terminates nearly opposite the extremity of the malleus handle, at the periphery of the membrane, and sometimes apparently merging itself into the meatus. In some instances it commences at a point above the short process, and in others somewhat below. This fold is frequently not sufficiently conspicuous to be seen except by a careful examination.

There is also a posterior fold, commencing near the short process and extending backward, upward, and finally downward, forming an obtuse angle with the malleus handle. It is ordinarily less conspicuous than the anterior fold. The explanation of the cause of these folds is easily understood by taking a handkerchief, drawing it tightly, then thrusting a knuckle into it, when folds will be produced resembling those of the membrana tympani. The inner extremity of the meatus terminates in a bony ring, into which the membrane is inserted. At its upper part a portion is lacking: this is called the Rivinian segment. This hiatus is filled in by cutis from the meatus and mucous membrane from the tympanum. The *Rivinian foramen* may occasionally be demonstrated passing from above somewhat downward, from a point just in front of the short process through the membrane, beneath the chorda tympani and tendon of the tensor. A small bristle may be passed into it. Occasionally air or tobacco smoke may be blown through it. The presence of this canal is denied by some, who claim that the appearance is due to arrest of development. *Shrapnell's membrane*, or, as Shrapnell has defined it, the membrana flaccida, is located as follows: Two bands extend from either extremity of the Rivinian segment and meet on the short process, thus forming a triangular space. This is filled in by cutis and mucous membrane. It is very yielding and readily moves to and fro, and serves to relieve the membrane of any sudden pressure which might otherwise rupture it. *Tröltsch's Pocket* may be thus described: A supplementary leaf of the drum membrane applied to its inner surface on the upper part of its posterior half, irregularly triangular in shape, about 3 or 4 mm. in diameter, and arises just behind the osseous border giving attachment to the membrana, and extends to the manubrium; it opens below and forms a "pocket" or cavity. The chorda tympani nerve runs along its posterior concave border.

DISEASES OF THE AURICLE.

PRELIMINARY OBSERVATIONS.

THE auricle, with the exception of the lobule, is chiefly composed of cartilage: this is complicated in form, in order to give its peculiar shape to the organ. It is covered by perichondrium, the fibres of which pass into the substance of the cartilage, form anastomoses, amongst which are found small cartilage cells (Stricker).

The cutis of the auricle covers the cartilage, and by its duplicature forms the lobule. Over the whole surface downy hairs are found, into the sheaths of which sebaceous glands enter. In the concha these glands are of greater size than elsewhere, and may be seen by the aid of a magnifier. Small sweat glands are found on the surface of the auricle, chiefly on the side next to the skull. On the concave surface of the auricle the subcutaneous connective tissue forms a thin layer firmly united with the perichondrium, causing the skin to be immovable; but on the convex side it is thick, and the skin is quite movable. The lobule is composed of skin, connective tissue, and fat cells. The cartilage of the auricle is continuous with the cartilaginous portion of the meatus externus. The auricle, with its helix and its antihelix, its tragus and antitragus, its fossa of the helix and its fossa of the antihelix, its incisura intertragica and its rounded and graceful lobulus, is indeed a beautiful appendage to the organ of hearing. I do not remember to have ever seen a portrait painting by even the most renowned master in which the auricle is even respectably delineated. Only a few persons possess auricles having the classical type. I cannot say that it has any special function other than to collect by its concave an-

terior surface the sonorous undulations for their passage to
the membrana tympani through the external auditory
canal. Many theories have been broached pointing to a
different function, but when we habitually observe people
who are hard of hearing, passing their hand behind the
auricle to prop it forward to catch the sound, we are reasonably certain that its function is, like the hearing-trumpet, to collect and convey sounds to the organ of hearing.

ECZEMA OF THE AURICLE.

This is a species of catarrhal inflammation of its integument. The parts seem often to be invaded by the eczema
of neighboring regions. Notably is this the case with
eczema of the face, when the auricle is frequently attacked
with the disease. The same is true of *E. capitis.* An acrid
discharge from the ear frequently induces an eczema of
the auricle and meatus. Exposure to cold is sufficient to
develop it, so also is uncleanliness. Any depressing influences may predispose to it. It is frequently met with in
those of strumous habit. Dentition seems occasionally to
cause it. Scarlet fever and measles have been known to
produce it.

Symptoms.—The auricle is intensely red, tender, hot, tense,
and considerably swollen; the vesicles characteristic of the
disease are well developed, and give rise to a fine discharge, which dries into crusts, falls off, and leaves a dull
red surface beneath. Frequently this is developed into eczema impetiginodes, when the ear becomes hypertrophied,
and small abscesses are formed. The acute variety is more
frequently seen in children. In the adult the chronic form
is more frequently seen, when few or no vesicles are visible, the auricle being dry, more or less scurfy, with fissures
here and there, and accompanied by considerable hypertrophy and distortion in the shape of the organ. In this
state the disease is very prone to invade the meatus, and
extend to the drum membrane.

Frequently considerable deafness results, accompanied
by tinnitus aurium and a stuffy feeling in the ear. This
condition is often very intractable to treatment.

Diagnosis.—However long eczema may have continued,
there will be found somewhere in the history evidence of
its having been a "moist" disease. Usually the vesicular

stage has passed before the physician is called. A redness
of the part will have been observed, which had a tendency
to "weep" or "discharge." The minute vesicles may often be seen on the edges of a patch of old eruption. We
often have a moist, swollen condition behind the ear,
and on the convex surface of the auricle at its union with
the side of the head, an *intertrigo*, which is the result of the
friction of the auricle against the head, and by its moist
surfaces somewhat resembles eczema, but there are no vesicles, and the cause of the trouble is evidently that of *intertrigo*. Herpes, with its small bullæ collected upon a red
base, which do not break as in eczema, but shrivel away,
with absence of light yellow crusts, will cause no error in
diagnosis.

Treatment.—The first indication in the treatment of the
acute form of eczema is the avoidance of all irritating applications to the parts, and to refrain from too much treatment. The excessive itching and burning must be relieved, and it is difficult to state what remedy will best
fulfil this indication until it is tested. When the part is
covered with scabs they must be removed by poulticing, as
with flaxseed, bread and milk, etc., but do not allow the
poultice to remain sufficiently long to macerate the parts,
which will add to, rather than diminish the tendency to
discharge. To allay the irritation of the parts, the following remedies are found to be serviceable: Wash the part
several times daily with bran infusions, or even cold water,
being careful, however, to keep the meatus stopped with
cotton for fear of the damaging effects of the cold upon
the tympanum. Acetate of lead and powdered opium, a
scruple of each to the pint of boiling water, is a very
soothing application; use by dipping lint into the lotion
and laying it on the parts, with frequent renewals. Five
grs. zinci sulph. with two grs. of morphine to an ounce of
water, is also good to allay irritation. Common whiting
mixed into a thin paste and brushed on the part does well.
After a little time we may commence to use remedies calculated to diminish the discharge; care, however, needs to
be taken not to do too much, for greater heat and inflammation may result from injudicious use of such remedies.
Bismuth, camphor, and starch, of each equal parts, may be
sprinkled on the auricle, but must be washed off once or
twice daily. Vaseline alone, or with finely powdered red
oxide of mercury, one or two grs. to the ℥i, may be ap-

plied several times daily. The linimentum calcis sometimes is successful. In washing off the parts frequently, a
very weak lotion of soda in tepid water may be used, or
even white Castile soap. In Vienna the soap treatment is
much in vogue, where common yellow soap is rubbed into
the part for the purpose of exciting a new action. This
latter idea, however, must be used somewhat cautiously.
Some recommend washing the ear with gruel.

When all vesicles are removed and the part is red and
not excessively hot, with infiltration and swelling of the
subcutaneous connective tissue, rather strong stimulants
may be used. For this purpose a solution of arg. nit., grs.
x to xxx to the ounce of water, may be painted on the
auricle every two or three days. Tr. iodine I have found
perhaps the best remedy for this purpose. If an excess of
it is applied considerable burning pain may result.

After the epidermis begins to exfoliate, Wilde is in
favor of pricking the part here and there with a sharp
lancet. I have no experience in this method, and feel a
little doubtful about its utility. As the parts need at all
times to be protected from the air, the suggestion to apply
collodion is a good one, especially in the latter stages of
the disease. It is like strapping parts, to diminish the tendency to relaxation, and give support. A little oiled silk
bag may be used to protect the part from the air during
the use of the remedies already suggested.

The constitutional treatment is very important. As we are
likely to deal with children more frequently than with
adults, the kind and quantity of food taken will be of great
importance. Nothing should be allowed which is indi-
gestible or innutritious. The periods of eating should
be regular. Frequent bathing and the most careful attention to cleanliness should be practised. The bowels need
attention, and also any acid or acrid condition of the
stomach.

Rhubarb and soda, or magnesia calcinat., hyd. c. creta,
or some other form of mercury, will readily occur to the
practitioner as being appropriate. Tonics, iron, and cod-
liver oil will frequently be required; nor must fresh air be
neglected, nor whatever tends to better the general constitution of the patient. In chronic cases Fowler's solution
must be used; iodide of potass. is sometimes indicated.

HERPES ZOSTER AURICULARIS.

Herpes zoster of the auricle, according to Tilbury Fox, is identical with that of the nose and mouth, and does not essentially differ from the disease in other parts of the body. It is characterized by the presence of vesicles larger than those of eczema, distinct from each other, and seated on an inflamed base, which as a rule do not rupture. The duration of the vesicles is seven or eight days. Severe neuralgic pain precedes their formation, lasting for one or two weeks, which usually subsides upon the appearance of the eruption. The latter is accompanied by heat, tension, and burning; fever usually ushers in the attack. Dr. C. H. Burnett, in his Treatise on the Ear (Phil., 1877), quotes Prof. Gruber, in " Die Bläschenfleschte am Ohre.," *Monatsschr. f. O.*, Mai, 1875, as having observed this disease in the form of an idiopathic affection, which is of very great rarity. He speaks of it as involving the skin of the auricle, meatus, membrana, and even the lining of the tympanum; for it is sufficiently well known that herpes involves mucous membranes as well as skin; when this is the case considerable deafness results. Dr. Burnett also quotes Dr. J. Orne Greene, in the *Transactions of the American Otological Society for* 1874, as alluding to a case of neuralgia of the ear, which resulted in an herpetic eruption "over the anterior surface of the helix." He in turn also quotes Dr. Anstie, in the *Practitioner*, as reporting a case of zoster of the nerves supplying the tragus and meatus. Dr. Burnett also speaks of a case of herpes zoster of the tragus in a young lady 18 years old, and also of a man who had herpes of the meatus and of the tragus. According to Gruber's article, the nerves involved in this affection are the auricularis magnus and the auriculo-temporal; and the eruption appears more frequently on the anterior and upper surface of the auricle, the vesicles being more numerous over tracts innervated by filaments of the auricular branch of the pneumogastric nerve.

Herpes of the auricle is developed in the same manner as in other parts of the body. There seems no doubt but that it is a disturbance in function of the sympathetic system of nerves, especially those presiding over the circulation, the vaso-motor.

As causes of this disturbance we may enumerate the following: The common phenomena of taking cold, with its accompanying herpes labialis, or "cold sore," will give a good idea of its nature. Inflammation of the deeper parts of the ear of a catarrhal nature, or a general catarrh, pneumonia or fevers, ague, direct exposure to cold, emotional disturbances—sometimes the administration of arsenic will cause it. The disease resembles somewhat an eruptive fever.

The diagnosis is not difficult; the neuralgic pain and fever preceding the eruption are very characteristic. The vesicles are larger than those of eczema and smaller than those of pemphigus, and are situated on an inflamed base. They do not rupture, except when in the meatus. The short duration and regular course of the disease sufficiently differentiates it from other similar affections.

The prognosis is altogether favorable, except in the very feeble or those who have been debilitated by extreme age ; then great effort needs to be made to keep up the strength of the patient. The vesicles after rupturing frequently leave large ulcers, which are a considerable time in healing. The hearing when affected recovers slowly but perfectly after a few weeks.

Treatment.—The parts simply require protection from irritation and exposure, as the disease naturally tends towards recovery. It is on the whole better not to rupture the vesicles. The auricle may be anointed with vaseline, then sprinkled with flour, the whole to be covered by absorbent cotton. Vaseline ʒi, plumbi acetat. gr. v ad x, is a good application. Coating with collodion is also serviceable. If the parts are painful, diachylon salve with morphine may be used, spread on lint and applied. Camphor and belladonna liniment, of each equal parts, may be used. In the meatus the vesicles are always ruptured, and a moist condition is the result. Some astringent will be needed on this part. Gruber, in the article referred to, recommends sulph. zinc. Arg. nit. painted on the canal will be serviceable. Plumb. acet., 5 to 10 grs. to the ounce of water, is a good remedy. If the neuralgic element is prominent it will be proper to use hypodermic injections of morphine, or any other means calculated to relieve pain. Stimulants and tonics may be necessary to meet any symptoms of depression. Careful attention to the patient's food will be necessary; a full and nutritious diet must be prescribed, and perhaps alcoholic stimulants.

ERYSIPELAS OF THE AURICLE.

The auricle is occasionally the seat of erysipelas. This is of the simple variety. It presents the same symptoms as that disease found elsewhere. It is ushered in by febrile symptoms more or less severe, lasting two or three days; on the appearance of the eruption the auricle has a burning or smarting sensation, with a feeling of tension; it is red, swollen, puffy, and shining. The edges of the patch are raised, the parts are tender and hot. In two or three days the redness increases, blebs form; these burst and 'dry into scabs. In five or six days convalescence commences, and the surface becomes yellowish and the cuticle becomes somewhat detached. The hearing may temporarily be diminished by closure of the meatus from swelling. This disease is caused by extension from other parts to the auricle, as when the face or scalp is involved. Cold is a sufficient cause. Any species of traumatism, especially when the patient is in a depraved state of health, may excite it. Leech bites sometimes result in erysipelatous inflammation, as I have in a few instances verified. In Bright's disease traumatism is peculiarly liable to induce it. Women are more subject to the disease than men.

Its duration is from eight to twelve days. The prognosis is altogether favorable, there being, as far as I know, no destructive processes, as in the phlegmonous varieties.

Diagnosis.—It is not likely to be confounded with any other disease except erythema, from which it may be distinguished by the absence in the latter of the tense, shining, smarting blush, and the implication of the subcutaneous connective tissue, so characteristic of erysipelas.

Treatment.—The pyrexia may be managed in the ordinary manner. It will be well to avoid any cold applications, on account of harm being likely to result to the hearing. In the earlier stages, a weak lead and opium wash may be used, or poppy fomentations. Protect the part from cold, and even apply heat if necessary. In a later stage, tr. of iodine painted upon the part is very serviceable; it may cause some pain, but subsequently it will afford much relief to the burning and smarting. The constitutional treatment is very important. Tr. ferri chlor. in large doses, four or five times a day, either with or without quinine, is strongly indicated. Stimulants may be needed to keep up the strength. The greatest care of the nutrition is necessary. .,

HORNY GROWTHS ON THE AURICLE.

These are occasionally seen; I have observed two or three cases. It seems to be an hypertrophy of the epidermis. The one I distinctly remember had a growth on the upper part of the helix, more than half an inch in length, tapering at its extremity, and of horn-like hardness. At its attachment on the auricle, it was moderately soft, and resembled cartilage. It occurred in a young man who first noticed it about seven months previously. It was. somewhat tender to the touch at its attachment. The color was yellowish, and it had the appearance of true horn. I removed it by an incision made somewhat deeply into the auricle, without difficulty; as far as I know, it never returned. Sometimes the same development may be found in other parts of the body. By a case reported by Dr. A. H. Buck, in *The Transactions of the American Otological Society for* 1871, page 18, it would seem that if the base of the tumor is not fully removed, it may be expected to return. Dr. Buck describes his case as "a blunted, horn-like protuberance, three fourths of an inch long, and nearly as broad at its base; it sprang from the upper and posterior part of the left helix. . . . At the extremity and middle portion it was hard like horn, but near the base it could be easily compressed, though yet comparatively hard. . . . The line of demarcation between the growth and the normal skin was very abrupt. There was no pain to patient, or tenderness on pressure. It was removed and did not return. The young man had been in the habit of paring it off with a razor. On the eyelid corresponding to this ear was a small sharp point resembling the aural growth at an earlier stage." Dr. Burnett, in his Treatise on the Ear, reports the following case which occurred in his practice : A horny growth was observed on the upper and outer portion of the helix of the left ear, in a large, strong man, forty-five years old ; it was of smaller size than in Dr. Buck's case. The patient picked it and it grew larger ; it caused no inconvenience. He disappeared from observation.

INTERTRIGO AURICULARIS.

This is a disease peculiar to children. It consists of an

excoriation behind the auricle, commencing in the sulcus corresponding to the point of attachment of the auricle to the side of the head. The part is always moist, except occasionally when it is covered with scabs. If left to itself, it may extend and even develop into an eczema, or a disease closely resembling it. It depends on a thin, irritable skin, found in the healthy or strumous subject. Moisture and uncleanliness of the part favor it, but the direct cause is friction between the posterior surface of the auricle and the side of the head, precisely as in the same disease when affecting the groins of little children or infants. Too much head covering, which is likely to press the ear against the head, with moistening of the part by prespiration and accompanied by uncleanliness, is probably the direct cause of the trouble.

The treatment consists in removing the conditions giving rise to the disease, and protecting the excoriated parts. The latter indication is fulfilled by sprinkling powdered starch on the denuded surface, previously cleansed by bathing with tepid water, aud a small addition of baking soda. Vaseline applied to a bit of lint or absorbent cotton may be sufficient to cure. Oxide of zinc, in powder, added to seven or eight parts of starch, is a good "drying" application. A bit of lint saturated with camphor water and borax will be of service.

R Aqua Camph... ʒ i.
 Sod. biborat...gr. x.
 M.

Vaseline and plumb. acet., one gr. of the latter to the drachm of the former, is a good application. When the excoriation is diminished, tr. iodine painted *lightly* over the part may be recommended. A coating of collodion will sometimes so protect the part as to cure by one application, but there must not be an excessive discharge, or the coating will be removed. In children of the lower class cleanliness must be insisted upon.

FIBROUS TUMORS OF THE AURICLE.

These are usually found in the lobulus, but may exist in other parts of the auricle. They are of fibrous hardness, somewhat paler in color than other parts of the ear, and more or less nodulated. They are also indifferently called fibro-sarcomatous or myxo-fibromatous tumors. The size

of these growths varies from that of a large pea to a hen's
egg; (Fig. 18) although in some cases they are found depending from the auricle of sufficient size to touch the
shoulder, presenting a huge and horrible appendage to the
ear.

The cause of these growths is an inflammatory proliferation
of the connective-tissue elements of the part, the result of
some form of irritation. This is generally dependent on
the practice of wearing ear-rings—
sometimes from their great weight,
as in those worn by many semi-bar-
barous races, as the African or Indian. In other instances the irritation
seems to result from the use of base
metal in the earring (Roosa), or
this factor combined with a possibly strumous diathesis, so often
found in mulattoes in this country,
in whom the affection is prevalent
(Turnbull).

In some cases, however, I have
seen inflammation and a tendency
to fibrous proliferation in the lobu-
les of the ears from simply piercing
them and drawing a thread through
the aperture preliminary to the insertion of the earring. Whether the
process commences by the formation of granulations at the borders
of the perforation, as suggested by
Gruber, or within the substance of

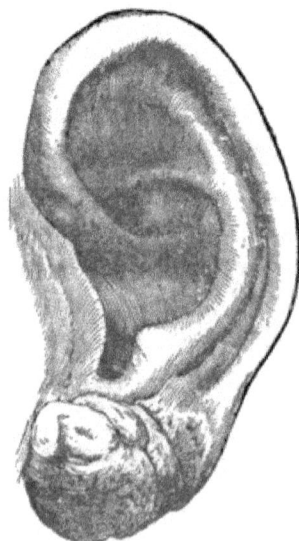

FIG. 18.

Fibrous Tumour of the Auricle.
(From Turnbull.)

the lobulus, is difficult to determine. A process almost or
quite identical with that found in ordinary fibromata of
the lobulus is sometimes seen in other parts of the auricle.
Thus Agnew reports a case of myxo-fibroma of the auricle
as follows:

C. S., æt. 11, received a scratch from a toilet pin upon the
left auricle when two years of age. The resulting scar at
the end of eighteen months reached the size of a buckshot.
It was removed by the knife. Almost immediately a tumor
returned in the old place, and at the end of two years was
about three times the size of the old tumor. This was removed. At the end of two years more, another tumor about
twice the size of the second one was removed. Some time

after this still another tumor had grown to such an extent as to involve most of the auricle except the lobulus, giving it a peculiar nodulated appearance with wavy outlines. The whole of this, with the cartilage of the auricle and most of that of the meatus, was removed. There remained a little skin and the lobule only. The wound was closed by sutures in such a manner as to form a rudimentary auricle. It returned again, however; but, as is the case with this class of tumors, it was benign in character, it was advised that it be again removed. This tendency to return is mentioned by many authors. Knapp, in the *Archives of Otology*, vol. iv., in speaking of Fibroma of the Lobule of the Ear, concludes that the recurrence of these tumors is mainly due to the fact of incomplete removal, as in one of his cases a portion of the tumor was left so as to make a better flap. The result was that the tumor soon grew again, but after a second and complete removal there was no recurrence. From the records of a large number of cases where the removal has been done with apparent thoroughness, but where the tumors have frequently recurred, it would seem that there was a tendency to recurrence of the growths not to be always prevented by thoroughness of removal. The microscopical appearances of these tumors are those of a condensed "connective tissue with intersecting yellow elastic fibres, with few but strongly developed blood-vessels running through it" (Bertholet, *Transactions of American Otological Society*, 1871). According to some authors (Knapp, l. c.), "the softer and semi-transparent interspaces between the trabecles (of fibrous tissue) contained a large quantity of roundish, spindle-shaped, and stellate cells, which anastomosed with one another, and were separated by an abundant homogeneous or finely striated intercellular substance." Little need be said about treatment. If the growths are small and not unsightly, it may be proper to refrain from operation, but otherwise removal should be accomplished at once. If the growth is in the lobulus, a V-shaped incision may be made, using a stout pair of scissors or a scalpel for the purpose. The scalpel will make an incision more likely to heal kindly than the scissors. Closing the wound with sutures is all that remains to be done. In the case herewith reported, it will be seen that the whole of the cartilaginous portion of the auricle and most of that of the meatus was removed in the extirpation of the tumor.

HÆMATOMA AURIS; OTHÆMATOMA.

The first symptom of this affection is a swelling of the auricle, with redness, burning pain, and sense of distension. After a few hours or a day or two the blood tumor makes its appearance. This has a somewhat livid red color, or it may be little changed from that of the normal auricle. It varies in size from a filbert to a hen's egg, and occupies the concha as a rule, but may occasionally make its appearance higher up. It never invades the lobule. The blood effusion is usually beneath the perichondrium, but sometimes it may be within the substance of the cartilage itself. When of considerable size, the tumor may convert the concave surface of the auricle to an even rounded convexity, occupying every part of the latter except the helix. This tumor is for a few days attended with signs of inflammation, as burning pain and sense of distension in the auricle; but after a week it becomes quite inactive and slowly subsides, and in from two to four weeks has disappeared. Sometimes, however, an abscess forms. It spontaneously ruptures in a certain proportion of cases. These usually have a more favorable termination than where incision has been made. A few end in complete absorption of the effusion, with very little resulting deformity.

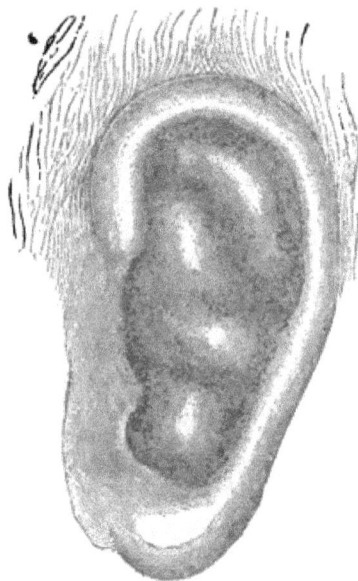

FIG. 19.
Othæmatoma (Gruber.)

The Etiology and Pathogenesis is not well understood. The disease certainly occurs in the sane as well as the insane, although without doubt it is more frequent in the latter. The characteristic symptom, hemorrhage, evidently does not occur spontaneously in perfectly sound ears, and where there is no brain disease, even though traumatism has been an element of causation.

The effect of traumatism in the perfectly healthy subject is worthy of note—there is no characteristic hemorrhage.

Virchow, Hun, and many others who have written on the subject, seem strongly of the opinion that the auricle is in some way prepared for the hemorrhage by softening of its cartilage into a viscid material, with the development of an excessive number of blood-vessels of the perichondrium, with increase in their size, which would tend to produce spontaneous hemorrhage, or one from very slight exciting causes. A depression of the vital energy of the part from any cause is a factor. Many cases evidently depend on some form of cerebral disease. Brown-Séquard, in a lecture in the University of Pennsylvania Oct. 10, 1872, (*Tr. Am. Otol. Soc.*, 1873), states that othæmatoma has a nervous origin, and is the result of disease of the base of the brain, and is usually found in the paralysis of the insane. It may be produced artificially in animals by irritation of the restiform body. In the insane he states that the tumor is on the side on which the brain is found to be diseased. Some form of irritation of the sympathetic nerve inducing hyperæmia of the cerebral vessels, together with those of the auricle, is regarded by some neurologists as a sufficient cause.

The Diagnosis may easily be made from the fact of there being a bloody tumor in the auricle, with fluctuation, which has made its appearance sooner after the first symptoms than would be possible for a fluid product of inflammation to develop. The amount of pain and inflammation is much less than in perichondritis. This, like the latter disease, never invades the lobulus, for there is no doubt but this is a disease of the cartilage and its perichondrium. Its duration is much shorter than that of perichondritis, and never commences in the tympanum and meatus, as the latter frequently does. In later stages, where suppuration occurs, there will be true perichondritis, but the symptoms in the first stages will indicate its true nature. In some cases of erysipelas of the auricle with great swelling there may be some resemblance to this affection, but the helix will be swollen at least as much or more than the other parts of the auricle, and the color will be intensely red. Moreover there will be swelling of the posterior surface of the auricle—an unusual symptom in hæmatoma. In the latter affection, too, the swelling will be quite round, even, and more bladder-like than in erysipelas, and also quite in contrast to the nodular swelling of perichondritis. Its duration being only a month or even less, is quite in contrast with that of perichondritis, which is often six months in recovery.

The Prognosis is favorable as far as the hearing is concerned. Few cases recover without deformity. No treatment so far devised seems to prevent this. In a few instances where the hemorrhage has been moderate in extent and no rupture has occurred, nor incision been made, very little deformity results; but these cases are quite exceptional. There seems little or no destruction of cartilage as in perichondritis, except in very rare cases of abscess.

With the view of adding to our knowledge of the nature of hæmatoma auris, I addressed a note to Dr. Ralph L. Parsons, of Greenmont, Sing Sing, N. Y., formerly physician-in-chief to the Lunatic Asylum on Blackwell's Island, N. Y., and also of the Flushing Asylum for Lunatics. The letter is as follows:

"Greenmont, Nov. 30, 1882.

" Dear Doctor Pomeroy:

"Your letter asking an expression of my personal views on hæmatoma auris has been received.

"My observation of hæmatoma auris has been entirely confined to cases occurring in the insane; and I am disposed to think that the disease occurs very rarely in the sane. My own experience has been to the effect that insane patients who have hæmatoma auris do not recover. My conclusions have been that the disease usually depends upon a grave deterioration of the nervous system as a cause. While the local disease may possibly be caused by a local injury in a healthy person, I have seen no such cases, and think they must be of exceeding rarity."

Treatment.—The propriety of incising the tumor and evacuating its bloody contents is somewhat doubtful. It would seem, at least on theoretical grounds, that if the tumor was destined to spontaneous rupture this should be anticipated by a free incision. Again, the difficulty confronts us, that it is not always possible to determine whether a given case is liable to spontaneous rupture or not. If the tumor is of moderate size, showing no tendency to rupture, and its disappearance by absorption is probable, it should not be interfered with. Where the extravasation of blood is extensive, and accompanied by considerable distension of the parts, and possibly by pain, with an undoubted tendency to still further separate the perichondrium from the cartilage, a free incision is certainly indicated, opening well the blood cavity so that it may receive proper treatment. This cavity may be thoroughly washed out with tepid water at least once daily. Sometimes a stimulant or astringent may be used in conjunction with the tepid water. Carbolic acid, one drachm to the pint of water, may be injected daily, or tr. of iodine may be used to swab out the cavity. Frequently these astringents irritate too much, and do not fully answer our expectations. Pressure by means of a roller bandage passed across the top of the head and underneath the chin, and enclosing the auricle, answers very well. Place a bit of absorbent cotton behind the auricle and another in front; then lay the pinna against the side of the head and apply the roller—just tight enough to be comfortable to the patient. Remove often enough to keep the parts clean and the bandage snugly applied. In the earlier phases of the disease, if it is decided not to incise, a variety of measures may be adopted to subdue any inflammation and promote absorption. Dr. Kirkbride, physician-in-chief to the Pennsylvania Hospital for the Insane, thus writes to Lawrence Turnbull: "The best treatment that I have tried has been the application of ice to the tumor from the very commencement." This, he thinks, "prevents the enlargement which otherwise is pretty sure to occur." Dr. John Curwin, of the Pennsylvania State Lunatic Asylum of Harrisburg, finds tr. of iodine painted on the auricle once or twice a day the best treatment for hæmatoma auris (Turnbull, "Diseases of the Ear," pp. 138 and 139). Very few neurologists are in favor of incisions in this affection, and most of them believe that, whatever is done, considerable deformity is likely to result.

PERICHONDRITIS AURICULÆ.

This affection may appear as a primary inflammation of the perichondrium, commencing usually in the concha, but it may develop from any portion of the front part of the auricle except the lobulus. In some cases the cartilage will be found softened at the point from which the inflammation first arises. Again, the perichondritis plainly depends on an inflammation of the meatus or tympanum, which involves the perichondrium of the auricle by extension outward. The swelling may be of a bright red color, although not always, and quickly extends to the whole anterior surface of the auricle except the lobule. Frequently, however, inflammation escapes the helix.

In the earlier stages the process is quite rapid, and is accompanied by much heat, pain, and sense of fulness in the auricle. In the subacute stage its progress is slow, being sometimes more than six months in duration, and often very painful.

In the more severe cases a large portion of the cartilage may be denuded of its perichondrium, and sometimes completely destroyed over a considerable area. This results in very great deformity to the auricle. In this class of cases the auricle may be the site of an immense abscess which subsequently may be partially filled with granulations. In other cases the cartilage is thickened by inflammatory proliferation and becomes much distorted in shape, and may obliterate the normal outlines of the auricle. In milder forms of the disease the inflammation may terminate in resolution after a duration of a few weeks only (three to six).

Symptoms.—In the severer forms of the disease an inflammatory swelling may be noticed in the meatus, which gradually extends outward, involving first the concha, then the greater part or the whole of the front of the auricle except the lobule, when it will be swollen so that the normal configuration of the parts will be obliterated and the pinna will be nearly flat or somewhat irregularly convex on its anterior surface. In some cases the posterior surface of the auricle may be involved. The color of the swelling will be red or reddish, in many instances inclining to a bright rather than a livid red; in other cases there may be little change in color. The burning, smarting, and in some instances throbbing pain, may be so great as to necessitate the use of anodynes for relief. Where pus forms there is no difficulty in detecting it by the sense of fluctuation, but sometimes there is no purulent formation, and the ear may

FIG. 20.

Auricle of J. S., showing the swelling posteriorly.

have some of the signs of fluctuation, without, however, proving the presence of pus. These signs will be more like a boggy or doughy sensation to the touch, similar to that which is observed in cellulitis, with considerable swelling over the mastoid. If the disease develops secondarily to an otitis media or otitis externa, there is likely to be a discharge from the meatus.

At an advanced stage of the inflammatory swelling, if the presence of pus be suspected and an incision made, a glairy, somewhat bloody fluid will be evacuated, which may or may not contain a small amount of pus.

This will indicate pretty certainly, according to Virchow and others, that we have to deal with an inflammation of the cartilage. In the course of the affection large quantities of this fluid may be evacuated. In some instances the swelling of the auricle will be enormous, when it may be seen to stand out at right angles to the side of the head. In later stages the anterior surface of the auricle may have a nodular appearance, simulating malignant disease. The inflammation may extend to neighboring parts, and result in abscesses, enlarged glands, etc. The resulting deformities are very unsightly, altering the shape of the auricle in some instances past recognition. The helix is less likely to be deformed,

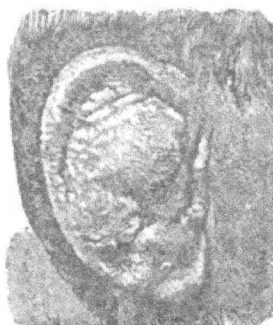

FIG. 21.
Shows the anterior surface of the auricle of J. S.

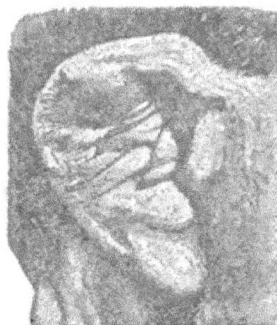

FIG. 22.
Auricle of J. S. after recovery, and showing the resulting deformity.

but it may be, as seen in the appended cut. As this is an affection of the cartilage, the lobulus, having no cartilage in its structure, will be unaffected—a point to which Dr. Knapp has called attention. The cut will also show this. One point to which Dr. Pooley calls attention is, that the auricle is peculiarly liable to perspiration as a consequence of this affection. As far as I know, the hearing is unaffected. It will be seen that this disease has some points of resemblance to hæmatoma auris or othæmatoma. The latter is also a disease of the cartilage in many instances; but the constant hemorrhages of this affection are not present in the former. Perichondritis may be secondary to disease of the tympanum or of the meatus. A furuncle in the latter may lead to the development of this affection, or even the rare instance of a furuncle in the auricle may do so. Often it is idiopathic, again it is traumatic. It has never been observed in the insane. The swelling in perichondritis is likely to be somewhat nodular, but in hæmatoma it is quite round and regular.

My own case, illustrative of this subject, is briefly as follows: J. S., aged 42, in May, 1874, had a violent pain in the right ear, which continued for eleven days with very little abatement, when the membrane ruptured and the ear discharged large quantities of thick creamy pus, with but little relief to the pain. After one month a polypus made its appearance in the meatus and was removed. Subsequently a very painful, circumscribed swelling of the meatus was incised, which gave great relief.

In July a swelling made its appearance in the region of the concha, rapidly extended to the whole of the auricle, increasing its size prodigiously. On September 1st I saw the patient for the first time, and found a large abscess occupying the region of the concha and extending upon the auricle in all directions so as to involve about half of its superficial area. As the engraving will show, it pointed both in front and behind, causing the auricle to stand out from the head at right angles. In front of the meatus, near the tragus, was a circumscribed swelling, with a fistulous opening in its apex. The swelling of the auricle was somewhat nodulated, giving it the appearance of carcinoma. Fluctuation was easily detected both in front and behind, as the walls of the abscess were very thin. An incision, made posteriorly, evacuated about five drachms of pus. A finger passed into the incision revealed the fact that the cartilage of the auricle, in the region of the pus cavity, had entirely disappeared in front and behind; there was nothing left except integument and connective tissue. Subsequent to this, small abscesses made their appearance as follows : three were incised just above the region of the lobulus, one in the tragus, and two on the upper portion of the helix. The principal pus cavity closed in three weeks without special treatment. Carbolic-acid solutions were injected at first, but they were not well borne, and were discontinued. Syringing out the cavity with warm water daily was practised. The discharge from the abscess in the helix was somewhat glairy in consistency. The incision in this part may be seen in the third cut.

The duration of the affection was a little more than five months. There was no insanity, present or past, in the patient or his family, and no history of previous traumatism was elicited. My opinion of the nature of this case, which coincided with that of Dr. Mathewson's, who previously had treated the patient, was that these abscesses depended on a perichondritis, and that this had been developed from an inflammation of the tympanum, which had passed outward along the meatus, and involved the auricle by extension.

From this case and subsequent ones, to be herewith referred to, I shall adhere to the opinion that this affection is not identical with hæmatoma auris, although such an authority as Buck regards them as identical, or at least closely allied affections. One of the most important cases reported, which also strongly resembles my own, is that of Dr. Knapp, in the *Arch. Otol.*, vol. ix., No. 3, p. 195 *et seq.*, and is as follows: H. A., of New York, 16 years of age, had always enjoyed good health. On August 29, 1879, he had a moderately painful swelling in the right ear, of a few days' duration. External meatus slightly red and swollen, so as to close the canal. If the auricle was drawn backward a small opening appeared in the canal; hearing good. On the anterior lower part of the canal fluctuation was observed. Diagnosis of furuncle was made, and the swelling was incised and a watery pus escaped from the opening. A week later the swelling of the canal diminished, but the lower part of the concha was red, swollen, and indistinctly fluctuating. After three days, the symptoms becoming more declared, the swelling was incised, and watery pus escaped. Four days later the swelling filled the concha, was dark red, and had a doughy feel. It was opened above the lobule, and a viscid fluid, containing dense yellowish flakes, was evacuated. A probe introduced through the opening could be passed more than a half-inch upward. The perichondrium was detached from the cartilage. A silver drainage-tube was introduced into the opening, and the ear was covered with picked lint, held in position by a roller bandage. The swelling crept slowly over the whole anterior surface of the auricle excepting the lobule. In some places it was diffuse, in others nodular and fluctuating. The latter were lanced, and a material was evacuated similar to that before described. At the beginning of the sixth week the swelling had attained its height. The helix

was swollen, and the fossa of the helix was preserved only
at its upper part. The remainder of the auricle, except the
lobule, presented one uneven, nodular, reddish swelling.
The tragus was swollen and misshapen, and in front and
below enlarged glands were felt. The whole pinna was
enlarged to about one and a half times its normal size. For
some weeks the meatus was found to be moist, with a scant
puriform discharge, which evidently came from an opening
communicating with that of the auricle. Some granulation
tissue was scraped out of the cavity of the auricle by means
of a sharp spoon. After the sixth week the swelling began
gradually to subside, the aperture leading to the abscess
cavity closed, discharge ceased, and the auricle shrunk and
remained irregularly corrugated, with an obliteration of its
normal configuration, except the somewhat atrophied helix,
which remained, only slightly changed. The pinna had
been reduced to about two thirds its normal size. It was
pressed against the side of the head, and was much disfig-
ured by nodules and ridges. The lobule was unaffected.
The process lasted ten weeks. On May 30th, 1880, the patient was found to have a perforation in the drum-head, the
result of an inflammatory process occurring subsequently
to the symptoms already described, and not in the least dependent on them (so I infer). Knapp also reports a second
case—that of a young lady aged about 17 years. There
was no apparent cause for the trouble. Her ear looked
exactly like the one already described during its fourth
week. The concha was filled up, and there was distinct
fluctuation. The patient declined treatment and disappeared. He quotes a third case from Dr. R. C. Brandeis,
of New York. The patient had a mild chronic aural catarrh, with pronounced pharyngitis. The same reddish
diffuse swelling developed in the cartilaginous portion of
the meatus and the adjacent portions of the concha which
was noted at the beginning of the first case. A large and
deep opening was made, watery pus escaped, and the
swelling subsided in a few weeks.

Dr. K. also quotes Dr. R. Chimani, in the *Arch. f. Ohrhk.*,
vol. ii. pp. 169–171, 1867, as follows: Without assignable
cause, the concave surface of the auricle (of a patient) became painful, red, and so much swollen that on the fourth
day all elevations and depressions had disappeared. Poulticed it for three days, when fluctuation manifested itself
over the whole of the swelling. A large vertical incision

resulted in the evacuation of a synovia-like fluid mixed with pus. The cartilage was laid bare, and on probing, its integument was found detached to the whole extent of the concavity of the auricle. The abscess cavity was syringed with lukewarm water twice daily, and dressed with lint and cotton wadding. In twenty-four days the patient completely recovered, without any deformity of the auricle. Dr. Pooley, of New York, also reports a case of perichondritis auriculæ in the *Tr. of Med. Soc. of the State of New York* for 1881, p. 212 *et seq.*, as follows (the patient consulted the doctor on September 24, 1877): Mrs. G., aged 21, born in New York, of German parents. Three weeks previous to consulting the doctor she had some pain and itching in the right ear. She said that a boil formed in her ear, broke, and gave exit to a greenish-looking core. She felt better for a time, but the ear again troubled her, and she sought advice. On the anterior lower wall of the meatus, near its outer portion, was a swelling resembling a furuncle ; on incising it, watery-looking pus escaped.

October 2d, the swelling subsided, but the lower part of the concha was red and swollen; no distinct fluctuation. The swelling rapidly increased, and became tender and painful. There was an ill-defined sensation of fluctuation. In a few days the swelling had obliterated the concavity of the concha; it was of a dark-red color and boggy to the touch. An incision gave exit to a thin, glairy-looking fluid, mixed with yellowish-white shreds. A probe could be passed along the anterior surface of the auricle as far as the antihelix. The cartilage was roughened, hard, and could easily be detached and raised, together with the skin, by means of a probe. A tent of charpie was introduced into the wound, the ear covered with absorbent cotton, and a flannel roller bandage was applied with pretty firm pressure. The swelling slowly extended to the whole of the front of the auricle except the lobule. This swelling was in some places diffuse and in others nodular, the whole being of a diffusely red color and painful to the touch. Two abscesses were formed—one in front of the lobule and the other behind—which were opened (probably suppurating glands). A probe passed into the fistulous opening, which remained for some time in the anterior incision, could be pushed into the concha and along the anterior surface of the auricle until its point could be felt under-

neath the upper margin of the helix. The acute inflammatory symptoms lasted for about two months, accompanied
by considerable pain, which was aggravated at night. Narcotics were often required for its relief. In consequence of
her suffering she lost flesh, and became nervous and irritable.

Eleven incisions were made in the auricle wherever points
of fluctuation were discovered. Carbolic-acid solutions,
and afterwards solutions of iodine, were injected into the
abscess cavities. The compressive bandage was kept quite
constantly applied: it seemed to allay the pain. Remedies
for supporting the strength were used. There was no discharge at any time from the meatus. The membrana was
normal and the hearing was but little affected. The case
was under treatment until March 27, 1879. The tragus
assumed nearly a normal appearance; but the calibre of
the meatus was diminished from permanent thickening
and enlargement of its cartilage; the deformed concha also
pushed forward and encroached on its calibre. The entire
auricle, except the lobule, had participated in the deformity, and was slightly diminished in size. January 18,
1881, the hearing is found normal; no change in the deformity, but the ear "sweats more or less all the while, and
in all temperatures," according to the patient's own statement. Dr. P. quotes from Dr. Born, who showed him a
case in which the disease seemed to be developing from an
external otitis and affecting the lower part of the concha,
as in his own case. He holds the opinion that this affection should not be confounded with hæmatoma auris. He
regrets having made so many incisions, as "they were followed by no relief to the pain, nor did they in any way
hasten the cure;" but he admits that it may be well to make
one incision into the most dependent part of the swelling to
gain access to the abscess cavity. Dr. Kipp, of Newark,
reports in the *Tr. Am. Otol. Soc.* for 1873, p. 79, a case of
what he designates as spurious othæmatoma, the result of
a burn. A boy, aged 5, was brought to the Infirmary on
account of a painful swelling of his ears. Four weeks previous his face and ears were burned by an explosion of gunpowder. Two weeks later appeared the swelling of the
ears. The auricles were pushed out from the head, and
the upper two thirds were pear-shaped. Auricles hot, and
sensitive to the touch; surfaces of the swellings convex and
smooth; skin of normal color, but looking as though it had

been blistered; fluctuation in both. Incision gave exit to
a yellowish glairy fluid. The swellings did not collapse.
In both was found a cavity the size of a hickory nut, lined
by a smooth, shining membrane. The anterior wall of
both cavities was about a line in thickness, and apparently
composed of thickened perichondrium; the posterior wall
was formed of cartilage. Tincture of iodine was applied
to the interior and exterior of the swelling, and in two
months recovery took place, without deformity, except a
wrinkled condition of the walls of the fossa helicis. He
remarks that not a single case like this is recorded; but
that spurious othæmatoma, the result of contusion of the
auricle, with or without fracture, are not unfrequently observed.

Pathogenesis.—Already a number of cases are on record
where inflammation of the tympanum has first extended to
the cartilage of the meatus externus and thence to the cartilage of the pinna. An otitis externa, with inflammation
of the cartilage of the meatus, has also involved the auricle
by extension. Again, a few cases have been recorded
where the perichondritis has commenced in the auricle,
generally in the concha, by a minute spot, where the cartilage seems to have undergone softening, by some process
of degeneration not precisely understood; it would, however, seem to be dependent on some disturbance in the nutrition of the part. A great variety of injuries to the auricle, where the cartilage has sustained violence, is sufficient
to provoke the disease. The statues of some of the renowned athletes in olden times exhibit deformed auricles,
undoubtedly the result of violence received in boxing en-
counters, and the perichondritis resulting therefrom. The
sword practice of the German university students in duel-
ling often results in perichondritis auriculæ, the ears seeming to be a favorite point of attack. Burns of the auricle
(see Kipp's case) or frost-bites (Tröltsch's case, where
a patient rested the auricle against a window-pane, to
which it froze) will develop this disease. Simple exposure
of the auricle to cold is often a sufficient provocation for
the disease. A furuncular inflammation of the auricle in
subjects lacking a normal amount of energy may lead to
the development of a perichondritis. The causes which
lead to the development of perichondritis in hæmatoma are
considered under the head of the latter.

The diagnosis is mainly from hæmatoma auris. The

latter affection is essentially a bloody tumor of the ear. As far as I have observed, no case of perichondritis has exhibited the symptom of blood effusion. It is true that the discharge from an abscess in perichondritis is often *tinged* with blood, but it never is pure blood or a blood-clot. The onset of the two diseases is also quite different; the collection in hæmatoma appears much sooner than in perichondritis, or sooner than pus is capable of forming in consequence of inflammation. It is not denied that later on a hæmatoma may result in a perichondritis, consequent on the presence of the blood effusion, but this does not place it in the same category.

There seems also no question but that a hæmatoma may run its course without resulting in a perichondritis. The swelling in hæmatoma is even, rounded, and utterly devoid of the irregular nodulations seen in perichondritis. As the lobule is unaffected in both hæmatoma and perichondritis, this fact is not of value in differentiating between these affections, but it does suffice to separate diffuse inflammation of the auricle from inflammation of its cartilage, as in the former the lobule is as readily attacked as the other portions of the ear.

The color of a hæmatoma is usually a darker red than that seen in perichondritis. In hæmatoma from injury the difference is not so well marked; indeed, in many instances it is as easy to make the diagnosis of perichondritis as of hæmatoma. If the injury has resulted in hemorrhage only, we may place it in one category; and, on the other hand, if the cartilage has received considerable injury there will be an inflammation of that tissue. The term "Spurious Othematoma," used by many writers of distinction, indicates that there has been embarrassment in making an exact diagnosis.

So far, no insane people have been observed with perichondritis from the outset. In the latter disease only one ear seems to be affected, while in the former the fellow ear may occasionally be involved; but usually this occurs subsequently to the attack on the first ear. Hæmatoma is not as painful as perichondritis. The febrile symptoms are much more prominent in perichondritis. Dr. Roosa (Textbook), p. 113, says: "Any inflammation of the integument, connective tissue, and cartilage of the auricle, leading to the effusion of serum, blood, or the formation of pus, will be apt to cause a deformity of the part; but such a case should be distinguished from an othematoma." This shows that as long ago as 1873 Dr. Roosa had doubts about placing all tumors of a character similar to othematoma in that category. He appends a cut representing " an auricle deformed by inflammation," showing a perfect lobulus, however, where an inflammation of the cartilage of the meatus had extended to the tissues of the auricle (I suppose including the cartilage). Dr. Pooley (loc. cit.) speaks of a symptom of perichondritis not mentioned by other authors, namely, a disposition of the auricle after a perichondritis to perspire much of the time, whether the temperature be high or low. This is not mentioned in any reported case of othæmatoma, as far as I know.

The prognosis as to the hearing is, I believe, always favorable. In the earlier stages the hearing may be lowered from swelling of the walls of the meatus or of the concha, which may encroach on the calibre of the meatus. The principal lesion in this affection is the distortion of the auricle, due to inflammatory changes in its cartilage. I believe my own case shows these changes as thoroughly as any I have heard of, the cartilage over a large area of the auricle being entirely destroyed by suppuration. When the cartilage is not destroyed, it sometimes takes on a proliferative action, and increases in size, becomes irregularly thickened and greatly distorted. The concha and its neighborhood is likely to suffer most, while the helix may often be unchanged. A few cases recover by resolution, without any distortion of the auricle whatever. If any considerable change has taken place in the shape of the auricle, it will also be diminished in size.

Treatment.—The cases herewith reported give valuable hints as to treatment. This being a disease of which little has been known until recently, there is still wanting a sufficient amount of experience to meet all the indications for treatment successfully. Inflammation of cartilaginous structures seems not to be as well understood as that of other tissues. In the earlier stages of the disease when the parts are swollen, hot, and tender, it is well enough to try the effect of cold applications, having previously stopped the meatus with cotton so as to prevent harm to the tympanum from the effects of the cold. A bladder or rubber bag filled with cold or iced water may be applied to the parts. Sometimes this will cause pain: then it will be well to reverse the temperature of the water, making it warm, and so regulating its temperature as to make it agreeable to the patient. We have in mind, however, the tendency of the cartilage to soften and break down with purulent formations, which may be encouraged by an excessive use of the warm applications. The irritation of the inflamed surface may be diminished by lead and opium washes, or even painting upon the part tincture of iodine. Sometimes the warm applications may be applied best by gently bathing the part. Early in the treatment it may be necessary to administer opiates internally or inject morphine hypodermically, for the prolonged sufferings of the patient may depress the vital energies to a considerable degree.

There are diverse opinions about the propriety of incisions. If the evidence of fluctuation is decided, and there appears to be a pus cavity of considerable size, there can be little doubt about the propriety of making a free incision. If, on the contrary, the fluctuation is obscure and attended by a boggy or doughy sensation imparted to the touch, and the suspected spot is not of large area, it is better not to use the knife. On this point, however, there are conflicting opinions. Sometimes it seems to do good to open these places, and again it accomplishes nothing. The affection often disappoints expectations, and seems to follow a law of its own, in its progress towards recovery.

After the incision has been made, some sort of a compress is needed, as the abscess cavity shows little disposition to close in consequence of the inelasticity of its walls; moreover, the compress often gives relief from pain. A tent of charpie may be used to keep the wound open, or, if the patient can be seen often enough, it is better to pass in a probe as frequently as the incision narrows or closes, and break up any adhesions which may have formed. This indication may also be met by the use of a drainage tube. I prefer one of soft rubber. In some instances the cases seem to do better if the abscess cavity is injected with tepid water only, and in others tincture of iodine injected in full strength or diluted does well. Carbolic acid, one drachm to the pint of water, or a ten to twenty gr. sol. of arg. nit. may do good service used as injections. If these seem to irritate, increase the pain, and add to the swelling, they should be diluted or even discontinued. Occasionally granulations may form within the abscess cavity; they may be removed by means of forceps, curette, or sharp spoon. In such cases nitrate-of-silver injections will very properly follow, or the caustic may be introduced on a piece of cotton in strong solution, to destroy any remains of the granulations.

In using the roller bandage for a compress, flannel will be found more elastic than muslin or linen, and may be preferred. A piece of cotton may be placed behind the auricle, and another in front; the piece covering an incision may be carbolized, and should be changed frequently to prevent the discharge from irritating the auricle. The roller may be passed over the top of the head and beneath the chin. After the acute stage has passed, I know of no better application to the auricle than tr. of iodine, applied at first twice a day, and after the parts become sensitive once every two or three days. Occasionally the oxide of zinc ointment will do good service, or even plain vaseline. Sometimes, however, greasy applications irritate. The auricle should always be protected from the cold. As the canal sometimes shows a tendency to narrow, it may be well in the absence of pain or inflammation to introduce a sponge-tent to resist this tendency.

Inasmuch as the affection is often tedious, and the patient may become depressed by the pain and the consequent disturbance to nutrition, it will be well in the outset to endeavor to resist this tendency by the careful regulation of the nutrition of the patient, attending to the stomach and bowels, not forgetting that the pain may depend in part upon general depression, when it will be better relieved by brandy than morphine in many instances. A variety of tonics will naturally be suggested, at the head of which stand bark and iron.

INJURIES OF THE AURICLE.

These for the most part belong to the domain of general surgery. The auricle may be bruised, frost-bitten, incised, its cartilage fractured, the lobulus divided into two equal portions by the cutting out of an ear-ring, etc. In any surgical operation for the repair of a damaged auricle from traumatism, it is well to remember that the cartilage is of low vitality, heals languidly, and does not bear the insertion of sutures well. The woodcut (Fig. 23) represents an auricle on which I performed the following operation :

It was in the case of a child who some years before had had a periostitis of the mastoid destroying the external table. This ulceration also extended to the posterior aspect of the auricle, and destroyed by ulceration so much of its cartilage as to produce a large gap, dividing the auricle into an upper and lower portion. I pared the edges of the auricle at these points, and brought them together by sutures introduced into the skin only. The engraving shows the somewhat imperfect union of the parts, and also the cicatrix over the mastoid. I have had occasion to close a lobulus which had been torn across in a vertical direction by an ear-ring. This operation is as simple as the previous one: pare the edges of the lobule where it was torn across and coaptate them by sutures. These may extend deeply into the parts, but should not remain longer than two days. Under the heading of Perichondritis Auriculæ is detailed the effect of burns upon the auricle.

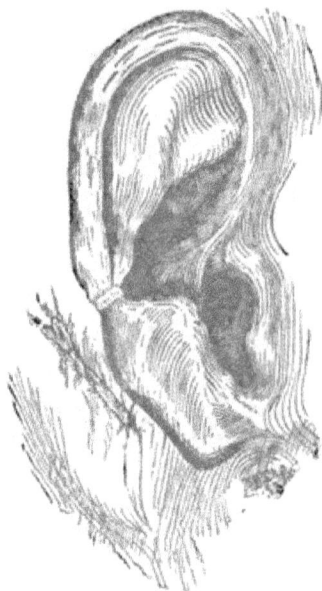

(FIG. 23.)

Injuries of the Auricle.

DISEASES OF THE EXTERNAL AUDITORY CANAL.

FOREIGN BODIES IN THE MEATUS, AUDITORIUS, EXTERNUS, AND TYMPANIC CAVITY.

Children have a strong propensity to thrust various foreign substances into their ears. Among these may be enumerated beans, Indian corn, peas, bits of chewing gum, buttons of various kinds, beads, sealing-wax, slate-pencils, nails, pebbles, gravel stones, pins, needles, shells, seeds of grain, pieces of glass, coffee beans, pieces of paper, fruit stones, shot, and a variety of other objects.

For the purpose of relieving pain in or about the ear various articles are thrust into the meatus, such as bits of camphor, cotton wool soaked in some form of anodyne, bits of tobacco, core of an onion, etc. Many objects are likely to get into the ear without the agency of the patient or his friends, as flies of various kinds; and if the ear contain any foreign substance favorable to the hatching of their eggs, the larvæ may subsequently be found in the ear acting as irritating and even painful foreign bodies. The cockroach, the earwig, bedbugs, or any other small animal that creeps or flies, may get into the ears. Sequestra of carious portions of the osseous meatus sometimes act as foreign bodies in the ear. A carious third molar tooth is reported by Politzer as resting within the meatus for forty years. Hairs are sometimes found doubled upon themselves within the meatus, and causing considerable irritation. Dr. R. F. Weir, of New York, reported a case where a hair growing from the tragus had passed inward, rested upon the membrana, and caused considerable irritation.

Foreign bodies are not by any means always likely to cause mischief in the meatus if they are not meddled with by injudicious and unsuccessful efforts at removal. Dr. Ludwig Mayer—*Monatsch. f. Ohrenheilk.*, Jahr. iv. No. 1 (Burnett's reference)—reported a case where a foreign body remained in the meatus for over sixty years without apparently doing serious harm. Where the foreign body is of such a nature as to swell, in consequence of imbibing moisture, it is likely to do harm. If the edges of the foreign body are sharp or abrupt, like a piece of rough stone, a bit of broken glass, a nail, a shell with sharp edges, or the like, there may be an inflammation occasioned even when it is not interfered with. In the larvæ of the muscidia sarcophaga and the muscidia lucilia, as described by Blake, —*Arch. Ophth. and Otol.*, vol. ii. No. 2 (Roosa's reference), the apparatus by which the grub attaches itself to the part is composed of a delicate framework of horny consistency, from which spring two hooks having the same structure as the framework. By means of these the larva burrows its way into the tissue on which it feeds, by alternately piercing and tearing. This results in agonizing pain and excites inflammation.

By far the most frequent cause of trouble from foreign bodies is the injudicious interference by ill-advised attempts at removal. As a rule the foreign body will be found not pushed beyond the middle, contracted portion of the canal, unless something has been done with a view to its removal.

In that case it may be pushed down to the membrana, and often through it, so that it rests within the tympanum itself. Some years since I had a little patient with a small pebble in the ear. Several attempts had been made to remove it, causing great pain. When I saw the child I found the stone crowded through the membrana and lying impacted in the inner tympanic wall. The results of attempts to remove a foreign body by forceps, spoons, hooks, etc., are sometimes surprisingly disappointing. From the nature of things it is apparent that any essay to remove in this manner tends to push the obstruction further in. I once attempted to dislodge a brass-headed nail from the meatus by instruments. The head of the nail was considerably wider than the meatus; it was only half-way down the canal, and I felt sure that all that was necessary was to catch it with a forceps; it was in plain sight, but on attempting to withdraw it I somehow failed to catch it properly, and each effort drove it further in, causing considerable pain to the patient. I then resorted to syringing, and to my great surprise it came out without any trouble whatever, and with little pain. I would like to formulate the following proposition: *If the foreign body has not been disturbed in its position by efforts to remove it, the syringe is nearly always the most painless, the least violent, and most effective and safe means of removal.* This proposition holds good without excluding those cases where the foreign body is swollen, or has sharp edges, like pieces of glass, or nails, or having the form of slate-pencils. It is simply surprising what power there is in a stream of water thrown into the ear for forcing out foreign bodies. *The choice of a syringe* is of great importance. There is little difference in the power of a Davidson soft rubber syringe and a good valve syringe. One point is of very great importance—to have the nozzle long, narrow, and provided with an aperture sufficiently small to throw a stream of the highest possible intensity. The syringe I have elsewhere described (see index) fulfils these indications, and the same kind of a nozzle on a Davidson syringe will do nearly as well. If no instrument-maker is near, a tip may be improvised for the Davidson syringe by drawing a glass mint-julep tube to a narrow point in a flame.

In removing a foreign body by means of the syringe the auricle should be grasped firmly and pulled upward, outward, and backward, so as to straighten the canal. Throw in a stream of pretty warm water, being careful to direct it on the outer edges of the obstruction. The nozzle of the syringe should approach very near to the object to be removed, and in some cases where the canal is swollen so as to somewhat envelop the foreign body the nozzle may even be gently insinuated at the side of the obstruction so as to be sure to allow the stream of water to pass behind it. Alternately, every side of the obstruction may be approached if necessary. When there is no doubt of there being something lodged in the ear, considerable force may safely be used. By attending carefully to these details there is not often any need to resort to other methods. I have not used such means for removing foreign bodies from the ear more than twice in a year. I am sure there is a great mistake made in using instruments for the removal of foreign bodies from the ear to the extent to which many surgeons of undoubted ability resort.

As a matter of simple justice to so high an authority as

G. TIEMANN & CO.

FIG. 24.

Gross' curette for the removal of foreign bodies from the ear.

Zaufal, I would state that in his clinique, during 1880, he had twenty cases of foreign bodies in the ear, thirteen of which were removed by syringing and seven by instruments. (Review in the *Am. Jour. Otol.*, 1882, No. 1.) Löwenberg, in the *Berlin Med. Woch.*, No. 9, 1872 (Blake's reference), relates that he extracted a foreign body from the meatus, which had resisted other efforts, by dipping a brush into joiner's glue, passing it in and allowing it to become adherent to the foreign body. After thoroughly drying, the obstruction was removed by using traction on the brush (some older authority, I have since found, has, however, used this method). Where the foreign body becomes impacted from any cause whatever, it may be necessary to resort to forceps, curettes, spoons, hooks, and the like, for its removal. The different varieties of polypus forceps of a somewhat small size may be used for this purpose. A probe, slightly bent at its extremity, and introduced at the side of the obstruction, acting as a vectis, often succeeds better than forceps. If it be grooved on its concave side it is more likely to engage with the obstruction. The curette of Prof. Gross of Philadelphia, which also has a tooth at the opposite extremity, placed at right angles to the shaft of the instrument, is well adapted to the purpose. (Fig. 24.) Sometimes Politzer's fleam-shaped knife for incising the meatus (Fig. 25), may be passed in flatwise at the margin of the foreign body, and when beyond it turned so as to impinge against it, and if it be of a material which can be incised, a cut may be made in it so as to facilitate its breaking up. It may then be removed by syringing, or it may be drawn out bodily by the knife. In all these efforts the size and direction of the canal must be recognized, namely, that the inner end of the canal is larger than the middle, that the anterior wall has a convex curvature looking backward, and that when the foreign body becomes impacted it is more likely to be found in the antero-inferior portion of the inner extremity of the canal, and on account of the greater diameter of this portion, a pocket is formed from which it is difficult to dislodge the obstruction. If a curette or hook is used it is better to pass it in the direction of the antero-inferior portion of the canal, as here the membrana is farthest from the external orifice, and the tendency of the foreign body to strike the convex curve of the anterior portion of the wall of the canal will be diminished. Where the canal is so much swollen at its outer extremity as to partly envelop the foreign body it would not be amiss to make a deep incision into it with Politzer's fleam-shaped meatus knife, or the canal may be dilated by a Gruber's bivalve speculum. When the foreign body can be penetrated the spiral end of one of Gross' instruments (Fig. 24) may be sufficiently inserted into it so as to allow of its extraction. Dr. E. D. Spear, Jr., in the *Am. Jour. Otol.*, July, 1881, recommends a novel mode of procedure in removing foreign bodies from the meatus. If the foreign body is in the outer portion of the cartilaginous meatus, "the fingers are now pressed firmly upon the skin close to and in front of the tragus, carried upward and around the meatus upon the auricle, and back again to their starting-

FIG. 25. Politzer's meatus knife.

point; then lift up, and, the manœuvre repeated several times, the foreign body will be seen to move outward, and will finally drop into the depression at the bottom of the concha. . . . . When the foreign body is lying beyond the centre of the cartilaginous portion of the canal, and has even been pressed partly into the osseous portion, the same movements of the canal will bring it nearer the entrance of the meatus, or will perhaps change its position so that other means, such as syringing, will more easily remove it." This idea seems based on the mobility of the canal, and the method is evidently a good one. I have, however, no experience with it. In removing larvæ from the ear, whether alive or dead, the hook-like appendage of the animal is so fastened into the tissues that syringing will not remove them, and it becomes necessary to pick them out with forceps. If it is desirable to kill them previous to extraction, pour in a little of the liquor sodæ chlorinat. (Roosa).

In some instances, where the canal is exceptionally narrow or excessively swollen, or where the foreign body is so large and so firmly impacted as to make it impossible to remove it by other means, the auricle may be partially separated from the canal at the junction of the osseous and cartilaginous meatus. This is an old operation, performed in various manners by Von Troeltsch, Langenbeck, Schwartze, and by Paulus Ægineta in 660. It has also been recommended by Hyrtle (J. O. Greene's references in the *Tr. Am. Otol. Soc.*, 1881). The operation is not difficult; no better method can be employed than that described by Greene (l. c.). The patient must be placed under ether. Make a semicircular incision above and behind the auricle at its insertion, which must divide the periosteum. This may be further extended so as to sever the upper and posterior portion of the meatus at the junction of the osseous and cartilaginous portions, when the osseous meatus is well uncovered. Sometimes, however, the canal does not after this operation present so decided and satisfactory an opening as might be expected or desired, but by means of some of the instruments we have already enumerated there will be no difficulty in removing the foreign body. In certain cases it might be necessary to open the mastoid antrum by the usual method and approach the meatus and tympanum from this direction. I have had considerable experience in opening into the meatus by an incision behind the auricle, which has been referred to under the head of suppurative otitis. There is no difficulty in the operation whatever. Dr. Ely, of New York, in performing an operation for deformity of the auricle, was enabled to demonstrate the feasibility of this operation. After the foreign body has been removed the auricle may be turned back into its proper position and fastened by sutures in the usual manner.

A foreign body may excite an intense degree of inflammation in the meatus and tympanum, extending perhaps to the mastoid cells, and even to the brain itself, sometimes with fatal results. This is not likely to occur, however, unless great violence is inflicted in the efforts made to remove it. Granulations and polypi are sometimes found in the meatus, as a consequence of the abrasion of the part from the presence of the foreign body. As is well known, any irritation of the meatus may produce a number of reflex symptoms. In the case of irritation dependent on the presence of a foreign body in the meatus, we may have a cough, vomiting, epileptic seizures, paralysis of the face of the same side (Meyer, quoted by Roosa), atrophy of the arm, anæsthesia of the whole of one side of the body, convulsions, etc.

IMPACTED CERUMEN.

The meatus externus, in a normal state, secretes a material of a light-yellow color, which, presenting some of the characteristics of wax, has been denominated cerumen. In some cases there is so little of this secretion that it is hardly visible to the eye; yet, on touching the wall of the meatus, the sensation communicated to the finger will leave no doubt of its presence. In other cases there will be sufficient to smear the wall of the meatus, but not enough to cause obstruction. This substance is the combined product of the ceruminous and sebaceous glands.

The cartilaginous portion of the meatus contains sebaceous glands with their downy hairs, and ceruminous glands. In the osseous portion the downy hairs with their glands are few in number, and the ceruminous glands are found only on the posterior upper wall, extending often to the membrana tympani. The ceruminous glands resemble the sweat glands in many particulars, but the secretion of the former contains coloring matter not found in that of the latter. The ceruminous and sebaceous glands, together, secrete a light yellow, rather fluid substance, which consists essentially of small and large fat globules, masses of corpuscles of coloring matter, and cells in which single globules of fat and coloring matter are imbedded. In addition, hairs and scales of epidermis from the lining of the meatus, and foreign bodies of different kinds, are also found.

When the cerumen has collected in considerable quantity and remains some time in the external passage, it changes its color, and forms from the loss of its fluid contents solid masses, (Stricker's " Manual of Histology," English translation, edited by Buck). It is, however, prevented ordinarily, from remaining unduly long in the meatus, so as to become inspissated, by the action of the outer layer of the epidermis lining the meatus. This has the property of constantly moving from within outward, thus carrying the cerumen with it out of the meatus. The time occupied in the removal of cerumen by this physiological process may be estimated at from one to three weeks. In some cases cerumen is secreted so rapidly, according to Sir William Wilde, as to simulate a discharge of pus or sero-pus from the ear. Buck also mentions this fact. This, however, is rare; for when cerumen is present in any considerable quantity, as a rule, it is of much greater consistency than when it first leaves the glands, having been secreted so slowly as to become inspissated by evaporation of its fluid contents. This will not however, always explain the phenomena of large collections of cerumen in the ear. I have in mind at present a lady who has had, perhaps, twenty collections of cerumen in her ears, and I have seen a small mass, which had collected within five or six days, which was friable, dry, and quite hard. Again, in the cases to which I shall hereafter refer, I found one where the cerumen had not been removed for eighteen or twenty years, yet was of a light color, and soft in consistency. I have a feeling which I cannot well explain, that in some diseased conditions of the glands the secretion is much thicker than normal when first secreted.

The conditions which give rise to augmented secretion in the meatus are somewhat difficult to determine. It seems well-nigh a truism, that some form of irritation exists, which excites the glands to increased secretory activity. It will be seen hereafter that in most of the cases of considerable collection of cerumen a diseased condition, accompanied by increased hyperæmia of the tympanum or canal, or both, exists; probably also with accompanying throat trouble, which I had stated in the *Trans. Amer. Otol. Soc. for* 1872, p. 62, as the probable cause of the secretion. Buck ("Diagnosis and Treatment of Ear Diseases") expresses similar views, but adds, "With regard, however, to the first step in the series of changes, I should rather favor the view which refers its origin to a reflex influence, than that which makes the irritation spread directly from the mucous membrane of the middle ear to the skin of the external auditory canal." Whether these explanations are correct or not, the fact seems to remain, that in most of the cases of considerable ceruminous collection there is disease in the throat, the tympanum, or meatus externus. In the article of my own, previously referred to, out of two hundred ears examined, most of them being impacted with cerumen, there were only seventy-seven with perfect hearing. The glands in the meatus seem to behave in quite a similar manner to those of the throat and tympanic cavity during a catarrh. In the earlier stages there is increased secretion of the mucous glands of the tympanum and throat, with more or less hypertrophy. Subsequently the opposite condition obtains: the glands shrivel, become atrophic, with greatly diminished or altogether interrupted function, as in pharyngitis sicca, or in what may be called the stage of dry catarrh of the tympanum. Usually in this stage of the disease the secretion of the glands of the meatus is just as much diminished as is that of the mucous membranes, and the result is, a meatus almost utterly devoid of cerumen, although it may have been in an earlier stage impacted with it. This latter condition has been noticed by Tröltsch (Diseases of the Ear, Roosa's translation, p. 77). In the one hundred ears actually containing cerumen, which were carefully examined, in the article above referred to, eight were associated with perforated drum membranes. This seems rather unusual, and is not easy of explanation. Roosa, in his Treatise on the Ear, explains that in some instances a hard mass of wax may so press on the membrana as to excite an inflammation sufficient to cause perforation. My impression is, that my own cases were those of old suppurative inflammation, and that the cerumen was an after symptom. Chronic eczema may also increase the secretion of cerumen. There is a certain class of patients, rather inclined to be fleshy, with a moist and perhaps greasy skin, who, over the whole body exhibit activity in the glandular secretions, being quite opposed to another class, who are thin, with a dry rough skin, and presenting little evidence of activity of the cutaneous secretions. The former type of patient is much more likely to have an excess in the ceruminous secretions, and the latter a diminution of the same, without presenting in either case any actual disease of the ear. I have, however, yet to see more than an occasional patient, whom I have watched for two or three years, with frequent collections of cerumen, who did not eventually develop some form of otitis. The means for the natural removal of the cerumen from the ears may be inadequate. Nothing is more common than to find a coal-heaver with ceruminous collections in his ears. Undoubtedly the presence of dust in the meatus, which would naturally adhere to the cerumen, is sufficient to interfere with the normal efforts at removal, and the more it accumulates the greater the difficulty of removal. Undoubtedly, lack of cleanliness has somewhat to do with these collections, but Dr. Roosa has much to say about the efforts at cleanliness sometimes preventing the removal of the cerumen, especially if the corner of a towel, or a bit of sponge attached to a holder (aurilave), is used for the purpose, thereby forcing inward any particles of cerumen about to drop from the meatus. I believe this to be frequently true. It is well to advise patients not to interfere with their ears, and if anything is amiss, to consult a surgeon. Another cause which prevents the normal removal of the cerumen, is the presence of foreign bodies in the ear. The most frequent of these are bits of cotton pushed into the meatus. The cerumen collects on the foreign body and so completely envelops it as to make it unrecognizable until it is removed. I once found a bug concealed within a mass of cerumen. Occasionally, when there is an excess of exfoliation of epidermis from the canal, the cerumen becomes entangled with it, and failing to be removed, collects in large quantities. Any narrowing of the canal, by whatever cause produced, prevents the proper escape of cerumen. Notably this is the case with elderly people, more especially women, whose tissues have become much relaxed, and the auditory canals somewhat collapsed, especially at the outer orifice, so that the cerumen becomes shut in, so to speak; and as there is a tendency of the mass to cause absorption of the soft parts, the cavity containing the cerumen grows larger, and thereby aggravates the existing condition.

Appearances of an excess of cerumen in the ear.—Ordinarily nothing is easier than to diagnosticate the presence of unusual collections of cerumen in the ear. The color is usually quite dark, and sometimes nearly black; in other instances it is gray or dark brown, and sometimes whitish in color (light gray). Rarely has it the clear, light-yellow color of the normal cerumen. *Its consistency* is almost always greater than the normal, sometimes being almost solid. A more usual condition, when it is removed *en masse*, is to be so hard as to crumble with some difficulty in the fingers. I have already alluded to the fact of its being sometimes fluid, and in one case, before referred to, where the cerumen had remained in the meatus for many years, it was quite soft. In Case 8 of my series, the cerumen was insoluble in water or any menstruum I could select, and utterly resisted the syringe. It occurred to me that a different quality of cerumen was secreted from the normal, but as it had a somewhat sandy consistency it may have been only an admixture of gritty material with the cerumen during its formation. Sometimes the mass of cerumen will be nodulated with a polished surface giving a light reflex, possibly simulating that on the membrana tympani. Occasionally a flake of dark-colored cerumen is seen covering the drum membrane, which may sometimes be mistaken for it. Occasionally a reflex is seen from the meatus, having a bright lustre like cholesterine crystals, which evidently depends on degeneration of the fatty element contained in the cerumen (Tröltsch). Before the cerumen has filled the meatus it may be seen as a dark mass closely adherent to the walls of the canal, and an opening more or less fissure-like may enable us to see some portion of the membrana. Buck (loc. cit.) points out that in some instances the cerumen only covers a mass of epidermis, consequent on desquamative inflammation of the meatus, and in others a dried and hardened mass of pus, due to suppurative inflammation of the tympanum, is encrusted with cerumen. Some of the cases I have reported, which were accompanied by suppurative disease, may have been of the latter character. The plug of cerumen, after removal, besides being mixed with hairs and epidermis, may actually be encased in the epidermic lining of the meatus, giving it a light-gray color. The appearances of the cerumen above detailed will be sufficient to settle the diagnosis.

Consequences of ceruminous collections in the ear.—The interference to the hearing from these collections depends on the following conditions: If the cerumen is not impacted nor resting on the drum-head no diminution to the hearing results. I have often seen a meatus nearly filled with cerumen without the patient's noticing anything amiss. Deafness often comes on suddenly, which may be accounted for in this manner: The canal may have been full of cerumen, except a little fissure leading to the membrana; the plug of cerumen, by a violent jolt or jar may have fallen inward, or so changed its position as to close up the fissure. In washing the ear in the morning's toilet the water may soak into the cerumen, causing it to swell and hermetically seal the meatus. A similar condition results from bathing. A thin flake of cerumen on the membrana always diminishes the hearing, and may diminish bone conduction as well.

When the cerumen presses upon the membrana, tinnitus aurium is very likely to occur, as well as vertigo, and possibly nausea and a feeling of faintness. A cough may be occasioned by impacted cerumen, or even if it smears the canal. Dr. Andrew H. Smith once brought a patient to me who had an obstinate cough which had no explanation from laryngeal or lung conditions. I found the auditory canals simply smeared with a mixture of cerumen, epidermis, and dust. On cleaning the ear carefully, the cough at once disappeared. The sympathetic relations between the ears, eyes, naso-pharynx, teeth, larynx, lungs, and stomach are well enough known. Some years since a patient at the N. Y. Eye and Ear Infirmary had an auditory canal so sensitive, that on touching a certain spot he would immediately vomit. He seemed amused at the interest he excited among the surgeons, and would smilingly allow the ear to be tickled whenever it pleased any one to test his peculiarity. One of the most remarkable consequences of ceruminous collections is related by J. Rudd Lesson, M.D., in the *London Lancet* for 1879, vol. xii. p. 833. A young woman had a bronchitis which had resisted treatment for several years. During the last two years she gradually grew hard of hearing, first in one, then in the other ear, until the deafness became complete. This resulted in an inspection of the ears, when the canals were found filled with cerumen. This was removed, and the cough suddenly ceased and the hearing became normal. In the *Tr. Am. Med. Assoc.* for 1880, Dr. S. D. Risley, of Philadelphia, contributes a couple of cases bearing upon this point. The first patient, with impacted cerumen, had some symptoms of locomotor ataxy. He described himself as feeling as though he were walking on a deeply-padded carpet, and otherwise being uncertain in his gait. The cerumen was removed, and in ten days all the symptoms disappeared. The other case was also one of ceruminous collection; the patient complained of tinnitus, loss of memory, and giddiness. So troublesome was the latter symptom that he could not continue his work. He was dull in his intellect, low-spirited, and dreaded suicide. He found by placing his finger or a piece of cotton in the ear that these symptoms would temporarily disappear. On examination a large dark mass of cerumen was found in the inner extremity of the canal. The latter had become enlarged, and the mass of cerumen was found to be loose, and would move about in every direction whenever the patient's head moved, striking the membrana with considerable violence. On removing the mass all the symptoms disappeared. I infer that the cause of the temporary relief resulting from thrusting the finger or a piece of cotton into the ear was the fact of its preventing the loose piece of cerumen from moving about and striking the drum membrane, which evidently caused all the symptoms. Drs. Roosa and Ely, in the *Archives of Otology*, N. Y., 1880, No. 1, p. 19, report a case of impacted cerumen in a student, who had the following symptoms: Great mental depression, and inability to concentrate his mind on a subject for more than a few minutes at a time. He was fearful of having to give up his studies. The right ear was affected; the hearing distance before removal of the mass was, for the watch, $\frac{9}{16}$; after the removal, $1\frac{4}{8}''$. In a few days all the symptoms disappeared, much to his delight. Epileptiform seizures have been traced to impacted cerumen.

From the record of my published cases, it would seem that suppurative inflammation of the tympanum and impacted cerumen had some connection with each other, for, out of one hundred cases of impaction eight cases of perforate membrane were found. What has been said in a previous portion of this article may explain the presence of cerumen in suppurative cases, but I think that a few of them were the result of the extication of an inflammation due to the presence of the cerumen, the latter acting as an irritating foreign body. The meatus is often damaged by the presence of the cerumen. Case 8 had not slept for three nights previous to applying for aid, on account of pain, the result of otitis externa. After the removal several large granulations were found in the canal. Case 33 exhibited redness of the meatus, as though the cerumen had caused considerable irritation. Case 51 had a reddened and excoriated meatus. In Case 70 there was redness and swelling of the auricle five days previous to the removal of the cerumen, but without pain. Case 16 had pain in the ear five days previous to the removal of the cerumen. In Case 26 the patient had caught cold and suffered from pain in the ear. Case 29 had had pain in the ear five days previous to the removal of the cerumen. It is easy to understand that a meatus filled with hardened cerumen would take on inflammation on slight provocation, and that a moderate degree of swelling would render the part more painful on account of the pressure which would rapidly ensue.

The prospect of recovery from the deafness accompanying the impaction depends upon the following considerations, to wit: If there is no disease of the ear, perfect recovery of the hearing results, as far as I have observed, or witnessed from the experience of others. One of the cases I reported in the above-mentioned series had an impaction of cerumen existing twenty years, but the patient recovered with good hearing. It must, however, be remembered that this was exceptionally soft, and less liable to do harm by pressure upon the delicate mechanism of the middle ear than hardened masses. Ordinarily there is only a little pressure on the drum membrane, which pushes it inward, but which may again be restored by Politzer's inflation; a few days at most suffices to correct the malposition. It has already been seen, that in the majority of the cases before referred to, other affections of the ear accompanied the ceruminous impaction, and the removal of the cerumen could scarcely improve the hearing because the deafness mostly depended on the disease accompanying the ceruminous collection. If the patient fails to hear a voice somewhat elevated, a foot or two from the ear before the removal of the cerumen, it is evidence that the deafness does not all depend on the cerumen, for an ear may be stopped as tightly as a finger may do it, and yet hear a voice two or three feet, and the cerumen hardly does more than this. There is one exception to this: if the membrana is pressed upon, a greater degree of deafness may result than would otherwise, and bone conduction may even be weak, contrary to the rule. A suspicion of labyrinth affection might also be raised. As far as I know, removal of the cerumen will relieve all the symptoms caused by its presence. It is stated by some that the cerumen, in its constant growth by accretion, may enlarge the meatus, by causing absorption of its soft parts, and in some instances even of the bone itself. I am extremely doubtful that bone has been absorbed by this pressure; leastways I have never seen a case where this could be verified. The membrana is often reddened after the removal of the cerumen. This is the result of two factors; simply syringing the ear with warm water is sufficient to congest the membrana. The violent removal of the mass of cerumen will cause it—and if the

G. TIEMANN & CO.

Fig. 26.—Hard rubber basin for use in syringing the ear.

plug rests on the drum head it cannot be removed without violence to the latter. The redness and erosion of the canal noticed in some of the cases I have reported may depend on the violence necessarily used in removing the impacted mass.

The treatment consists in the removal of the mass, in the gentlest manner possible.

The best method of doing this is by syringing with warm water. The latter should be a little too warm for the hand, for that temperature which is warm to the hand is cool to the ear. It is desirable to use two basins or bowls—one as a receptacle for clean water, and the other to receive what returns from the ear in the syringing (Fig. 26). Our success will depend on very careful attention to every detail of

the process. The syringe to which I have affixed a flange
and long narrow tip, I believe to be as good for the purpose
as any other (FIG. 27). A Davidson rubber syringe may,

FIG. 27.—Pomeroy's Flange Ear Syringe.

however, be used (FIG. 28), but it should be provided with
a narrower tip than that usually found upon it. A bit of
glass tubing drawn to a point in a spirit lamp will answer
this purpose very well—any surgeon can
make this for himself. To prevent the
return current from impinging on the
surgeon's person, a circular disk of
leather, with a central perforation for allowing it to pass over the nozzle of the
syringe, will answer very well. (I believe this is Dr. J. S. Prout's suggestion.)
The objection to this syringe is, that the
operator cannot as well hold the auricle while syringing—a
manœuvre of great importance. To accomplish our purpose
in the best manner, the auricle should be grasped firmly and
drawn upward, backward, and outward, so as to straighten
the meatus, and overcome a somewhat relaxed condition of
the canal, whereby it tends to collapse, and to actually
open a canal which may have wholly collapsed. The

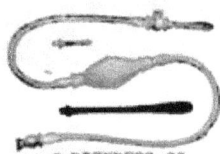

FIG. 28.—Davidson's Syringe.

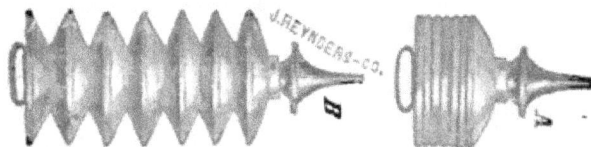

FIG. 29.—Dr. Wilson's Bag Syringe. Is also used to perform Politzer's inflation.

stream of water may be thrown in very forcibly at first. As
long as it is known that cerumen remains in the ear, considerable force may with perfect safety be used. The stream
of water should be thrown at the sides of the cerumen, and

on the canal—successively on every side. I have long practised this plan, but I have found the same description in a
book by Sir William Wilde, a man who knew so many things
about ear surgery, so long ago. My experience is, for the
cerumen to come out *en masse* after a few minutes' syringing, and often with so much force as to strike the tip of the
syringe, so as to make itself felt. In other instances the
mass becomes dislodged, and comes out to the base of the
concha, where it may be caught by forceps, curette, or
spoon. In still other instances it comes out piecemeal,
staining the water, and producing few cerumen masses. I
state my experience here very concisely, for I know how
far it diverges from that of many surgeons I cannot but
respect. For several years—at least four or five—I have
used no means but these described, and have not in a single
instance occupied more than one sitting of five minutes
or less in removing cerumen from the ear. I remember
some years since, that two or three sittings were occupied by me in accomplishing the same purpose. At the
Manhattan Eye and Ear Hospital I occasionally have
an assistant come to me with the statement that a given
patient had better be sent away with some softening
menstruum poured into the ear, and another sitting ap-
pointed. I have answered this suggestion by going myself
and syringing out the offending mass. In my 100 cases
of cerumen, one case only resisted the syringe, and recourse was had to forceps, spoons, etc., to break up the
cerumen, which was subsequently removed by the syringe.
Politzer of Vienna speaks of the syringe as being adequate
to the removal of cerumen from the meatus, in his Lehrbuch lately issued. Wilde, in his book on the ear, American
edition, Philadelphia, 1853, says: "I have, however, seldom
met a case in which, with a little care and patience, I could
not remove the wax at one sitting." I feel sure that the
power of the syringe for removing cerumen and a great
variety of other foreign bodies from the ear is not sufficiently appreciated by many of the profession. About its
harmlessness: as long as a considerable amount of cerumen remains in the ear to protect the membrane, I have
not the slightest fear of doing harm by the syringe. It is
already seen that I do not use agents to soften the cerumen
ordinarily—I believe it to be utterly unnecessary, except in
very rare cases indeed. Saturated solutions of bicarbonate
of soda are ordinarily used for this purpose, poured into

the ear, the latter being plugged with cotton to keep the fluid in contact with the cerumen. Dr. Roosa in one instance used strong nitric acid to dissolve the cerumen. Dr. Blake, of Boston, sometimes applies a saturated solution of caustic potash or liquor potassæ to a limited central portion of the cerumen, so as to saponify the fatty portion of the mass. Olive-oil and glycerine have been used. Some are in the habit of syringing out the ear with a solution of soda, a drachm or two to the pint of water. This may exert some influence on the cerumen, but much more in dissolving the oil in the packing of the valve of the syringe and rendering it unfit for use. In the very exceptional cases, picks, ear-spoons, or curettes may be used to break up the ceruminous mass, which may afterwards be removed by syringing. A sharp-pointed wire sometimes does well in breaking up the surface of the mass, which is often much harder than the interior. As I have done so little of this kind of work, I feel that my experience is of but little value. I confess I cannot possibly understand how many aural surgeons of undoubted ability can habitually use instruments for removing cerumen, which act by spooning, or mining it out. I feel that I cannot seriously argue the question. In syringing it makes a great difference whether the small and rather pointed nozzle of the syringe is pushed far enough in to nearly or quite reach the cerumen, or if there is a fissure in the mass, the nozzle should be inserted into it. Naturally the surgeon will avoid throwing the stream directly upon the mass, which would have the effect to drive it further in. After the removal of the cerumen, the patient should wear cotton wool in the ears for a few days to protect from atmospheric changes. The ear is also sensitive to loud sounds, and the cotton is an agreeable protection from these. Whatever diseased condition is found, not dependent on the presence of the cerumen, may be treated as detailed under the proper heading in other portions of this book.

Inasmuch as the cerumen in many cases persists in filling the ear again and again, something should be done to prevent this. Naturally we study out the condition on which the secretion depends, and appeal to that. I am forced to confess that I occasionally meet with a patient whose ears fill up frequently in spite of treatment. Painting tr. of iodine on the canal is as good as anything. Arg. nit. solutions are serviceable. I have a patient under my care now, a lady in middle life, who has nearly perfect hearing—as perfect as others of her age. The cerumen will collect in one or two weeks, sometimes sufficiently to be annoying to her. Nothing that I have done has succeeded in checking it more than very slightly. I am expecting her to grow hard of hearing from catarrhal otitis, but so far the hearing is perfect.

DESQUAMATIVE INFLAMMATION OF THE MEATUS AUDITORIUS EXTERNUS AND MEMBRANA TYMPANI.

This seems hardly worthy to dignify by a separate name, but inasmuch as it is frequently met with in literature, it can hardly be overlooked. It consists of accumulations of the epidermis of the canal or membrana, sometimes mixed with masses of dried pus, cerumen, or fatty particles; the latter being the result of degenerations in the collected mass (cholesteatomata, or pearly tumor; Lucæ, Buck, Kipp, and others). Some form of inflammation or at least irritation must have preceded this formation. According to Buck (*Med. Record,* Dec. 15, 1877), a condition of the canal similar to eczema or identical with it is the starting-point of the affection. According to Wette—*Monatsch. f. Ohrenheilk.*, No. 2, 1882 (*Am. Jour. Otol.*, vol. iv.),—an acute purulent inflammation of the membrana may be the starting-point of the collection. According to Kipp (*Arch. Ophthal. and Otol.*, vol. iv.), the affection seems to follow at a somewhat remote period, an acute purulent scarlatinous inflammation of the drum cavity. According to F. Graf,—*Monatsch. f. Ohrenheilk.*, No. 12, 1881 (*Arch. Otol.*, vol. xi.),—it may follow abscess of the meatus. The collection may completely fill the meatus. It has the appearance of a brownish-gray or whitish mass, with a glistening appearance, when fatty particles are visible. Where the membrana only is involved "inspection shows a grayish white appearance in the place of the membrana tympani" (Wette, l. c.). Granulations and polypi sometimes are found in the meatus. The affection may not be painful unless the epidermis accumulates sufficiently to cause pressure, when the symptoms may resemble those of impacted cerumen; as vertigo, perhaps nausea, pain in the head, tinnitus aurium, etc. Naturally the symptoms of acute inflammation of the meatus or drum membrane, when present, will be added to those already mentioned.

One important fact, specially mentioned by Buck (l. c.), is the tendency of the affection to recur, and frequently without apparent explanation. One case he mentions, in which the mass had been reproduced from every few months to a year, for four years. It then had just as strong a tendency to recurrence as at first.

The microscopic examination of these masses reveals a lamellated structure composed of flattened epithelial cells, crystals of cholesterine, fatty acids, and minute shining bodies, some round, others irregular (Kipp, l. c.). Buck's description is quite similar, but he mentions the fact that the mass contains dried pus and cerumen. Wette (l. c.) simply says, "A microscopical examination of these masses showed them to be undoubted epidermoid formations, polyhedral cells destitute of a nucleus."

The diagnosis is sometimes difficult. Many cases of impacted cerumen show an epidermic covering to the mass, or at least epidermis is mixed with the surface of the cerumen at the point at which it is visible. It is also admitted that the epidermic collections are sometimes composed in part of cerumen. One fact is valuable: it cannot be removed by syringing, as is the case with cerumen. Where the epidermis is collected on the membrana, or on the walls of the meatus, in the form of a thin whitish covering, it may resemble the macerated epidermis covering the parts in acute inflammation, or it may even resemble the lardaceous material of aspergillus. In the last two instances, however, the parts are moist, and the material is easily separated from contact with the skin. I recently saw a case which was apparently one of impacted cerumen. The ordinary means were used for removing cerumen (syringing, etc.), but failed to dislodge the mass. It had a brownish-gray appearance, was evidently composed of flakes of epidermis mixed with cerumen. It filled the canal. There was considerable pain, which had continued for some days. After a rather energetic effort to dislodge the mass which only partly succeeded, the patient disappeared, and has not again presented himself. I have no doubt but this was a true case of desquamative inflammation of the meatus auditorius externus.

Treatment.—The first indication is removal of the mass. Inasmuch as a part of the obstruction consists of fatty matter, it will be proper to effect solution of this by means of tolerably strong solutions of the bicarbonate of soda (Buck).

After this the mass must be removed by mining it out as best we may, by picks, breaking up the mass and removing it by syringing. Bent wires may be passed in, and by a vectis-like movement dislodge the mass. Curettes are useful. Gross' instrument for the removal of cerumen will be serviceable (Fig. 24, p. 75). Where a thin flake of epidermis lines the meatus a small spatula is of service to scrape it off. The same instrument bent at right angles will serve to remove the collection on the membrana. After the removal of this material the parts beneath showing signs of inflammation or irritation require attention. Buck seems to place very little reliance on treatment of this kind, but Weite (l. c.) used in one case pulverized boracic acid and cured his patient in eleven days; but it must be remembered that this was an acute affection. Painting the canal and perhaps the membrane with tr. iodine every two or three days will be as effective as anything. Strong solutions of arg. nit. promise good results.

ACUTE CIRCUMSCRIBED INFLAMMATION OF THE MEATUS AUDITORIUS EXTERNUS.—FURUNCLE.

This is often an extremely painful affection, and presents many analogies to the ordinary boil or furuncle. The inflamed point is in the cartilaginous portion of the meatus as a rule, although it occasionally may be located in the osseous portion. Furuncles are usually found in the neighborhood of the ceruminous glands, the sweat glands, or the hair follicles. It is generally believed that one of these glands becoming inflamed constitutes the trouble in question. Löwenberg, in the *Progrès Medicale*, Nos. 27 to 36, 1881, has devoted much study to the germ theory of Pasteur, as applied to the causation of furuncles in the ear. He believes that a microbe peculiar to the furuncle, and belonging to the family of the micrococci, is introduced into the orifices of these glands and causes the inflammatory attack. The boil often has a "core," which Thomas Barr, in the Glasgow *Med. Journal*, describes as being "composed of a slough of connective tissue, or a necrosed follicle or gland, around which there is more or less purulent formation." I have not often seen this "core," which most authors describe as being a pretty constant characteristic. As a rule, in the outset there is but a single furuncle, and though a considerable number may be subsequently developed, there will not be present more than one at a time. *The subjective symptoms* are at first only a little itching or a feeling of irritation in the auditory canal. Subsequently, however, this sensation develops into a pain, which may increase to an unbearable degree. As a rule, the throbbing character peculiar to tympanal disease is wanting, unless the boil is near the inner extremity of the canal, when the tympanum may become congested, and be accompanied by the characteristic pain of middle-ear disease. In some instances the side of the head and region of the jaws may be painful; any effort at deglutition causes pain. Pulling at the auricle, or any movement of the jaws causing pressure on the inflamed part, is painful. The hearing is not usually affected unless from the mechanical obstruction of the canal due to the swollen furuncle; or when the deeper parts of the canal are involved, deafness may result from the tympanal complication. *Tinnitus aurium* may sometimes be present, but is not likely to exist unless there is considerable pressure of the inflamed tissues on deeper parts or pressure from confined pus in the meatus, or from some possible tympanal complication. Sometimes the *systemic disturbances* may reach the point of high fever, delirium, and even convulsions, the patient being very ill indeed.

The appearance on inspection is not as characteristic as one could desire when attempting to make a diagnosis. Sometimes the canal is swollen so that nothing can be seen : there may be a furuncle, tender to the touch of a probe, but the excessive swelling prevents this observation from being made. More frequently, however, some point of greater swelling calls attention to the fact that a furuncle exists; if this be touched with a probe, exquisite pain may be elicited, localized swelling, tenderness, or soreness being the characteristic condition. The color of these swellings is frequently similar to that of the surrounding meatus, but may sometimes be pale, and at other times somewhat reddened. After from one to several days (according to its depth), the boil will rupture by a small opening, discharging only from one to several drops of thick cheesy-looking pus, which escapes slowly and not always giving the expected relief to pain. The site of this rupture may be bounded by rather jagged edges, from which granulations sometimes spring. The abscess cavity may present a considerable hollow, as the inelastic quality of the parts seems to prevent collapse of the walls after the evacuation of the pus. The discharge from the furuncle may so irritate the canal as to excite diffuse inflammation. When the inflammation extends deeply into the cartilaginous region of the canal, the auricle may be involved in a chondritis or perichondritis, and much mischief result. Abscess of the parotid gland sometimes results from the extension of the furuncular inflammation. The mastoid region may sometimes be involved, although this is not frequent.

The duration of the disease in favorable cases may be a little more than a week; in others, especially if furuncles recur, the patient may be tormented with the affection for months. Tröltsch, in Roosa's translation, relates the case of a man, whom for twelve years he treated for furuncles of the meatus. The boils regularly recurred, first in one, then in the other ear, at intervals of from two weeks to two months. At each attack there was febrile disturbance, and the patient was obliged to spend some days in bed.

The causes of furuncle are not always easy to determine. Facts adduced by Löwenberg and others seem to afford substantial support to the view that bacteria act as its exciting cause; this, however, can scarcely be considered as definitively established. Age and sex have an influence, adults being much more subject to the affection than the extremely youthful or the aged. Females seem to be afflicted oftener than males, especially at the climacteric period. Those of strumous habit, or where any influence has acted depressingly, are more liable to the disease. I have seen a much greater number of cases in warm weather than in cold. Again, I have seen so many cases at a given period that I could not resist the impression that an epidemic influence was at work. This opinion has been held by others. Any form of traumatism may bring on a furunculosis, as picking or scratching the ear when an itchy feeling is experienced; or an awkward attempt made to clean the ear with a bit of stick or a swab. Ear-washes of various kinds have been known to excite the affection. I am certain that a strong solution of arg. nit. will do it, and Tröltsch (loc. cit.) states that a solution of alum will also. Eczema of the meatus will sometimes result in furuncular disease. An irritating discharge from a suppurating ear may act as its exciting cause. Collections of cerumen and epidermis may cause furuncular inflammation. The general application of cold to the ear, whether of air or water, has been known to cause furuncle.

The diagnosis is not usually difficult. The absence of deafness when the canal is open and there is no discharge, with no carious teeth to explain the pain, and a localized swelling, which is tender to the touch, are conditions which make the nature of the affection certain. If there is discharge from the suspected part it must be wiped away, so as to enable the hearing to be estimated accurately. If the swelling of the walls of the canal closes it, there is no method of certainly diagnosticating the nature of the affection, unless a probe touches some tender point. When the swelling is near the membrana there will be much trouble in making a diagnosis. Sometimes there may be a localized swelling in the canal which communicates with the mastoid. If on opening it a deep fistula is found, the evidence will point to its being a mastoid complication. The finding of a minute spot in the canal which gives signs of fluctuation when touched by two probes is well-nigh diagnostic of the presence of furuncle. Sometimes a reddened and tender exostosis may simulate a furuncle, but it is of stony hardness, although it may be tender to the touch. Moreover, the fellow ear may have a similar swelling, which is not often the case in furuncle.

The prognosis is favorable. If the hearing, by reason of the furuncles developing at a point near the membrana, is somewhat lowered, it subsequently returns. The recurrence of the affection is what is most to be dreaded, and it cannot always be prevented. Witness the case previously reported from Tröltsch. Sometimes a diffuse inflammation of the meatus follows the affection, which is often of considerable duration, and may affect the hearing. If the granulations which sometimes spring from the edges of the rupture of the furuncle are allowed to grow without hindrance, the canal and the membrane become macerated from the discharge, and an inflammation of these parts may be the consequence. Occasionally the cavity of the abscess fails to close promptly, owing to the non-elasticity of its walls. When the furuncle has extended deeply into the cartilage of the meatus there is some danger of destruction of a portion of it, and even of extension to the cartilage of the auricle.

Treatment.—Before the inflammation has advanced sufficiently to result in the formation of pus, the treatment should be directed towards relieving pain. Very few furuncles result in resolution. The weight of authority seems

to be in favor of making early incisions, whether pus be
evacuated or not, the relief of tension and the depletion
resulting from the incision being the principal explanation
of the relief to the pain which frequently ensues. I do not
often make incisions of furuncles before some signs of
fluctuation appear, unless there is considerable localized
swelling. The operation is very painful, the pain continuing after the incision has been made, especially if no pus
has been evacuated. Painting the furuncle (or the tender
part of the meatus) with tr. of iodine often relieves pain,
and in some instances aborts the inflammation. Arg. nit.
in saturated solutions has more power in aborting furuncles,
but it leaves a black crust which acts as a foreign body and
may subsequently give trouble. Warmth, moist or dry, is
almost always soothing to the part. If the ear be soaked
too long with moist warmth, the canal is prepared for a
chronic diffuse inflammation, and the tendency to granula-
lations, which always exists in this affection, is greatly increased, and in some instances the moist applications aggravate pain. Dry warmth is ordinarily preferable. The
most elegant mode of employing it is to place a rubber bag
filled with warm water upon the auricle ; the temperature
of the water must not be too high ; be guided in respect
to this by the sensations of the patient—make it agreeable.
Common salt heated and placed in a bag is a good application. A readier method is the use of hot water in a bottle ;
lay the ear on the bottle, the latter being wrapped in a napkin. The leech, which is so valuable in middle-ear disease,
often fails here—it is not always easy to explain why.
From one to four may be applied as near the painful point
as possible—this will be within the canal, or near to it. If
relief is to follow their use, it is likely to occur in an hour
or two ; if not, do not repeat them. Magendie's solution of
morphine dropped into the ear may sometimes do good,
or a solution of atropine, four grs. to the ounce of water,
may be used. In the latter instance watch the pupil for
evidences of the systemic effect of the drug. The last-
named remedies are not very reliable. Often we are compelled to use morphine internally to relieve the pain. After
pus has formed there is little question of the propriety of
evacuating it. By introducing a couple of probes and
pressing on the suspected swelling, the presence of pus
may be indicated by fluctuation, or a single probe touching a point where pus is present, the tissues will sometimes be

found softer or more yielding than neighboring parts. The apex of this swelling will sometimes be found of a lighter color than adjacent parts, and possibly the skin may be macerated. My practice is to open these boils with a puncture rather than an incision. Nothing can be more painful than a long deep incision of a furuncle, and I feel that the indication for the incision is the evacuation of pus rather than the relief of pressure by extensive divisions of tissues, which is so successful in whitlows. The instrument I use is a Graefe cataract knife, which makes only a narrow puncture. The pus being thick, it frequently requires pressure with a probe to cause its evacuation. It is also a good manœuvre to pass in a small probe and break up the thick contents of the pus cavity, when its extrusion becomes much easier. This operation may need to be repeated at first every day or sometimes oftener, especially if the abscess again fills and causes a return of the pain. Another objection to a long incision is that it heals very slowly, although it does allow of the free evacuation of pus. After the incision the ear needs to be syringed frequently to keep it clean. Often it will be necessary to wipe out the canal carefully with cotton on a probe. Indeed, when the canal is full of pus, possibly exerting some pressure, a thorough cleansing with the cotton will relieve pain and add to the comfort of the patient. To do this properly it will often be necessary to draw the auricle upward, backward, and outward, when the canal, which previously may have been in a state of collapse, will gape sufficiently to enable us to cleanse it thoroughly. After a few days, if the abscess cavity fails to close, it will facilitate matters to rub well into it a strong solution of nitrate of silver by means of cotton on a holder. As to the propriety of opening the abscesses at all, the following observations made by Dr. Buck, in the *Am. Jour. of Otol.*, vol. ii., 1880, p. 28, are in point. In 17 out of 28 cases the furuncles were incised. In 8 cases, or nearly one half, decided relief was afforded. In one third of the remainder no relief was experienced, or it was at least temporary. In the 11 cases where no incision was made the stage of pain was the same as in those which were incised, namely, five to seven days. Buck explains that owing to the tubular shape of the canal there is a mechanical difficulty preventing the proper gaping of the incision. The granulations which, with some frequency, spring from the sides or bottom of the abscess cavity, consequent on languid repair, may be treated exactly as those found in the tympanum. These polyps are identical in structure with those found in the tympanum, except in their epithelial covering, which is of the laminated, flat, or trabecular variety, resembling somewhat the epidermis. I remove these by forceps if necessary, and cauterize their bases. A saturated solution of arg. nit., or the crystals of the same fused on a probe, do excellent service. It may be necessary to bend the probe at right angles, the better to properly reach the part. Fuming nitric acid may be used with the cotton on a holder. Care needs to be taken to wipe off any excess of acid, for it should never be allowed to flow upon neighboring parts where healthy tissue would be destroyed. After the patient has fairly recovered from the furuncle, the canal may still remain more or less diseased, being in a condition of diffuse inflammation. These symptoms are met by ear-washes similar to those used in suppurative otitis.

Tr. of iodine painted on the canal is very often useful, although it sometimes is rather painful. Arg. nit. is also of value applied in the same manner. The ears may be filled with a bit of cotton wool to protect them from changes in the weather for one or two weeks after recovery. Inasmuch as it is usually admitted that the constitutional condition is at the bottom of the frequent relapses, much may be done by appropriate treatment directed to this condition. A great variety of tonics are indicated, at the head of which stand quinine and iron. Any faulty condition in the system must be looked after. Often great exhaustion from overwork must be recovered from. The stomach and bowels require attention. Change of air and scene are often indicated. Sulphide of calcium may be tried with the hope of arresting the purulent formation, given in doses of from $\frac{1}{10}$th to $\frac{1}{4}$th of a grain every three or four hours. I cannot say that I have seen any great results from its use, but many physicians of the highest scientific attainments believe firmly in its efficacy. Small alterative doses of sulphur often act well.

Von Tröltsch has great confidence in Fowler's sol. of arsenic. It certainly does well in other forms of suppuration, as impetigo for instance. Iodide of potass. seems sometimes to act well. Inasmuch as the antiseptic treatment is much thought of by many, it has been thought

advisable to insert Politzer's antiseptic treatment of furuncle. It is from his *Lehrbuch der Ohrenheilkunde*, vol. ii.
p. 682.

" The introduction of the antiseptic treatment marks a
real advance in the therapeutics of furuncular disease.
While under the former methods of treatment cases of
multiple furuncle were much more frequent, and they were
not able to prevent fresh eruptions; but under the antiseptic
treatment, however, recurrences of fresh eruptions are much
more rarely observed.

" The most effective applications are carbolic acid in the
form of carbolized glycerine, 1 part to 30, to be applied with
a brush, and boracic acid in powder (Morpurgo), or alcoholic
solution, 1 part to 20 (Löwenberg), to be instilled into the
meatus.   These applications may be made either before or
after the opening of the abscess.   I have seen resolution of
a furuncle occur, after brushing it over repeatedly by these
agents, without its opening into the meatus.

" Löwenberg obtained the same result in a case where
incision could not be practised, by the instillation of an
alcoholic solution of boracic acid.   This method is, at all
events, to be preferred to the cauterization with nitrate of
silver, proposed by Wilde, or the very painful injection into
the furuncle of a five-per-cent solution of carbolic acid, recommended by Weber-Liel.

" After division of the furuncle, the incision should be
immediately brushed over with the carbolized glycerine or
boracic-acid solution, to act on the parasites and prevent
the emigration of bacteria to the neighboring follicles.
The instillation of the boracic-acid solution should be continued until the opening in the furuncle is closed by cicatrization."

DIFFUSE INFLAMMATION OF THE MEATUS AUDITORIUS EX-
TERNUS.

This affection consists of an inflammation of the cuticular lining of the meatus and subcutaneous connective tissue, and of the outer or dermoid layer of the membrana
tympani.   The inflammation in many instances, however,
extends much deeper, involving the periosteum and perichondrium of the meatus, and often attacks all the layers
of the drum membrane.

It varies in degree from the most violent suppurative inflammation of these parts with destruction of tissue, to the mildest form of inflammation of the skin, with scarcely other symptoms than a slight itching in the ear and partial desquamation of the epidermis of the canal.

The affection is very frequently secondary to inflammations of neighboring parts—as of the tympanum, auricle, various cutaneous affections of the scalp and face, etc.

The subjective symptoms, as might be expected, are extremely variable. If the disease is of the acute variety, with great swelling, especially if it is consequent on tympanal disease, excessive pain, sense of fulness and throbbing, will be present. Notably there will be pain during any movement of the jaws, or when the auricle is handled somewhat harshly. The same symptoms are also likely to accompany those varieties of the disease which depend on the exanthemata.

Deafness or tinnitus are not likely to occur unless the inflammation also involves the tympanum ; then these symptoms will be in accordance with the kind and degree of existing tympanal trouble. If the canal alone is involved, but is swollen so as to mechanically obstruct the passage of sonorous undulations to the tympanum, there will be deafness.

In milder cases, or in those which are in the sub-acute or chronic stage, there may be only an itching, and possibly a hot or burning sensation in the canal. In these there is great danger to the patient from his inclination, which is often irresistible, to introduce a variety of articles into the ear for the purpose of relieving the unpleasant sensation.

In a subsequent division of this subject it will be seen how mischievous this practice is.

Objective Symptoms.—In some instances little or nothing will be seen, because the canal is closed. Where inspection is possible, we may see the following peculiarities: In acute or sub-acute cases a discharge of pus, blood, mucus, or serum, as in suppurative otitis media, will be observed, which on being removed reveals a macerated condition of the canal and membrana, as in suppurative otitis. Here and there the epidermis may be removed so as to exhibit a reddened surface. Occasionally small polypi may be observed, but these are infrequent rather than otherwise. Sometimes the whole canal may be denuded of epidermis, and the surface exposed may be reddened and somewhat rough, with here and there minute elevations, as though granulations were about to spring up. In a later stage, when the discharge has ceased, or in those cases in which a discharge has never existed, the canal will present a scurfy appearance, with more or less redness of the surface, when the epidermis has been removed. In those cases, associated with eczema of the auricle, the canal may present an excessively scurfy appearance, with here and there fissures, which, on being roughly handled, will bleed. The appearances described in the canal are often repeated on the drum-head.

In the diffuse otitis accompanying suppurative inflammation of the tympanum, deep ulcerated grooves may frequently be seen in the lower part of the meatus, consequent on the corroding action of the discharge.

In old cases of catarrhal or proliferative otitis media the canal will look reasonably normal, except there be little or no cerumen present; and the inner extremity of the meatus may be smooth, shining, and red, with few or no signs of loosened epidermis. These reddened portions are sometimes tender to the touch, and sensitive to cold. Another phase of the disease exhibits a canal smeared with an excess of cerumen, which on removal reveals a surface more or less reddened, and somewhat tender. In a very few instances the inflammation extends outward so far as to cause some redness and œdematous swelling in the mastoid region.

The etiology may be stated as follows: Inasmuch as the disease is rarely primary, but depends on inflammation of adjacent structures, we must look to them to account for the presence of the inflammation.

A large percentage of cases of diffuse inflammation of the meatus depends on tympanal trouble, the inflammation having involved the whole of the tympanum, including the dermoid layer of the membrana; and it requires only a glance to see that the inflammation having once involved the dermoid layer of the membrana, it easily spreads to the lining of the contiguous derma of the meatus. It is well-nigh impossible to have a decided inflammation of the tympanum, whether acute or chronic, without the meatus becoming involved. It is true that in many of these cases only the inner extremity of the meatus may be inflamed. In high degrees of inflammation of the tympanum it is not uncommon to see the meatus closed by the inflammatory swelling, and in old catarrhal cases nothing is more com-

mon than to see the inner end of the meatus reddened, and
perhaps tender from a chronic diffuse inflammation, dependent wholly on the middle-ear disease. In mastoid disease it is not uncommon to see complications of diffuse
inflammation of the meatus; but these are, as a rule, rather
of the circumscribed variety, and are also often associated
with bone destruction. In furuncle, if the abscesses are
numerous and the repair is languid, a diffuse inflammation
of the canal often results.

All inflammatory affections of the auricle are liable to
involve the meatus, as erysipelas, eczema, herpes, impetigo,
etc. All the exanthematous affections are liable to pass
inward and involve the meatus. Eczema of the scalp, face,
or auricle is prone to attack the meatus, as also is herpes
of the auricle. Causes that result in inflammation of the
drum cavity often operate to produce otitis externa diffusa; as drafts of cold air upon the ear, cold water entering
the meatus in bathing or while using cold-water compresses
on the head. Usually there is no danger of cold drafts
upon the ear when in the open air, unless there is exposure
to a very high wind—for instance in riding, so that it is not
usually necessary to cover the ears for protection against
the cold. There are many irritating ear-washes which are
liable to inflame the meatus. The subject of aspergillus
will be considered in another place (see Index). This
often leaves the canal in a state of obstinate diffuse inflammation, more difficult to cure than the aspergillus itself.
Traumatism in a great variety of forms is productive of
inflammation of the meatus: familiar examples are the
practice of picking the ear with a hair-pin or a tooth-pick,
ear-spoons, aurilaves, or even thrusting the finger into the
ear to allay any itching or sense of discomfort. Collections
of cerumen and detached epidermic scales excite an otitis
externa. Irritating discharges from the middle ear also
very frequently excite inflammation of the meatus. Excessive syringing of the ears, especially if the water is very
warm, will, by macerating the part, produce otitis externa.
Poultices act in the same manner. It has long been believed that infants often have a discharge from the ear, the
consequence of otitis externa excited by dentition. I believe that in some cases this is true, being clearly a case of
reflex influence. Tröltsch states (Roosa's translation) that
specific condylomatous patches in the meatus, as well as
those of pemphigus, result in diffuse otitis externa. Most

authors agree that patients predisposed to struma are more subject to otitis externa than others.

Diagnosis and Prognosis.—It is important to determine in a given case, whether tympanal or mastoidal complications may not be of greater importance than the otitis externa, which is so frequently secondary to other and graver affections of the ear. If the major part of the disease is tympanal, the deafness, tinnitus, obstructed Eustachian tube and catarrh of the pharynx will reveal that fact. With considerable mastoid disease, a simple inspection will determine which affection is the more important. In the exanthematous variety we must not overlook the fact that serious disease of the tympanum almost always accompanies this form of the disease.

Under the heading of circumscribed otitis externa, the diagnosis has been given between that disease and the one under consideration. It is worthy of remark, that a prominent and characteristic symptom of this affection in the acute variety is tenderness in the meatus, whenever the jaws are moved or the auricle is pulled upon.

The prognosis in uncomplicated cases is always favorable. Whatever grave prognoses may be made depend altogether on the complications which may exist. It is true that the meatus is sometimes destroyed by inflammatory processes, but in such cases the disease is one commencing elsewhere, as in the tympanum or mastoid region, and is not properly a disease of the meatus. Specific affections of the meatus are promptly recovered from under appropriate treatment. The cases of this disease caused by aspergillus are quite obstinate to treatment; those arising from eczema are probably the most obstinate of all the varieties. Sometimes, though rarely, the disease may extend to the periosteum of the mastoid process, but the complication is not serious. Again, it occasionally gives rise to furuncles of the meatus. One of the annoying features of the disease is its tendency to relapse, especially on exposure to cold, or to sudden-atmospheric changes.

Treatment.—When the inflammation is acute, with great swelling and pain, especially if there is a feeling of fulness in the part, and possibly throbbing, with pain on manipulating the auricle or moving the jaws, leeching is certainly indicated. The leeches should be placed within the meatus, or as near to it as possible. If the symptoms are not urgent, douching with warm water is a good method of treatment. Clarke's douche (FIG. 30) or a substitute may be used. A competent person may douche the ear very successfully with a Davidson syringe. With this a stream may be thrown into the ear with the greatest possible gentleness. Warm water may be poured into the meatus and allowed to remain there for a time. If the inflammation threatens to involve the bone, free incisions through the inflamed tissues will be proper.

In making the incision select any narrow knife—a tenotomy knife or a Graefe's cataract knife of the wider variety, or the fleam-shaped knife suggested by Politzer (FIG. 25, p. 74)—is very serviceable. The incision should

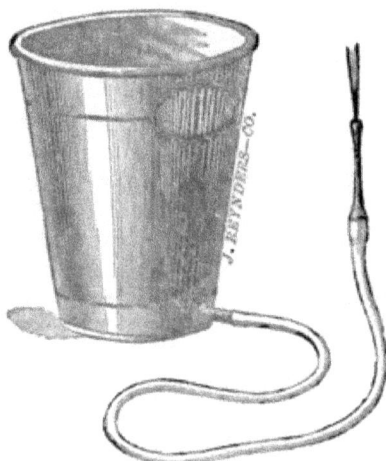

FIG. 30.—Clarke's Aural Douche.

penetrate nearly to the bone, and be of such a length as is indicated by the extent of the swelling. Afterwards, bathing or douching with warm water will encourage the bleeding and aid in dispersing the swelling. Some surgeons make two or more incisions in the canal, but ordinarily one is sufficient.

Magendie's solution of morphine may be dropped into the ear every half hour until pain subsides. It is quite safe to pour into the ear a larger quantity than would be administered internally. If the pain should resist all other modes of treatment, the morphine may be administered hypodermically or internally. If poultices are used they are admissible only for a short time. Longer use is likely to soften the canal and predispose to polypi. Anodynes may be added, if necessary, to relieve pain. Sometimes a 4-gr. solution of sulph. atrop. will act well as an anodyne. Pour a few drops into the ear every half hour, watching the pupil in the mean time for evidences of atropine poisoning. (If the pupil dilates, stop the application.) If there is considerable discharge, remove it by syringing and a little cotton wound upon a holder. Burnette, in his Treatise on the Ear, recommends cleaning the ear by aspiration, using a Siéglè's otoscope for the purpose. He has seen minute points of pus exude from the canal walls during its use, and claims that the plan is quite superior to other methods for cleansing the ear. It will be seen that the treatment of the acute stage of this affection is quite similar to that of otitis media acuta, and under that topic will be found further details of treatment bearing upon the subject. After the pain has subsided there may be considerable discharge. The astringents mentioned under the heading of acute suppurative otitis may be employed. It is, however, advisable to give considerable personal attention to cleansing the ear and making the applications. If a remedy is applied to the wall of the canal by means of cotton on a holder, especially if it is thoroughly rubbed in, it becomes much more effective than when used by instillation. The canal wall is almost always covered with detached and loosened epidermis, which effectually protects the diseased surface from the action of the remedy unless applied as directed. I believe the best remedy for arresting the discharge and subduing the inflammation to be nitrate of silver in strong solutions—from 20 to 100 grs. to the ounce of water. If it is carefully applied every second day the effect will be sufficient. If the dermoid layer of the membrana tympanis is involved, the application may also be made directly to that part. Sometimes the canal will have a soft, red, roughened look, with here and there perhaps, a nodule of granulation. In this class of cases the arg. nit. will act very satisfactorily indeed. The method of using arg. nit. solutions in this connection, as described by Buck in his "Diagnosis and Treatment of the Diseases of the Ear," N. Y., 1880, is worthy of insertion. He uses the silver nitrate by instillation, and of a strength of 60 grs. to the ounce, and even stronger; cleanse the ear, drop in the solution, and allow it to remain until warmth or throbbing is produced, then syringe it out with warm water. A single application often arrests the discharge. The stronger solution will produce warmth or throbbing sooner than the weaker solution.

I have not as much experience in the use of boracic acid in this disease as in suppurative inflammation of the tympanum. But where the meatus is denuded of its lining, and there is a tendency to granulations, it acts satisfactorily. Pack the canal with it, as directed in chronic suppurative otitis. Chloride of zinc, sulphate of zinc, acetate of lead, etc., may be brushed on the canal daily, in solutions of from 2 to 5 grs. to the ounce of water. The chloride should be used a little weaker than the others. Iodoform blown into the meatus sometimes acts well in arresting the discharge. A little balsam of peru mixed with it diminishes its unpleasant odor.

In the cases of diffuse inflammation, seen in connection with " dry catarrh," where the inner end of the meatus is reddened and perhaps tender, with few or no loosened epidermic scales, I know of nothing so serviceable as painting the canal with tincture of iodine. At first apply it very lightly, for fear of causing an intense burning pain. This may be repeated every day or two. The same treatment is also very effective in the eczematous form of this disease. I do not propose to treat of eczema of the meatus as a separate disease, having said what I desire on that subject under the heading of " Eczema of the Auricle." When the latter disease has extended to the meatus, it then practically assumes the character of diffuse inflammation of the meatus. Sometimes the ung. plumb. carb. acts well, smeared upon the meatus, although as a rule greasy substances are objectionable. They are used much less than formerly. Occasionally nitrate of silver acts well in this class of cases, but it is unpopular on account of its forming upon the meatus a black crust difficult of removal. In all of the scurfy forms of otitis externa, Fowler's or Donovan's solutions will be found valuable. Protecting the part from atmospheric vicissitudes is always indicated. This is best done by filling the meatus with cotton.

Where the otitis depends on a fungus, I shall detail the treatment under the heading of " Aspergillus." Hebra's diachylon ointment sometimes is very serviceable. Where granulations or polypi are found, treat them as detailed under the head of " Management of Aural Polypi." Specific ulcers in the canal or gummous tumors may be cauterized with strong nitric acid or acid nitrate of mercury, followed

by sub-nitrate of bismuth sprinkled on the part.   Iodide of
potass. will be indicated, and possibly mercury.   The bichlor-
ide and potass. iodid. will almost certainly be useful.   What
has been said about constitutional treatment and prophylaxis in tympanal disease is applicable in this connection.

MYRINGOMYCOSIS ASPERGILLINA (WREDEN).

This is a disease of the ear resulting from the development of a fungus upon the wall of the meatus and membrana tympani.   It resembles in most of its varieties the
blue mould found on bread and other articles of food
when exposed for a certain period in a damp atmosphere,
especially if it be in a dark place.   The principal varieties
are the aspergillus nigricans, the aspergillus flavescens,
the aspergillus glaucus, and the otomyces purpureus.
The fungus is more frequently found in a diseased ear, or
one recently affected, except where there is a purulent discharge, when it is not often seen.   Somewhat rarely it is
observed in an ear otherwise healthy.   Comparatively few
cases are reported in this country, but in some foreign regions, especially Russia, the affection seems to be of frequent occurrence.   In and about St. Petersburg, Wreden
has, during a few years, reported seventy-two cases.

The fungus occupies the meatus externus, preferably its
inner extremity, and the membrana tympani—that is, those
portions more excluded from the light, and where no hindrance to the formation of the fungus exists.   The appearance of the aspergillus nigricans is usually described as
that of a whitish lardaceous material, moist looking, a little
like macerated epidermis, and interspersed with numerous
black dots.   It adheres quite closely to the part, and often
requires to be forcibly removed by forceps or a curette, when
it leaves a surface somewhat red, but not often bleeding.
A very constant characteristic of every variety is its tendency to reappear.   From two to four days is often quite
sufficient for the fungus to again make its appearance.   In
a case of my own I found that two days after the removal
of the fungus minute hair-like stems appeared in the meatus, having brownish tops, somewhat resembling onion
stalks.   At this stage there were few or no signs of the
whitish lardaceous material usually described.   The fungus
was examined and found to be A. nigricans.

Weber, quoted by Wreden, describes the aspergillus glaucus as presenting dark-green velvet-like patches (Wreden, in the *Arch. Otol.*, vol. iv., C. H. Burnett's translation). The otomyces purpureus of Wreden has the appearance of a very red blood-clot. When removed it quickly returns, showing minute red spots, which might be mistaken for hemorrhages. Swan M. Burnett, in the *Arch. Otol. for* 1881, No. 4, p. 319, reports another case of otomyces purpureus, with a lengthy description of the microscopic appearances. J. O. Greene reports a case which he denominates aspergillus rubens, and which is evidently the same as Wreden's otomyces purpureus.

The microscopical appearances of aspergillus nigricans, accord-

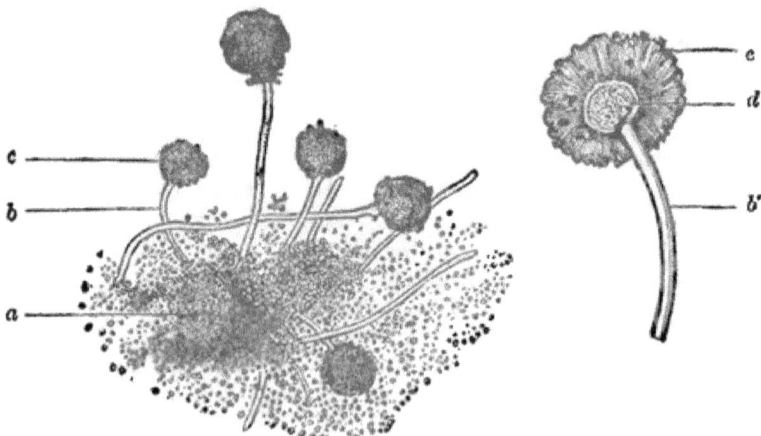

FIG. 31.

Microscopical Appearances of Aspergillus Nigricans.

ing to Politzer, in his " Lehrbuch der Ohrenheilkunde," p. 694, is as follows : " A compact felt-like mass, composed of epidermis and meshes of mycelium fibres, out of which arise perpendicularly cylindrical, tubular stems (hyphæ) *b, b'* (Fig. 31), with firm walls, which frequently have a double contour. These stems sustain the head of the fungus, which is called the sporangium (*c*). This consists of a central bladder-like expansion, the receptaculum (*d*), from which radiate long cells, sterigmata or basidia (*e*), having round spores at their free extremities. The color of different varieties of fungi depends in part on the color of the spores; those of A. nigricans being dark brown, those of

A. flavescens and A. glaucus yellowish or greenish, and those of A. fumigatus dark gray."

Wreden (loc. cit.) thus describes the microscopic appearances of the otomyces purpureus: " The mycelial layer consists of delicate, very transparent, colorless, branching, and septate rootlets, from which the stronger fructiferous hyphens arise. These fertile hyphens, with double contour, manifest at different places, like the fructiferous hyphens in the varieties of aspergillus already found in the ear, transverse septa. The width of the broadest of them is 0.00572 mm. to 0.00715 mm. The wall of the fungus—that is, the double contour or outline—is of a bright yellowish-red color, and is 0.00143 mm. thick. The fruit end of the hyphen is composed of a comparatively very large, red, round, vesicular sporangium, which consists of a thick-walled capsule and a number of round spores which completely fill the cavity of the capsule. The diameter of the larger sporangia is 0.0572 mm. to 0.06435 mm.; that of the smaller ones is 0.0014 mm. to 0.0429 mm. The thickness of the capsule wall = 0.00143 mm. to 0.00214 mm. Upon the younger, less developed sporangia we are able to distinguish, between them and their hyphens, a separation by means of a plain or somewhat arched septum. These unripe sporangia are of a brighter, yellowish-red color, have a thick wall, and are filled with a finely granular protoplasm, from which, as development advances, the round-celled spores are developed. The spores are small, bright-red (by transmitted light), round cells of a diameter varying from 0.00286 mm. to 0.00429 mm., which before germination show only a simple smooth contour and bright-red homogeneous contents. The germinating spores, on the other hand, show a distinctly double contour or outline, and a dark, eccentric nucleus, the spores having now attained a diameter of 0.00715 mm. to 0.00858 mm. After the rupture of the ripe sporangium the spores pass into the open air and distribute themselves over the surrounding neighborhood."

Subjective Symptoms.—These are usually stated to be itching, smarting, burning, a sense of fulness, pain, tinnitus aurium, vertigo, and diminution of the hearing, and after a day or two a serous discharge. Out of seventy-four cases, Wreden (loc. cit.) found twelve in which there was no pain nor itching, and four in which there was no pain nor tinnitus aurium. In the *Tr. Am. Otol. Society for* 1869, Dr. J. O. Greene graphically describes some of the symptoms of aspergillus as it appeared in his own ears. He says: "In August I felt some ill-defined irritation in the ears, attended by a slight serous discharge, just enough to be felt with the fingers; this, however, soon ceased without any treatment, and was only recalled to mind on a return of the trouble in the next November, when I noticed in both ears the same slight serous discharge, with pricking, itching, and occasional slight pain and feeling of fulness in the ears. The hearing was found by the watch to be somewhat impaired." The pain is not liable to continue more than a few hours, or a day or two, when the discharge makes its appearance, and the membrane is not so closely adherent to the parts as at first. In some cases, however, especially when the canal is nearly or quite full of the fungus, the pain may continue until it is removed. If the symptoms all depend on the presence of the fungus, removal of the latter disposes of them, but very frequently the hardness of hearing and tinnitus will depend on some antecedent disease.

The consequences of aspergillus in the ear are ordinarily not serious. It often induces an inflammation of the dermal lining of the meatus and membrana. Infrequently it extends to the subcutaneous connective tissue. The diffuse otitis externa resulting from the fungus is sometimes long in disappearing. Occasionally, according to Wreden, an otitis externa circumscripta may be a consequence of the fungus. Politzer, quoted by Wreden, has observed the fungus to involve the tympanum itself, and result in serious complications. It is the expressed belief of Wreden that the fungus cannot remain long in the meatus without doing harm.

The diagnosis is made almost altogether by the microscope.

The causes of the formation of the fungus in the ear are twofold: sporules of the aspergillus may be floating in the air and come in contact with the meatus, or this contact may be accomplished by using a speculum which previously has been inserted into an ear affected with aspergillus; and secondly, the ear must be in a condition to provide a lodgment for these spores. Dried or decomposed products of inflammation in the canal and on the membrana, or the desquamated epidermis accompanying eczema, furnish favorable conditions for the development of the fungus. Oils, ointments, salves, etc., when used in the ear, may decompose and furnish a favorable nidus for the reception of the fungi. If the ear is frequently cleansed there is less liability of the parasite to effect a lodgment. Dampness favors their development; this may account for the fact that many cases have been reported from Russia, where so large a number of the people live during a long winter in damp, often filthy and ill-ventilated houses. Steudener and De Barry, quoted by Wreden (l. c.), state that "aural fungi will be developed on any dead organic substance which accumulates in the auditory meatus or upon the membrana tympani, and that, finally, when they are thus developed upon the membrana tympani they will act like foreign bodies and excite an inflammation in this delicate structure."

The influence of age and sex in the development of the fungus is shown in Wreden's tables. Out of his seventy-four cases fifty-one were males and twenty-three were females. The youngest patient was 13 years old; the oldest was 69. He has never found myringomycosis in children, and rarely in the very aged. Fifty-two out of the seventy-four cases were affected in one ear only, and eighteen in both. Of the latter, only six had the fungus in both ears at first; the remaining twelve had the fungus in only one ear, and not until from two to four weeks did it appear in the other. He has never found but a single form of fungus in both ears of a given patient.

Treatment.—The first thing to be accomplished in a case of aspergillus, is the removal of the fungus. If it has recently formed, it is likely to adhere so closely to the part that it requires forceps or a curette to remove it. It is not always indicated to make special efforts, however, to do so, for there is more likely to be pain consequent on the removal than later on; and the forcible removal of the fungus may aggravate any existing inflammatory symptoms. A little later on the ear begins to discharge a thin serous fluid, and the membrane has become somewhat loosened, when it may frequently be removed by syringing. I do not on the whole believe that syringing the ear with warm water and removing the fungus will always result in cure, although many high authorities assert to the contrary. Indeed, too much soaking of the ear may furnish a favorable soil, on which future fungi may flourish. The remedy first recommended by Wreden still holds its ground—the hypochlorate of lime. It may be used in the strength of two grains to the ounce of water, two or three times a day, by instillation. It is better to make the solution at the time it is used, on account of the risk of decomposition. It is un-irritating, and is very agreeable to the patient. Carbolic acid may be also used as a parasiticide. Twelve to fifteen grains to the ounce of water may be used by instillation several times a day. Dr. J. O. Greene found it quite irritating in his own case. Dr. C. H. Burnett, in his text-book on the Ear, expresses great confidence in alcohol, either absolute or diluted, to destroy the fungi.

My own practice has been to paint the canal and membrane, after the fungus has been removed, with a strong solution of arg. nit.—from 30 to 60 grains to the ounce. When applied to the canal, a saturated solution may sometimes be used. I believe it to be a good parasiticide, and what is better, the most effective remedy for the diffuse inflammation of the canal, which is such an important element in most cases. Apply it every two or three days. Iodoform freely sprinkled on the canal and membrane frequently acts well. Tröltsch (J. O. Green's translation) recommends permanganate of potash, in, from a 2 to 4 grain solution to the ounce of water. It is better to make it fresh. One of the objections to this remedy is that it is somewhat irritating, and may add to a pre-existing inflammation of the canal. Tr. of iodine painted lightly on the canal after the fungus is removed is a good remedy, and acts well on any diffuse inflammation of the canal. Apply every day or two; if it causes pain, syringe with tepid water.

Wreden has used Fowler's solution locally with good effect.

After making an application to the canal, the ear should be protected from the air by carbolized or plain cotton wool. It may also prevent any spores of the fungus, which may be floating in the atmosphere, from depositing on the canal. Buck, in his "Diagnosis and Treatment of Ear Diseases," objects to syringing the ear for some time after an application has been made, preferring that a scab or scurf should form, and remain a few days to protect the part from further attacks of the fungus. He thinks the syringing often causes pain and discharge. After the fungus has been completely destroyed, there still may remain much to be done; the canal and membrane may be reddened and tender, and accompanied by a discharge, or there may be an amount of hyperæmia and pain sufficient to require leeches and hot applications.

The Eustachian tube may be more or less impervious, and naso-pharyngeal catarrh, so frequently accompanying this affection, may require attention. This class of cases may be referred to their appropriate heading, in the Index of this book, for further details in treatment. If the fungus seems to be developed from the surroundings of the patient, let him be removed to a more salubrious locality, or let his room, bedding, and furniture be thoroughly cleaned, and the place put in such a state that no more spores of the fungi may be floating in the atmosphere of his apartment.

EXOSTOSIS AND HYPEROSTOSIS OF THE MEATUS AUDITORIUS EXTERNUS.

These growths in the osseous meatus are variously described as circumscribed bony tumors of the meatus (exostoses), or as simply hypertrophies of the same, with a more or less regular narrowing of the meatus (hyperostoses). The exostoses are usually described as having a base as narrow as the rest of the growth, or often much narrower, being decidedly pedunculated. Of this form there may be from one to three or four in the meatus. Mr. Field, in the London *Lancet* for April 1, 1882, notices that where three or four of these nearly occlude the meatus a somewhat triangular fissure is left near the centre of the obstruction. In the case of simple hypertrophy of the bony walls, the narrowing of the meatus may be quite regular. The hyperostoses seem oftener to be found on the upper wall of the osseous meatus. Very frequently a bony growth will be found in both ears, but more developed in one than in the other. The surface of these tumors is covered with a very thin skin, and often this is surmounted by epidermis, which may have sufficiently accumulated to become obstructive. Ordinarily the tumors are described as being quite free from sensibility, but there are many exceptions to this rule—redness and tenderness being often noticed. These growths, it is true, cannot by any means be accounted for in numerous instances; frequently, however, nothing is easier than to recognize them as consequent on a previous inflammation of the middle ear, either purulent or catarrhal. In some instances this inflammatory proliferation seems to result in the formation of condensed connective tissue, previous to the ultimate osseous development. My own case of polypus of the meatus having a cartilaginous and osseous base, published in the *Tr. Am. Otol. Soc. for* 1874, p. 541, will illustrate this. The case is as follows: Miss. T., æt. 15 years; chron. supp. otitis in both ears from scarlatina ten years previous. Discharge continued with brief intermissions until the present time. A mucous polypus in the right ear was removed and did not return. The left meatus was filled to within three or four lines of the concha with a polypoid material, which, however, was of greater firmness deep in the meatus, as became evident on touching it with a probe. The soft superficial surface was removed by forceps, when a cartilaginous material was found. This, together with cancellous bone, was removed by bone forceps to about half the depth of the meatus. Five days after the operation there was great swelling of the meatus and of the mastoid region, with some constitutional disturbance. After ten years the condition remains about as it was two months after the operation. The hearing was greatly improved by the operation.

Another case (to be hereafter alluded to), reported by A. Hedinger, M.D., of Stuttgart, also illustrates this variety of osseous tumor. The patient was a switchman; always had been healthy, and formerly had good hearing; for the last eighteen months has had suppurative otitis, which latterly has been very painful. On examination, diffuse swelling of the parts around the ear and rise in temperature were observed. Hearing and bone conduction almost gone. In the right ear there was a new formation, which the patient states has made its appearance since the suppuration commenced. This fully occluded the meatus, and was of bony hardness to the touch of the probe. "A dense fibrous tissue closely enveloped the osseous nucleus, and was connected with its surroundings by cords." Chiselling aided in removing a part of this tumor, as also did the laminaria, to be referred to hereafter. It would seem almost probable that some of these hyperostoses dependent on suppuration of the ear, may have commenced as granulations or polypi. In Dr. Roosa's "Treatise on the Diseases of the Ear" appear two cases, under the head of Exostoses, which further illustrate this subject. In Case III., p. 406, is this statement: "A gelatinous growth from the meatus was removed by torsion. It was found to have its origin from a general bony expansion of the meatus." In Case V., p.

407, appears this statement: "A gelatinous polypus was found attached to the hypertrophic posterior wall of the auditory canal." Dr. Cox, at a meeting of the New York Ophthalmological Society, presented several rounded polyp-shaped bodies of bony hardness, which had been removed from the meatus of an ear, with a history of previous suppuration. These were found on examination to be of the nature of true bone. Schwartze, in his " Pathological Anatomy of the Ear," (Greene's translation), says: " Exostoses, congenital or acquired, pedunculated or with a broad base, spongy or eburnated, are found" (in the meatus). "The eburnated may be developed from the spongy variety, and perhaps *vice versâ*. Both are only different stages of development of the same process." He explains the presence of polypi or granulations in a manner somewhat different from ours, as follows: " From the pressure of the exostoses against the opposite wall, painful inflammation of the meatus with the formation of granulations may take place." Dr. James Patterson Cassells, of Glasgow, in the *Brit. Med. Jour.*, December 15, 1877, seems to coincide with an opinion already expressed in these pages. He speaks of the origin of exostoses in this wise: A sub-periosteal abscess of the mastoid bursts into the meatus; granulations sprout from the opening of the abscess, and by the gradual conversion of their cells into bone cells, develop the true exostoses. This process seems analogous to the ossific degeneration of inflammatory proliferations occurring within the eye.

Etiology.—Numerous causes have been assigned to account for the presence of bony growths in the auditory meatus externus. Gout, rheumatism, rachitis, and syphilis have been stated as causes. More recently, purulent inflammations of the middle ear have been ascribed as the cause in a majority of instances. Roosa, in his " Treatise on the Diseases of the Ear," has very ably presented this phase of the question. I make no doubt but that in many instances the tumors may be traced with some directness to a previous purulent or catarrhal inflammation of the tympanum. In a case (alluded to under the heading of treatment) reported by Dr. Hedinger, he states that " it is evident that the tumor was a consecutive one, and arose in the course of an inflammation of the mucous membrane of the middle ear," although he states in the opening of the article that the " pathogenesis of these growths is as yet entirely unknown, and scarcely rises above the level of hypothesis." He, however, regards the case under consideration as somewhat exceptional. I am strongly of the opinion that in a large number of instances no possible cause can be ascribed for the presence of these tumors. Dr. Knapp in a discussion at the International Medical Congress in London, in 1881, spoke of one case where syphilis probably produced an exostosis. On the whole, the profession is inclined to discard syphilis as a cause of exostoses, together with rheumatism, gout, and rachitis. Much has been written on the presence of exostoses in the auditory canals of prehistoric races and of the American Indians. These occur with some frequency, and various hypotheses have been constructed to explain the phenomenon, but, on the whole, rather unsuccessfully.

Treatment.—Ordinarily, little needs to be done. If a suppuration of the ear is keeping up the irritation on which the existence and growth of the exostosis depends, it should be attended to. Usually the osseous growth is devoid of great sensibility, but occasionally it will be tender, somewhat painful, and the skin covering it will be red. For this condition it may be well to paint tincture of iodine or nitrate of silver solution upon it. *An attempt at removal*, or making an opening, may be made when both ears are so involved as to seriously impair the hearing. If there is one ear with good hearing, it would not be proper to operate on the other. Where there is pain, tinnitus, and vertigo, with any symptom of cerebral irritation or inflammation, an operation may be undertaken. If there is a discharge which fails to find a ready exit from the ear, with or without symptoms of intra-tympanal or intra-cranial pressure, an operation is justifiable. An opening into the mastoid antrum and effecting a communication with the tympanum, may be made, if it is not practicable to operate on the bony growths. It is true that an Eustachian tube of normal size may succeed, by frequent Politzerizations, in evacuating the tympanum of any collection, if it is not of too great consistency. But the tubes very frequently, in this class of cases, become obstructed, and then fail to evacuate the tympanum. A great variety of operations for the removal of the whole or a part of these growths has been devised.

Dr. Arthur Mathewson, in "The Transactions of the First International Otological Congress," refers to some of the means which have been used to remove exostoses, as follows: A case reported by Tröltsch had a laminaria bougie introduced for the purpose of dilating the meatus. It could not be removed until two months had expired, when small sequestra came away, enlarging the canal sufficiently to restore the hearing. He quotes a case of Dr. Roosa's, in which Toynbee attempted dilatation with bougies, but they caused great pain and effected nothing. The patient subsequently died from retention of pus. The first reported case of surgical operation on exostosis of the auditory meatus was by Bonnafont in *L'Union Médicale*, May, 1868. The growth filled the meatus and obstructed the hearing. The soft parts over the tumor were destroyed by five or six daily applications of nitrate of silver, and a fine rattailed file was used to bore through the bone. It was only after the fourth sitting, that he succeeded in gaining a starting-point for the end of the file. The boring was continued for ten days. After each sitting a whalebone probe was introduced to maintain the opening gained. The opening finally made remained for some years after, with satisfactory improvement to the hearing. Dr. L. B. Hamburg, in the *Archiv. für Ohrenheilkunde*, vol. x. p. 110, relates his own case as follows : He had deafness and tinnitus from exostoses in the meatus. A drill was used by Dr. Knorre daily for four days, which caused much pain from the slipping of the instrument during the operation, and inflammatory reaction of the meatus. Two days from the last attempt with the drill a chisel and hammer were used, which caused severe headache. After this, forceps and drills were used for ten more sittings, bringing away small pieces of bone of cancellated structure; the operation was attended with great pain. Muriatic and sulphuric acids were next applied to the tumor two or three times a day for eight weeks, with small effect. The actual cautery was applied several times. There was then a cessation of treatment for a time, and after the swelling of the soft parts had subsided a probe could be passed between the wall of the meatus and the tumor. Soon after this the doctor continued the operation on himself by filing away the growth with a small blunt file, roughened on one side only. He succeeded in making an opening sufficient to restore the hearing and relieve the tinnitus.

Prof. Heinicke operated on two exostoses in the same patient by means of a gouge three lines in breadth, propelled by blows from a hammer. In one of the cases there was great pain in the ear and a sense of pressure in the occiput on the third or fourth day after the operation. Both cases, however, were successful.

Dr. Mathewson's own case was briefly this: Miss M., aged 25—the case was previously reported by Dr. Roosa (loc. cit.). The tumor arose from the posterior portion of the osseous canal of the right ear, and nearly occluded it. A No. 2 Bowman's probe was passed between the wall of the meatus and the tumor into the cavity of the tympanum. The growth was slightly movable. Dr. Loring passed in a scissors blade, and by a boring motion somewhat enlarged the opening, which resulted in improved hearing.

The patient went on without serious symptoms for nearly three years, when she experienced a sense of pressure in the head, attacks of loss of consciousness, and other cerebral symptoms, which recurred at intervals for three or four months. Dr. Loring saw her at intervals, and proposed the operation which was subsequently performed. On account of her removal from New York to Brooklyn, Dr. Loring referred her case to Dr. M., who thus describes her condition: "I found the meatus nearly occluded by the exostosis; . . . the growth had become immovable, and had evidently increased somewhat since the time of Dr. Roosa's report." No discharge; hearing much impaired. Fearing a fatal termination, the symptoms seeming threatening, an operation was determined upon. Elliott's suspension dental engine was employed to propel the drills used. The patient was placed under the influence of ether. The integument over the tumor was removed by a dental instrument known as a scaler, by being circumscribed and scraped off with it. The bony growth was perforated at several points near its centre with the smallest of the drills—about one and a half mm. in diameter. The growth was eburnated and excessively hard. Drills of nearly twice the diameter of the first were then used to enlarge the perforations already made, and cause them to run together. Lateral pressure was made to ream out the meatus. The bleeding was so excessive that probes were used to guide the direction of the drilling. The operation was continued until a drill of about 3 mm. in diameter was used, making an opening through the depth of the bone somewhat larger than its diameter. Syringing and swabbing out with styptic cotton were practiced during the operation. The duration of the latter was from twenty to thirty minutes. The hemorrhage was not

troublesome. The pain subsequently experienced was sub-
dued by the warm-water douche and moderate doses of
opiates.

A purulent discharge soon came on. For weeks after the
operation the meatus was so nearly filled with swollen and
granulating soft tissue that the membrana could not be
seen. Gradually the swelling disappeared, and an opening
nearly the size of the meatus appeared, except that at one
point, where there was a thin remnant of the exostosis projecting from the anterior upper wall of the meatus, a portion of the membrana became visible. The discharge ceased
after using astringents and nitrate of silver; no irritation
nor unpleasant symptoms; hearing nearly normal. Dr. M.
takes occasion to remark on the peculiar action of the
drills—and I can verify it by my own experience—that they
are perfectly manageable, require very little pressure to
cause them to operate with considerable rapidity, and the
hardest bone is quickly cut away without violence.

This is in strong contrast to the chisel, gouge, and hammer method, and seems altogether admirable. The engine
herewith alluded to may be obtained of Johnston Brothers,
812 Broadway, New York, together with a great variety of
drills for operating. Dr. George P. Field, M.R.C.S., has
operated according to the Mathewson method many times.
In the *Lancet* for April 1, 1882, as well as in other numbers,
he has reported cases. In one he operated five times, and
in another six. In two cases, where the growths had a distinct pedicle, they were broken off by a pair of stump forceps, such as dentists use for the upper jaw. Mr. Field
gives some valuable details of his operations. At least
three assistants are needed. The hemorrhage is a source
of great embarrassment, and requires one assistant for that
alone. He uses an iron guard, which he passes behind the
tumor, so as to give a hint as to how far the work has pro-
gressed, and to protect the tympanum. This has to be
made to fit any given tumor. He first selects a thin piece
of copper, and when the proper shape is determined upon,
duplicates it in iron; he formerly used steel, but found it
liable to break. He does not cut through the skin, but
penetrates it by means of the drill. He uses a small drill
at first, and enlarges this opening by one of greater size.
The operation is regarded by him as one of great difficulty
and some gravity, and he is in the habit of working very
slowly and cautiously. In one instance he evidently

wounded the facial nerve in the hiatus fallopii, but the resulting paresis was soon recovered from.   All of his cases
seem to have done well.   In a case of exostosis, which evidently depended on suppurative inflammation of the middle
ear, Dr. A. Hedinger, of Stuttgart, in the *Arch. Otol.*, New
York, 1881, No. 1, translation by Furst, used the laminaria
tent, made from the root of the plant, to dilate the canal.
It was introduced daily.   This ostosis was, however, not
wholly composed of bone, the diagnosis of the growth being
an "*inflammatory proliferation of the papillæ and of the connective tissue, with deposition of lime within it (osteoid metamorphosis).*"   In this case, also, the growth was accompanied by
purulent inflammation in its neighborhood, and a considerable swelling in the mastoid region was incised, with
evacuation of pus.   Granulation tissue was also present.
A. E. Cumberbatch, in *St. Barthol. Hosp. Reports for* 1880,
relates two cases of exostosis in which the tumors scarcely
filled the meatus, but an increase in the thickness of their
cuticular covering resulted in complete occlusion.   Both
cases were relieved by destroying this thickened covering
of the tumor by means of nitric acid.   This rendered
the meatus sufficiently pervious.   He raises the question
that exostoses rarely completely close the canal, except
from the increase in thickness of the integument covering
them.   In the case of Dr. Mathewson, it will be remembered that Dr. Loring had previously introduced a blade of
a pair of scissors, and had succeeded by this means in removing a sufficiency of soft tissue covering the exostosis to
open the canal.   Subsequent growth of the tumor, however, again obliterated it.

INSTRUMENTS FOR THE EXAMINATION OF THE THROAT AND NARES.

A tongue-depressor or spatula is the first requisite.   By
the following wood-cuts it will be seen that a great variety
are in use.   When the posterior wall of the pharynx, uvula,
and lateral half arches are to be inspected, a spatula resembling the handle of a dessert spoon is a very convenient
form.   A perfectly straight spatula is often well adapted to
this work.   The pocket folding spatula (Fig. 32), fulfils this
indication very well; so also does the hard rubber hinged
spatula (Fig. 33).   While examining the throat, if the cheek

is drawn out laterally, a better view may be obtained of the pharynx. Those spatulæ made with the handle at right angles to the depressor have many advantages,

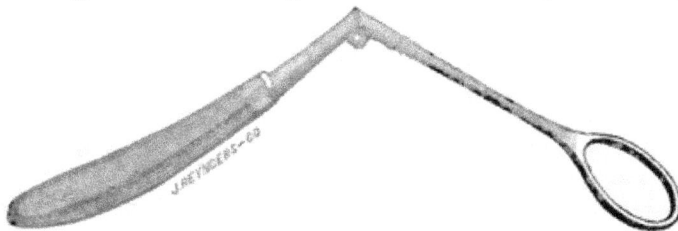

FIG. 32.—The Pocket Folding Spatula.

especially in rhinoscopic work. One objection to them is that the patient often throws the chin downward, when the

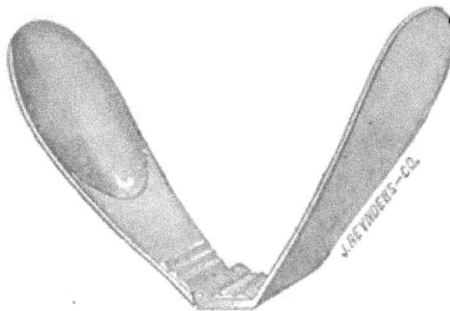

FIG. 33.—Hard Rubber Hinged Spatula.

handle will fall against the breast and embarrass the examiner. While doing posterior rhinoscopy they are a real

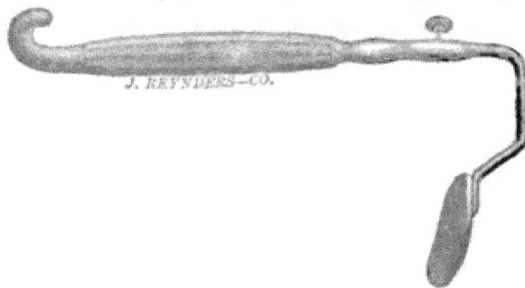

FIG. 34.—Tuerck's Tongue Spatula.

advantage, as the patient may depress his own tongue by their aid. The best instrument for this purpose is Tuerck's (Fig. 34), which is provided with three tongue-depressors of different sizes, which fit into the handle. By the upward bending of the stem attached to the depressor the instrument is lifted from the teeth, against which the ordinary spatula is so likely to impinge. Fig. 35 represents the right-angled folding spatula, which may easily be carried in the pocket. Fig. 36 represents Sass's spatula, which is a very elegant instrument indeed. For posterior rhinoscopy a small laryngeal mirror is used. The mirror here figured (Fig. 37), which may be denominated a rhinoscopic mirror, is of full size for the purpose. Fig. 38 represents two groups of mirrors useful for examination. In the division containing circular mirrors, Nos. 1, 2, and 3 may be selected for rhinoscopy. The oblong mirrors may be used in some

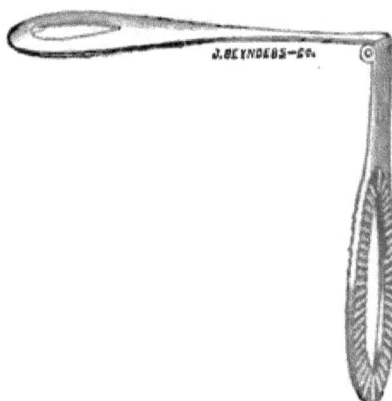

FIG. 35.—Right-Angled Folding Spatula.

cases where the pharyngeal orifice is narrowed by encroaching tonsils, or from other causes. For rhinoscopic examinations it is not absolutely essential to use other than a good gas-burner or coal-oil light; but if some form of condenser is needed, the one here figured, and known as McKenzie's, is as good as any (Fig. 39). Instead of using the large mirror after the manner of Tobold, as here represented, it is much better to use it placed upon the forehead, as previously described. Where gas is used, the universal bracket of Mitchell, Vance & Co., of New York, is by all comparison the best. The German student's lamp is the best arrangement where oil is used. In Vienna, Schrötter uses the ordinary argand gas-burner, without a condensing lens of any kind.

In using the rhinoscope, first warm it in the lamp; then

place it against the examiner's cheek to determine whether
it be of the proper temperature (sufficiently warm to prevent moisture from collecting on it); then, with the patient's

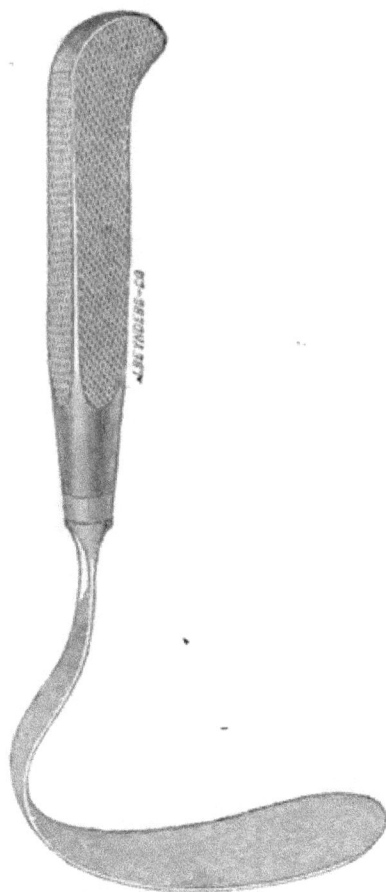

FIG. 36.—Sass's Spatula.   FIG. 37.—Rhinoscopic Mirror.

tongue depressed with a spatula, perhaps held by himself,
or drawn out of the mouth by the patient's catching the end
of it in the corner of a towel, pass the mirror into the posterior pharynx, if possible, not touching any portion of it
for fear of exciting reflex contractions of the parts, when a

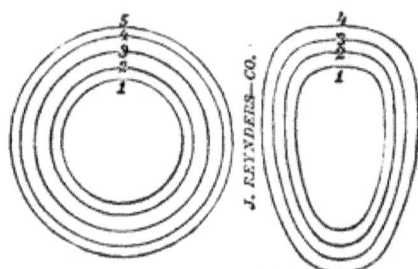

FIG. 38.—Rhinoscopic Mirrors of different Sizes and Shapes.

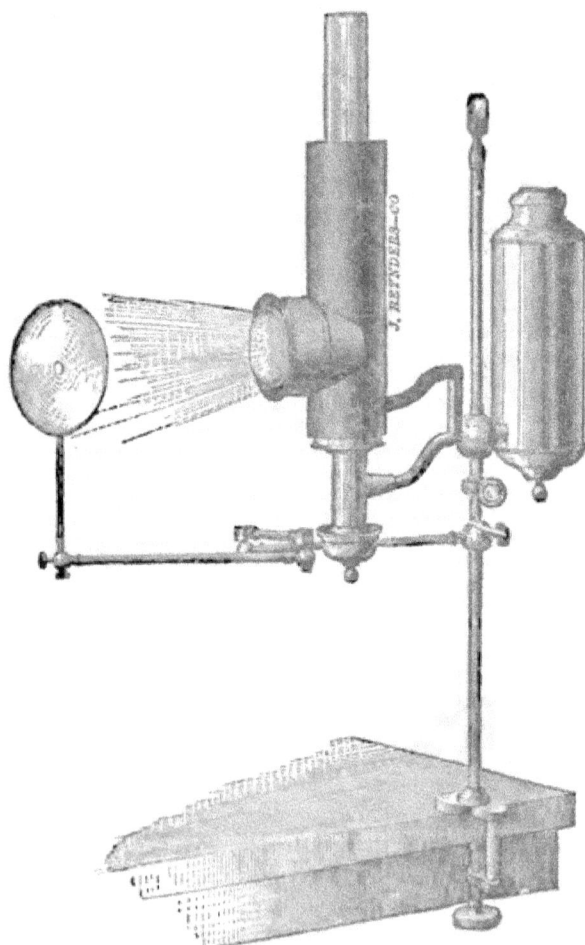

FIG. 39.—McKenzie's Condenser.

view of the upper pharynx, region of the Eustachian tubes, posterior nares, etc., will be gained. In some instances the mirror may gently rest on the base of the tongue, without provoking undue irritation. When the velum does not fall into proper position, direct the patient to say Ah in a prolonged manner. If the velum falls spasmodically against the posterior wall of the pharynx, a good procedure is to cause the patient to breathe through the nostrils. This may be practised by the patient previous to the attempt at examination, and should be done with the mouth wide open. It is not often necessary to use hooks to keep the uvula out of the way. A rhinoscopic examination posteriorly may be made easily without hooks, etc., to manage the velum, or it can scarcely be done well at all where

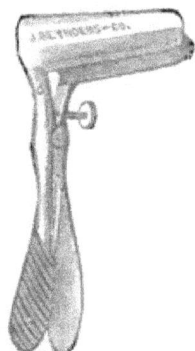

FIG. 40.—Roth's Bivalve Speculum.

violent coercive measures are necessary. The silvering on the mirror may soon be destroyed by carelessly wetting it in cleansing and insufficient drying, or by an excess of heating of the mirror in the lamp flame. It is a good rule to make the examination as quickly as possible, for the patient soon becomes fatigued and loses self-control, and consequently does not respond properly to directions given him.

In Anterior Rhinoscopy the following instruments are more frequently used. The old Gruber bivalve speculum is a valuable means of dilating the nostrils. Theoretically the dilatation should be made horizontally, but in practice we frequently find that the opposite direction of the dilatation—that is, the vertical—will expose a larger surface to view. Roth's bivalve speculum (Fig. 40) is similar to Gruber's, and answers a good purpose. Some years since I picked up in a shop in Berlin a speculum I have used very much since, with great satisfaction. It is Fraenkel's (Fig. 41). The screw seen at the end of the instru-

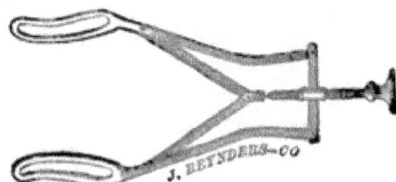

FIG. 41.—Fraenkel's Nasal Speculum.

ment operates with great rapidity, so that when inserted into the nostril it may be removed almost instantly by a very moderate turn of the screw. Goodwillie's speculum (Fig. 42) is a very convenient and inexpensive speculum. If

FIG. 42.—Goodwillie's Speculum.

a hair-pin is bent in the shape of a double hook, as suggested by Bosworth, it is very convenient for hooking into the nostril, drawing the outer wall away from the septum, and exposing the nasal cavity. In examining this cavity it is very important that it be thoroughly cleansed, and it will often be necessary to dry the part with cotton-wool on a holder. If the patient makes strong expirations and inspirations through the nostrils, many facts may be elicited in the examination, especially the presence of a polypus with a narrow base, allowing the growth to move in and out with the respiration, etc. It will generally be necessary to push the tip of the nose upward and backward, so as to place the nares in a position for better inspection.

DISEASES OF THE MIDDLE EAR.

ACUTE CATARRHAL INFLAMMATION OF THE MIDDLE EAR.

This affection is an inflammation of the lining of the drum cavity and Eustachian tube, of an essentially catarrhal nature. That is, it is of the milder order of inflammations of mucous membranes, the products of which are only serum or mucus. The same inflammation intensified so as to produce a purulent discharge, is placed in the category of purulent inflammations.

Acute catarrh of the tympanum shows comparatively little tendency to destroy tissue, as is the case in the purulent disease, and as a result, the membrana tympani is frequently not perforated, as is the rule in the other form. If the membrane is perforated, it is not likely to be the result of ulceration, which might produce a large opening, as in the purulent form, but rather of pressure from inflammatory products in the tympanum, and a fissure-like perforation, which rapidly heals, is likely to be the result. Inflammatory proliferation is much more likely to result than disorganization.

This affection is usually developed from some form of throat trouble, but it may also depend on causes located about the external ear. It may attack both ears simultaneously, but is more likely to attack the second ear one or several days after the first. It may again attack one ear alone, but as the condition of the throat is usually at the bottom of the matter, sooner or later both ears are likely to be involved.

The attack generally comes on towards evening or in the night, the symptoms subsiding by morning, to recur again on the next evening unless interrupted by treatment. Its tendency to return whenever there is exposure to cold, or when the patient has been irregular in his habits, is well enough known. This is one of its most serious features, and often leaves the patient in a state of incurable deafness, unless great care is taken to treat symptoms promptly as they arise.

The subjective symptoms are as follows: In a given case there may be a sore throat that has existed for a few hours or days, when a painful sensation is felt in one side of the throat, on which there may have been the greatest amount of trouble, which passes up the Eustachian tube towards the ear. After a little time the ear may have a full or stuffed feeling, which may be converted into a violent throbbing pain. The voice of the patient will sound hollow and out of pitch, the patient *feeling* it unpleasantly in the ear, for precisely the same reason that the tuning-fork on the teeth is heard better in the ear affected with middle-ear disease. Dr. Sexton, of New York, has called this symptom autophony.

The pain continues, in a severe case, until secretion is inaugurated or the membrane is perforated, when, with many premonitory loud crackling sounds, which may send sharp, darting pains through the ear, relief is experienced. I infer that the crackling sound depends on interchange of air between the throat and the tympanum. In very mild cases, where there is scarcely any actual pain, this crackling is the most noticeable symptom of all. The Eustachian tube normally is nearly or quite closed in a state of quiescence, and by its valve-like faucial extremity resists to some extent the passage of air from the throat to the tympanum. When in a catarrhal condition, if the inflammation be not too intense, there seems to be a relaxation of the tissue about the mouth of the tube, or the swelling of the lips of the tube prevents exact coaptation, and the consequence is that air is forced in or drawn out of the tympanum, causing this crackling sound. In other words, the tube is easily *forced*, and it will be found that Valsalva's operation is more readily performed than even in the normal state. Anything that condenses the air in the upper pharyngeal space will produce this symptom, as sneezing, coughing, blowing the nose, rapid expiration, especially if done through nostrils of narrowed calibre—which is likely to occur from the rhinitis so often accompanying this disease —eructations from the stomach, etc. When air is forced in with violence, much pain may be occasioned. On the other hand, rapid inspirations, especially if made through narrowed nostrils, rarefies the air in the upper pharynx, and air is drawn from the tympanum into the throat.

When the auricle is pulled upon there is likely to be little or no pain, as the meatus and region about the concha are less frequently involved than in the severer forms of suppurative disease. Movements of the jaws, deglutition, sneezing, coughing, or talking, are often painful. If there are severe throat complications, the swallowing of cold water may cause exquisite pain. Movements of the head from side to side are often painful. There may be tenderness in the scalp, and pain in the side or back part of the head. Occasionally, in the severer forms of the disease, there may be pain and tenderness in the mastoid region. The pain in the ear may extend forward to the teeth, and indeed may be more intense there than in the ear, just as in carious and painful teeth the. sensation may often be only in the ear.   It is due to the fact that filaments of the fifth nerve are distributed both to the ear and teeth.   In milder forms of the disease there may scarcely be any disagreeable symptoms except itching, deep in the meatus, which the patient often affects to relieve by scratching or thrusting his finger deeply into the ear, which frequently occasions throbbing and often results in much pain, with great aggravation of what were, mild symptoms.

Sometimes the pain in the ear depends solely on the pressure of air upon the outer surface of the membrane, the result of an insufficient pressure upon the opposite side. This is brought about by closure of the Eustachian tube, the consequences of the disease, the residual air in the tympanum soon becoming absorbed.   There is great rarefaction of the air, amounting almost to a vacuum, and the membrane pushes the ossicles sharply inward, and the labyrinth waters are compressed by the settling of the base of the stapes into the oval window.   I once heard that Dr. T. G. Thomas was so interested in the result of an inflation of the tympanum in the person of his coachman, that he related the case before his class.   The man had a mild form of tubal catarrh, which caused great pain.   This was instantaneously relieved by Politzer's inflation.   This pressure, incident to a collapsed membrana tympani, sometimes produces a feeling of fulness in the ear, difficult to diagnosticate from that of hyperæmia.

Deafness.—In the outset, before hyperæmia becomes excessive, and the parts are absolutely free from secretion, the ear being in a state in which irritation seems to predominate over inflammation, the hearing may be morbidly acute.   Very soon a slight diminution of hearing is observed, or it becomes so profound that loudly spoken words cannot be distinguished.   The deafness results from the following conditions: hyperæmia of the tympanum, producing pressure upon every portion of the contents of the tympanum, impeding the vibrations of the membrana and ossicula. This congestion may also extend to the labyrinth and interfere with the function of the acousticus, and for the time cause true nervous deafness, which may be recognized by the tuning-fork.

Again, after the stage of secretion commences, the tympanum becomes more or less filled with the products of inflammation, which obstruct the hearing by the retardation

of vibrations.   Naturally, filling the tympanum full, with
bulging of the membrane, would produce profound deafness; but it has been found that a small amount of secretion
may much diminish hearing.   I once removed one or two
drops of serum from the tympanum by paracentesis and
inflation, and the hearing was astonishingly improved.   I
believe that the smallest amount of mucus resting on the
round window, will interfere noticeably with the hearing,
or any clogging of the movements of the base of the stapes
by the same process will also much interrupt its function.
Even if mucus covers the inner surface of the membrana
tympani, and clings to the ossicles, there is interruption
to free vibrations sufficient to cause deafness.   There is
another cause of deafness, more important than any I have
mentioned—that is, the sinking of the membrana tympani,
in consequence of closure of the Eustachian tube.   As a
result of this collapse, the membrana tympani is put upon
the stretch, the ossicles are crowded together, the base of
the stirrup is driven into the oval window, the labyrinth
waters are pressed upon, and the membrane of the round
window may be pressed towards the tympanum.

Here are two causes of deafness: first, the interference of
free vibrations of the apparatus of the middle ear by this
pressure, and second, actual pressure upon the labyrinth.
Whether this latter acts by pressure directly upon the
nerve of hearing, or by interference to the conveying of the
sonorous undulations incident to increased pressure upon
the labyrinth waters, it is hard to say.   In my observations
on the tuning-fork and bone conduction, it will be seen
that I incline mainly to the opinion that interference in the
vibrations of the middle-ear mechanism is the principal
factor in the causation of the deafness.   The latter generally comes on suddenly, the patient describing the ear as
having all at once closed up.

From the conditions just described, we have certain extremely disagreeable symptoms, namely, tinnitus aurium,
hearing one's own voice too distinctly, which may also have
a hollow reverberating sound, a stuffed feeling, or a feeling
as though the ear was filled with some foreign substance.
The latter symptoms may point indifferently to engorgement of the tympanum or a sunken membrana tympani.
Another symptom dependent on pressure is vertigo.   It
may be so great that the patient has fears of falling out of
bed.

Frequently there is no fever, and again there may be a violent febrile movement. Sometimes delirium may be noticed, especially in young children. Many of the symptoms simulate brain trouble.

Objective Symptoms.—In the first instance, there is likely to be signs of catarrh of the pharynx, with or without pharyngeal or laryngeal symptoms. The upper pharyngeal space is the one to critically examine; for the description of which see naso-pharyngeal conditions giving rise to ear trouble. Naturally the rhinoscope will be needed here, and other instruments for examining the Eustachian tube. Externally, the meatus may show few signs of disease, except being reddened perhaps, and scurfy. The speculum auris may elicit some tenderness.

The membrana tympani is the important point of observation. It is true there may be a catarrh of the tympanum, without redness of the membrana tympani, provided it be sufficiently mild, but the membrane may show undue brightness of the light spot, from opacity of the mucous layer; hence the augmented reflection at the umbo. Besides, the membrana tympani will not present the normal pearly-gray translucency, but will look flat or opaque, and of a darkish hue. Where the membrana tympani has become sunken, there will be the following changes: the malleus handle will be too vertical, or too horizontal; it may be drawn inward, giving it a *foreshortened* appearance, not approaching to the centre of the membrane, but to a point upward and generally backward. In consequence of this movement the short process will be thrown outward, appearing too prominent. The light spot when present, will usually be too small; a better diagnostic, however, is a malposition of the reflex; it will present an acute angle with the anterior border of the malleus handle, or its angle may be too obtuse; it may be too near the centre of the membrana tympani, or be displaced towards the periphery. The folds of the membrana tympani may become greatly exaggerated, especially the anterior fold. The membrane may sometimes not be reddened at all. Again, where the mucous layer is intensely reddened, the membrana tympani will be more opaque, and when it still retains some translucency in its outer layers, this color may shine through as a faint dusky red.

The whole membrane may be congested so as to be completely reddened, but not of the intense red color described

in suppurative otitis. More likely, however, the redness
may be seen about the short process; after a little, a streak
of redness may be seen coursing down the malleus handle,
from congestion of the manubrial plexus of vessels. With a
greater degree of hyperæmia, vessels may be seen shooting
from the periphery to join those of the manubrium, when the
whole of the membrane may become reddened. It is quite
easy to discern individual vessels. Be careful to discriminate between the hyperæmia which is the result of disease,
and that which is a consequence of traumatism. Rough
handling of the ear in fixing the speculum or violent inflation of the ear will cause considerable redness of the
membrane. If a number of persons examine an ear with a
pretty hot artificial light, some congestion probably will
result; where a collection exists in the tympanum, the
membrana tympani may bulge; if so, the protrusion will be
generally found in the posterior superior quadrant and will
seem to merge into the meatus; occasionally it may bulge
in front as well, and the manubrium will seem to divide the
membrane into two portions. If there is not too much
congestion, and the outer layer is intact, a light reflex may
be seen on the summit of the ectasia.

Again, if there is a collection in the tympanum, with no
change in the position of the membrana tympani, the light
spot being opposite to an opaque fluid, will appear excessively bright and glassy. I have often diagnosticated fluid
in the tympanum by this symptom alone. If the subject is
not too old, nor has an excessively opaque membrana tympani, the fluid may be *seen* through the membrane. It will
give the latter a darker hue, perhaps of a yellowish or greenish cast. If there is considerable mobility of the fluid, its
level may be changed by tilting the head forward or backward or cause it to disappear altogether by turning the head
to the opposite side. If inflation is performed, the level of
the fluid is changed, and often air-bubbles may be seen
about the top of the fluid.

By the use of Siéglè's otoscope (see Index), a hint may be
gained as to the presence of fluid in the tympanum. If the
membrana tympani is pushed sharply in, the fluid will rise,
or it will recede if the opposite movement is made. The
malleus handle fails to make the normal lever-like movement when any considerable collection is present. In rare
cases it may be possible to assert that the contents of the
tympanum is mucus or blood, by its color, but none except

experts can determine this. Sometimes the fact of fluid in the tympanum may be elicited by auscultation with the diagnosis tube; a moist rattle or click is heard instead of a dry one, and the bursting of air-bubbles in the tympanic fluid will yield a peculiar crepitating râle, which the surgeon, after much practice, will be able to distinguish.

The diagnosis of acute catarrh of the tympanum is sometimes difficult. Very great pain, with speedy rupture of the membrana tympani and a purulent discharge, proves that we have the suppurative form. In the mild cases where there is little or no redness of the membrana tympani, with the presence possibly of carious teeth, there will be real difficulty in determining whether the trouble comes from the ear or the teeth; sometimes it comes from both. The use of the tuning-fork, testing the hearing by other means, and contrasting the amount of hearing before and after inflation, making careful observation as to the presence of fluid in the tympanum, etc., will determine the diagnosis. Where there has been an old catarrh with impairment of the hearing, but which is improved by inflation, it is very difficult to tell whether it is the teeth or the ear which causes the disturbance. I have before now been in doubt, and have caused the teeth to be attended to, and the pain has disappeared. The pain from the ear is often of a neuralgic character, and may be indifferently in the head or ear, and will sometimes mislead us. In such cases it may be found that the pain does not follow the laws of a neuralgia, and some disturbance in the function of the ear may be noted.

In the adult there need be no trouble about mistaking brain symptoms for ear symptoms. In children there is much greater difficulty in diagnosis on account of absence of rational signs. I have, in the person of my own children, made this observation. A child would be restless most of the night, and no cause could be assigned for the trouble, when in the morning a discharge from the ear revealed the correct explanation. I think we are too loth to examine carefully the membranæ of children. In a case of considerable pain in a young subject, naturally we look to the bowels or head, or possibly the teeth, for the explanation; but if we always have in our mind a suspicion that it may be the ear which causes it, we shall be sufficiently on our guard. In a suspected case, if we do not object to doing it, a hint may be gained by smelling the ear; if there be the slightest discharge, its peculiar odor will at once direct attention to a diseased condition. A little harsh handling of the ear may elicit tenderness, pointing to disease. If the patient has throat trouble of any kind, or presents the conditions which may give rise to ear trouble, we are put upon our guard. A little sufferer from ear pain may put the hand to the head and indicate the location of the trouble.

In the pneumonia of children, ear complications frequently arise, which may give us a hint to inspect the ear when necessary. In examining a suspected ear in a child, even though we find no redness, there may be retraction of the malleus handle, which in children is a more conspicuous symptom than in the adult. We inflate by Politzer's method, and the membrana tympani is restored to a more normal position, and the patient may be relieved from suffering, when the diagnosis is pretty well determined. Other means for the relief of pain in the ear may be used, and if they succeed, the fact points to the probability of ear trouble as the cause of the pain.

The causes of acute catarrh of the tympanum are so similar to those of the purulent form that I hardly need go into detail.

The first important factor is the inflamed throat. This may be excited by any catarrhal influence whatever.

The exanthemata are always liable to result in ear trouble consequent on the throat complication. Croup and diphtheria produce conditions in the throat liable to cause ear involvement. Some forms of fever may be accompanied by throat symptoms, which occasionally result in ear complications.

Pneumonia and pleurisy are often accompanied by acute ear troubles, which depend, first, on the accompanying throat disease, and second, upon the violent interchange of air between the throat and ear cavity, consequent on the rapid respiration incident to these diseases. This has been already explained. Besides producing labyrinth trouble, syphilis may, by its somewhat constant throat symptoms, develop middle-ear disease, without necessarily going deeper. Bright's disease, through defective bloodvessels, may result in middle-ear trouble; but hemorrhagic disease through rupture of the vessels is more likely to occur. Hooping-cough, by violent inflation of the ear during a paroxysm of coughing, sometimes ruptures the membrana tympani; but the violent inflation itself is sufficient to excite inflammation. Politzer's inflation may sometimes excite a catarrh of the tympanum. Tuberculous patients are strongly inclined to catarrh of the tympanum, in consequence of the extension of some form of bronchial or naso-pharyngeal catarrh, to which they are so subject, to the ear, through the Eustachian tube. Tuberculous matter (a cheesy material) in the tympanum often results in middle-ear catarrh.

Sea-bathing, as in the case of suppurative trouble, excites this affection in several ways. Breakers striking the ear may rupture the membrane, or even if it is not ruptured, the cold water enters the meatus, which is enough to excite inflammation. Again, the patient may take the water into the mouth, and in the agitation and excitement he may attempt imperfectly to swallow it, by which means some of it is forced into the tympanum and excites inflammation. Sometimes, in swimming, a wave dashes the water into the nostrils, which, in the strangling efforts to remove it, is often forced into the ears. The nasal douche has inflicted many aural catarrhs upon those who use it, and I have abandoned it almost altogether. Even snuffing up water into the nostrils from the palm of the hand, has been known to pass into the ears and do harm. In my own practice of sending a few drops of salt and water into the nostrils by means of an atropia-dropper, it has occasionally passed into the ears, with deleterious effects.

The irritation of carious teeth, or even the advent of the wisdom tooth, has been known to excite an otitis. Teething in children has long been asserted to be a cause of otitis, and I do not doubt that it is—less frequently, however, than it is asserted to be. The cold pack in water cures, or the Turkish and Russian baths, have been known to excite this disease. As aural catarrh is more prevalent in the spring and fall, when the air is changeable, we readily infer that catarrhal influences are at work. Exposing the ear to a draught of air in any manner may provoke a catarrhal attack. Taking cold, even though it be very slight, may, in those so predisposed, go to the ear. Another fact is of great interest—that the patient is much more likely to " take cold " when depressed, from any cause whatever. Many patients require cotton in the ears during the winter to keep them from becoming inflamed and painful.

Prognosis.—As a rule, patients make a good recovery. As has been previously hinted, there is little tendency to destruction of tissue. If the membrana tympani ruptures, there is usually no loss of substance : only a fissure, which rapidly closes. I have seen such fissures heal in one or two days, although as many weeks may be occupied in repair. The favorable prognosis often turns on the short duration of the disease; for, if it is continued a certain time, changes of a proliferous nature are likely to occur. If the cavity has been completely filled with secretion and there has been great pressure, not relieved by rupture of the membrana tympani or evacuation from the Eustachian tube, the membranes of the round or oval windows may rupture, and serious labyrinthine disease result. The mastoid cells may be filled with the tympanal secretion, and take on true mastoid cell disease. Mastoid periostitis is not as likely to occur. The result to be greatly feared in this disease is the obstinate catarrh of the pharynx, which so frequently accompanies the affection. Especially is this the case with those having struma or tuberculosis, where it sometimes seems impossible to overcome the difficulty. The same is true of many weakly, or at least sensitive, people, whose mucous membranes are constantly irritated by the harshness of the air in many parts of this country, and the only means of relief seems to be the selection of an atmosphere congenial to the patient. I have seen a patient with pain in the ear, tinnitus, etc., changed to a better atmosphere, when in a single day all the symptoms disappeared.

I infer that while a patient has a catarrh in the neighborhood of the Eustachian tubes he is always liable to ear complications. One great danger is, that in the frequent relapses, which do not present very noticeable symptoms and consequently the attention of the patient is not sufficiently drawn towards them, that treatment will be neglected, at a period when great deafness may result. I once had a catarrh patient go into the country for two weeks in midwinter, when he had a slight relapse of his symptoms, which were not treated until his return. The hearing had been fifteen inches for the watch, and it went down to three or four inches, from which I could never raise it.

In consequence of the proliferating changes in the Eustachian tube the latter is less under the control of the tubal muscles, and the patient may complain of crackling in the ears, due to the fact that the tube is frequently forced. Again, there is less of that opening and closing of the tube which in a normal state maintains exactly the proper amount of air in the tympanum. If the tube be permanently narrowed, on the recurrence of a subsequent catarrh it is more readily closed than it would have been had it been of normal calibre. This symptom may exhibit itself on very slight changes in the weather. In children, who have an ample Eustachian tube, it is less likely to occur. This condition may even be a source of safety where there are collections in the tympanum, which may have no other means of outlet, thereby possibly preventing grave complications from the effects of confined matter in the tympanum. The altered and often sclerosed condition of the tympanal lining, I have often thought, acted as an exciting cause to future attacks, somewhat as the cicatricial changes in granular conjunctiva predispose to relapses.

The hearing ought to be perfectly restored. If the disease has continued long enough to thicken the tympanal lining, there will be some diminution in the sharpness of the hearing. A more important state is the collapsed membrana tympani. If the latter has remained in this condition for a few weeks only, it may never return to a normal position. It seems as though the membrana tympani, by its collapsed state, had become stretched, so as to increase its superficial area, when a sunken or cup-shaped condition was the more natural. I have seen patients with an abnormally patulous tube and sunken membrana tympani, who could restore the membrane by inflation, with improvement to the hearing, momentarily, but on the first act of swallowing it was thrown back to the old condition, the membrane apparently remaining in proper position only when actually pushed outward by the intra-tympanal air-pressure.

The fact that patients subject to catarrhal otitis often have an ancestry with the same tendency points to heredity. I am sure that the catarrhal and the rheumatic diatheses often seem to be a matter of heredity, and these affections are quite sufficient to cause ear trouble. Tinnitis aurium usually disappears, but it is not certain to do so, the catarrh setting in motion a tinnitus which may continue a part or the whole of the time during the remainder of the patient's life.

Treatment.—Although the *hyperæmia* is not as intense as in the suppurative variety, it is still quite decided in some cases, and results in much pain and throbbing and a sense of fulness in the ears. From one to three leeches may be applied to the orifice of the meatus, preferably to the posterior face of the tragus. The objection to applying them to the concha is, that they sometimes cause abscesses and

erysipelas.   There is no question, from my own experience,
and from that of many others, that in this location, depletion accomplishes much more than when applied elsewhere.
After the leech has fallen off, encourage bleeding for one
or two hours by bathing with warm water.   In properly
selected cases the relief to pain and the feeling of fulness
is most extraordinary.   In other cases the pain is aggravated by the leech, a dull, dragging pain being substituted
for the former kind of pain.   This being the case, I at once
give large doses of morphia either hypodermically or otherwise.

It is worthy of remark, that where the pain depends on a
distended tympanum from inflammatory products the leech
will not relieve.   In milder cases warm or hot applications
to the ear are very serviceable.   The most elegant method
is to place warm or hot water into a rubber bag and lay it
on the ear.   The temperature should be regulated so as to
give the patient the greatest relief.   Very warm water will
sometimes aggravate, so will too cold water.   Latterly, it
has been the practice of some to use cold or iced water.   I
can imagine it to do good in some cases of very excessive inflammation, of the sthenic variety, but I am, on the whole,
timid about recommending cold to the ear.   If it relieves
the pain and throbbing, and there is not directly afterwards
a violent accession of inflammation, it is likely to be successful.   I may be superstitious about the action of cold
on the ear.   Common salt heated and placed in a bag is a
good application to the ear, or any dry agent susceptible of
retaining heat is serviceable.   A bottle of hot water wrapped
in a napkin, with the ear laid against it, is an inexpensive
and easily applied method.

Moist applications, from their macerating tendency,
should be objected to, unless used for only a limited period.
Even a hot flaxseed poultice may be applied for a brief
period.   The old-fashioned mode of applying the core of a
roasted onion by its conical extremity to the meatus may
properly be recommended.   A better plan for moist applications is the direct use of warm water to the meatus: fill
the latter with warm water, turn the head to the opposite
side, and let it remain for half an hour.   A better plan still
is to allow warm water to run into the ear from Clark's
aural douche.   (See Index.)   This may be continued twenty
or thirty minutes at a time, desisting when it relieves or no
longer feels pleasant to the patient.   This method is very

highly thought of by many surgeons of the best standing. I, on the whole, prefer the method of dry warmth. If the ear is excessively soaked with moist applications, it becomes at last very painful. Vapor of water gently blown into the ear often affords relief.

Sometimes chloroform vapor does well. Dr. Theobald, of Baltimore, instils a four-grain solution of atropia into the meatus, even in children, to relieve pain. In the latter, however, this should not be used if the membrana is perforated. Magendie's solution of morphia may be used in the same manner. In young children a small pinch of black pepper wrapped in cotton and inserted into the meatus will often relieve pain. It will be well for an adult to apply it to his own ear, and, if it burns excessively, diminish the amount of pepper or wrap it more deeply in the cotton.

Where there is a *collection* in the tympanum, the indication is to evacuate it. I do not always wait until the membrane bulges before I puncture. There are other signs of collection in the tympanum, besides this, which justify puncture and which have been already referred to. The operation is not specially painful; nay, it may be done without pain at all in some instances. An instrument that the ophthalmic surgeon calls a broad needle, with an extra long shank may be selected. Use a speculum which is large enough, on trial, for the needle to operate in, then with the forehead mirror carefully illuminate. Pass the instrument, held very lightly in the fingers, down close to the membrana tympani, for there is sometimes danger of pricking the canal. The puncture may be made at the protrusion, but if there is none, then select the posterior superior portion of the membrana tympani, for there, will be the greatest depth to the tympanum. In children, owing to the excessive obliquity of the membrana tympani, the puncture may be directed upward, so as not to allow the point of the instrument to glance off from the membrane. While the instrument is in the act of puncturing, the tactile sensibility should be appealed to, and the surgeon should *feel* when his needle has punctured the membrane. In any event, it is best to touch the inner wall of the tympanum— touch bottom, so to speak. Do not place the patient against a solid substance, lest he move his head towards the operator and do himself harm. I do not usually use ether in this operation, but with many the anæsthetic is the favorite method. I never operate as well with the patient

in a prone position: one seems to be out of his reckoning
—the relations are not natural.   I believe a quick, delicate,
and gentle thrust is the best manœuvre.

If the operation takes on the form of a somewhat lengthy
incision, a knife like Gruber's (Fig. 42 *a*), with a handle
affixed at an angle, is the best instrument, for every step
of the operation may then carefully be observed.   It is
proper to state, however, that an instrument of this kind is
much more difficult to operate with than a straight one.
If, after the puncture, no fluid is evacuated, then perform
Politzer's inflation, and the sero-mucus may be seen to flow
out.   If the head is turned towards the affected ear, then
the whole of the tympanum may be emptied.   The inflation should be repeated until the perforation whistle is
elicited.   The paracentesis may be repeated as often as the
tympanum refills, or, by a strong inflation the puncture
may be reopened, or, if this does not succeed, pass in a

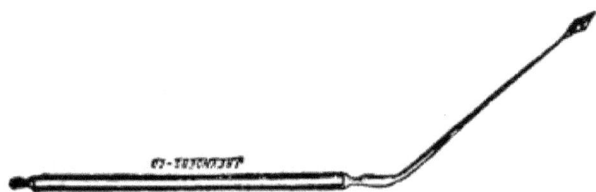

FIG. 42 *a*.—Gruber's Paracentesis Knife.

No. 1 Bowman probe.   This operation not only relieves
pain in a most wonderful manner, but has an equally
astonishing effect on the hearing.   I remember a case
where I suspected an accumulation in the tympanum, with
a collapsed membrana tympani, and I punctured and evacuated not more than two drops of serum, with very great
improvement to the hearing.

The profession is indebted to Prof. Schwartze, of Halle,
for this operation of paracentesis, which has contributed
so much to our surgical resources.

Another condition of great importance and requiring attention is the *collapsed membrana tympani*, consequent on the
closure of the Eustachian tube.   It has already been seen
that this is sufficient to cause considerable pain to the
patient and it often needs prompt attention.   At the outset,
I am inclined not to inflate if there is considerable hyperæmia, as it sometimes adds greatly to the patient's suffering;
but after a day or two it certainly would be proper to make

the attempt, but let it be done gently. At first it may occasion sharp pain, but usually it gives great relief. It may be done by Politzer's method rather than by the catheter, as the latter may irritate or lacerate the inflamed throat and do mischief. The rule, then, is, inflate very gently at first, and if it acts well continue it, and if there is too much pain, desist. The inflation should be done sufficiently to elevate the hearing to the highest attainable point. If this point is passed, and the hearing diminished, and tinnitus and pain results, then direct the patient to hold his nose and swallow several times; this will empty the tympanum of excess of air, and again restore the hearing, provided the tube is not too swollen to open upon muscular contraction. The latter not being the case, air will be confined in the tympanum, and may for a while aggravate the symptoms. This is one of the objectionable features of inflation. It

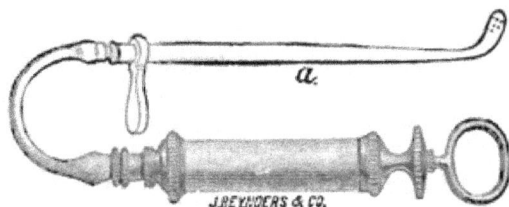

FIG. 42 b.—Pomeroy's Post Nasal Syringe.

is a good rule to do nothing to the patient which may occasion any throbbing or pain in the ear. Manipulations which are not altogether gentle, or even the mildest syringing, may do harm by adding to the existing hyperæmia. Even the leeching must be done by a careful hand, for rough manipulations have sometimes caused a return of the pain which the leech had previously relieved.

The inflation for the restoration of the collapsed drumhead is only a temporary measure, and the *tubal catarrh*, which is the cause of the trouble, should be treated. Sometimes a case may be so mild, and show no symptoms of disease except in the Eustachian tube, that the diagnosis of tubal catarrh may very properly be made. This condition may be met in the following manner: simple salt and water warmed, and of the strength of a drachm to the pint, may in mild cases suffice. It may be thrown behind the velum by means of a spray-producer or the posterior nasal syringe. I prefer the one devised by myself, which has a tip separated from the syringe by a rubber tube (see Fig.

42 b). I use my own hard-rubber spray instrument; any other may be used; the glass instruments are the best, but they often break. Not more than half an ounce or an ounce of the salt and water should be injected; do it very gently, instructing the patient to remove it at once by expectoration. The dropping tube may be used for this purpose, the fluid being thrown into the anterior nares; but in all these applications do not allow the patient to be frightened or agitated, for he may cough, sneeze, or swallow, so as to allow the fluid to enter the tympanum.

Some years since, while in Vienna, I saw Dr. Adam Politzer using for this purpose a little earthenware vessel having a nozzle which fitted the nostril, similar to our sick-feeder, which he used by pouring the fluid into the throat. When this medication is insufficient, chlorate of potash in saturated solution may be used, or carbolic acid, one drachm to the pint of water. It may be necessary to give the patient general treatment for naso-pharyngeal catarrh.

Where more energetic measures are needed, I employ the nitrate of silver. It may be used with the posterior nasal syringe or spray instrument, the Eustachian catheter, or my own faucial catheter. The strength may be from a few grains to the ounce, to a one-hundred-grain solution. When a drop or two is used by either of the catheters, or the spray instrument is used, a much stronger solution will be borne. In using the posterior nasal syringe it is a good plan to draw up five or ten drops of a ten or twenty grain solution, and force it in strongly enough to cause a coarse spray to be produced. If much pain results, throw in two or three drachms of salt water, which frequently, but not always, gives surprising relief. If the solution has been too strong, the tube may swell, and for two or three days become somewhat impervious. If a prolonged soreness is left in the throat, the remedy has been too severe. There ought to be a feeling of relief from the application after a few minutes or an hour or two. Tincture of iodine may sometimes be used, but only one to three or four drops should be thrown up at a time. It may be diluted with alcohol if necessary; it is sometimes painful. Chloride of zinc acts somewhat like nitrate of silver, but it is occasionally a harsh remedy. Many other astringents may be used, being careful to reject any that act badly. It is not possible to determine beforehand which remedy will be the most beneficial. Often sore and painful tracts of redness with pain on deglutition may be observed in the pharynx. The best plan is to touch these with a twenty to forty grain solution of nitrate of silver, using cotton on a holder for the purpose. Although it is a harsh remedy, I know of nothing which will so soon abolish the pain and inflammation.

Often at the outset of an acute catarrh with pain in one side of the throat which shows a tendency to travel up the Eustachian tube to the ear, the whole process may be arrested at once. A patient under my care was treated by mild remedies without effect. On returning home after one

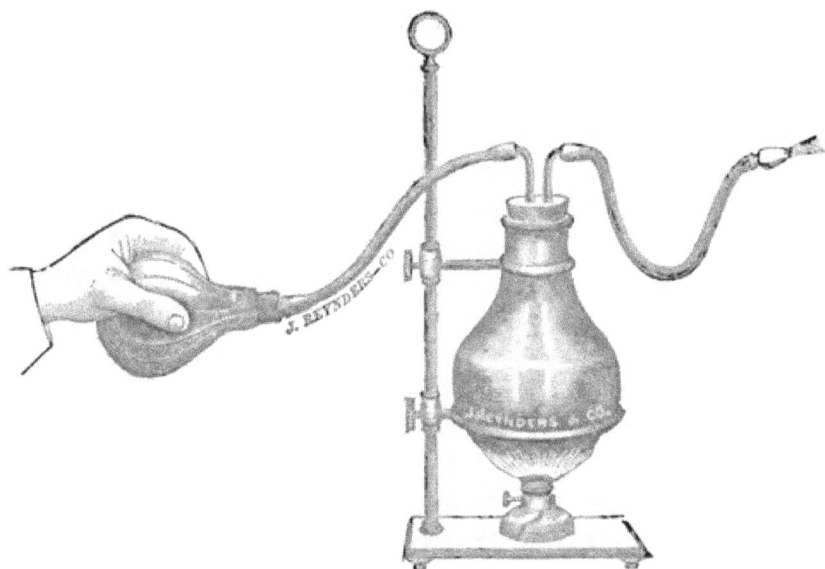

FIG. 43.—Apparatus for Generating Steam for Intra-Tympanal Injection.

of these applications, the pain was aggravated; it could be followed up the Eustachian tube to the tympanum; the latter began to throb painfully, and in desperation I poured a freshly made forty-grain solution of nitrate of silver into a hard-rubber spray instrument, and threw up an abundant spray for two or three seconds behind the velum, pointing it towards the Eustachian tube. It caused considerable pain for fifteen or twenty minutes, but the symptoms wholly disappeared, and no further treatment was necessary. A somewhat threatening attack was evidently wholly aborted.

I am aware that nitrate of silver often acts injuriously,

exciting severe pain, and causing much soreness of the parts, which may remain for days or weeks afterwards. The question is to select the cases in which it is likely to act favorably. The above indication of *pain and localized soreness* I believe to be favorable to the remedial action of nitrate of silver. I have in some instances applied a strong solution of the arg. nit. to a painful and hot locality in the upper part of the pharynx, the result of inflammation, where it produced no pain whatever, but actually felt cool, and afforded the most grateful relief. The principal trouble seems to be that the patient does not in some cases appear to react successfully from the application. Sometimes steam injected against the mouth of the tube, and often into the tympanum, acts well in reducing the engorgement of the Eustachian tube (see Fig. 43). The hard-

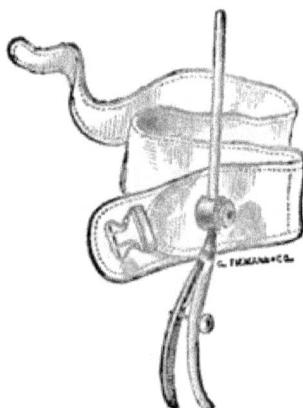

Fig. 44.—Pomeroy's Catheter Holder.

rubber catheter should be used for this purpose, as the metallic instrument becomes too much heated; it requires to be fixed in position by the catheter holder; I believe my own instrument is a sufficiently good one for the purpose (see Fig. 44). Do not prolong the sitting beyond four or five minutes, giving a few puffs, and then waiting a half minute, and then repeat. The first effect will be to congest the lining of the tube, but subsequently the hyperæmia will be diminished. I would not make applications to the throat oftener than once in two days, unless it be some remedy of the milder variety, like carbolic acid, a drachm to the pint; this may be used, perhaps a little diluted, in the upper pharnyx with a Davidson spray instrument (see Fig. 45) by the patient himself two or three times a day, but avoid any unpleasant irritation of the part.

Prophylaxis.—Inasmuch as this disease depends largely upon the habits of the patient and climatic influences, much may be done to prevent recurring attacks. Everything that tends to excite a "cold" should be scrupulously avoided. It takes years of experience to live wisely in this respect; it is only those who become somewhat venerable in years who may be seen carrying an umbrella almost habitually, with an overcoat on the arm, or actually wearing it, and with properly shod feet, who regulate the temperature of the body as carefully as a chemist would a compound. Patients subject to aural catarrh become surprisingly sensitive to any lowering of the temperature. I have known such to observe the direction of a draught by a pain, or an unpleasant feeling in the ear, when no other signs of its presence existed. They seem to be as sensitive to atmospheric changes as an habitually rheumatic patient. Rid-

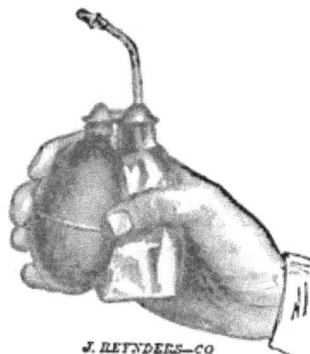

FIG. 45.—Davidson Spray Instrument.

ing in the wind is often extremely pernicious, and the patient should have the ears protected by cotton-wool in the meatus, or some other mode of protection. Wet or damp feet or damp clothing must be avoided; or on returning from a ride or walk everything damp should be removed and dry clothes substituted. The action of peculiar atmospheres has already been alluded to.

I believe *nothing* exerts so strong an influence on the well-being of a catarrhal patient as congenial air. It would seem almost as important as in the case of an asthmatic patient. If one remains in an atmosphere which constantly congests the lining of his pharynx and Eustachian tubes, it is well-nigh impossible for him to do well; he is likely to go on adding to his deafness by successive attacks, until great hardness of hearing may result. Another important indication is to cultivate as vigorous a state of the general health as is possible, thus enabling the patient to resist the ill effects of our climate.

I do not believe in avoiding the out-door air to prevent taking cold. I have noticed that our stage-drivers rarely suffer from catarrhal affections, notwithstanding they are on their boxes driving, sixteen or eighteen hours a day, and exposed to all weathers. Their vigorous physiques resist the catarrhal influences, and this state has been brought about by constantly being in the open air. Careful regulation of all the habits is a more important matter than is usually supposed.

Tuberculous or strumous patients may require, in addition to a nourishing diet, stimulants and cod-liver oil, the latter in some form of emulsion. Excessive fatigue or exhaustion should be avoided, as diminishing the power of the constitution to resist the effects of cold. I know a gentleman who will never admit that he has overworked or become exhausted, but he suffers frequently from "a cold" which seems always to attack him in times of depression. Even in midsummer the inclement atmosphere exists. I have observed that those who go into the country suffer especially. The middays are excessively warm, but the evenings often are cold, the temperature after sunset going down very suddenly, and blankets may be in requisition.

It is advisable to direct the patient to always breathe through the nostrils when possible. One of the best plans for keeping the nostrils free is to constantly breathe through them, besides the favorable influence resulting from a current of air passing across the faucial orifices of the Eustachian tubes. This matter will be developed more fully under the head of naso-pharyngeal catarrh, etc. If the mouth only is used in breathing, air is insufficiently warmed, and it irritates the pharynx. Sometimes I advise a respirator to be worn over the mouth so that the air may be rendered less irritating to the throat. From what has been said it would seem to be almost a fine art to protect the system properly from our variable and inclement atmosphere.

Constitutional treatment has already been hinted at in the previous topic. To those who accuse us of treating disease locally and depending on it, I would say that I am most heartily in favor of doing anything that benefits the general health of the patient. Some of our therapeutic friends will tell us that a large dose of quinia acts favorably in a commencing sore throat—may break it up, in short. Tr. aconit. rad. is also given for the same purpose. Others take pleasure in aborting threatening symptoms by a large dose of spts. vin. gal., or something of the kind. In the outset of an attack it may be broken up by somewhat prolonged and excessive perspiration in bed, with or without a cathartic dose. A warm foot-bath is a very harmless and effective mode of inducing perspiration, or bottles of hot water may be placed at the feet and on either side of the patient to accomplish the same purpose. If there be no undue obstruction in the bowels, a large dose of morphia may suddenly abort a threatening attack. A full dose of tr. of guaiac may sometimes be very serviceable.

CHRONIC CATARRHAL INFLAMMATION OF THE TYMPANUM.

Most of the works on diseases of the ear have a chapter on sub-acute catarrh of the tympanum. As this affection is one intermediate between the acute and chronic affections, it seems hardly necessary to elevate it into a distinct topic. Neither shall I make a separate heading for "proliferous" inflammation, "dry catarrh" or "moist catarrh," holding firmly to the opinion that these and all other manifestations of this disease are but the results of an inflammation which at different stages, and modified possibly by constitutional peculiarities, produces the peculiar products of inflammation characterizing this disease in its manifold developments.

I believe for the most part that the peculiar quality of the inflammation found in the upper pharyngeal space is repeated in the ear: if there is a moist throat the tympanum is likely to be moist, and where there is a pharyngitis sicca, with a dry shining mucous membrane, a similar condition exists in the middle ear. It will be inferred from this that I hold to the view that most of these affections of the middle ear depend on a throat trouble, whether past or present.

It may be stated here, that the ear affection is not usually as far advanced in its development as that of the throat.

Course.—The most frequent manifestation of this disease is seen in subjects who have had an acute affection of the tympanum, be it catarrhal or suppurative, which has failed to undergo resolution, owing to neglect of treatment or constitutional peculiarities. This statement is true in a much larger sense than would at first appear. The slightest cold that obstructs the Eustachian tubes and "stops up" the ears is, strictly speaking, an acute catarrh, and it is from these attacks that the foundation possibly may be laid for a serious chronic affection of the middle ear. I feel some doubt as to whether this affection ever commences as a chronic catarrh. I am well aware that many elderly people are found to have a considerable degree of deafness, which must have been of long standing, but who will stoutly persist that they never have had symptoms pointing to acute disease of the ear. This we may readily believe to be a conscientious statement, but not necessarily a true one.

The normal hearing is quite in excess of actual needs, and much of it may be lost without exciting the notice of the patient, unless he be unusually observing. It has been stated that many of these cases were unconnected with throat troubles, present or past. This is a statement impossible to prove, and the probabilities are against it.

When we reflect that a catarrh of the upper pharynx may be sufficient to obstruct the Eustachian tubes, and exhibit so few symptoms of throat trouble that the most expert surgeon may overlook it, and the patient himself be unconscious of its presence, or that a catarrh may have been cured before serious ear trouble made its appearance, it will be seen how difficult it would be to substantiate such statements. I do not forget that a few patients take on ear trouble from influences acting in the direction of the meatus auditorious externus, and naturally throat complications would be likely to be absent.

This disease is perhaps the most dangerous to the hearing of any, except certain forms of labyrinth disease. As its insidious beginnings escape the notice of the patient, or at least fail to excite his apprehensions, much valuable time may be lost before treatment is commenced. It will be subsequently seen that in the later developments of this disease changes have taken place which are partially or wholly irreparable.

I do not know of any instance where a more unreasonable demand is made on a surgeon than when he is expected to cure chronic catarrh of many years' duration. As well might you expect to cure a man with disorganized kidneys, liver, or lungs. The deafness varies from very slight lowering of the hearing to that which is so profound as not to admit of hearing the loudest conversation. In the latter instance the disease has extended to the labyrinth, and bone conduction may be weakened or abolished. The most distressing and discouraging form of tinnitus aurium frequently accompanies this form of otitis. When this disease has produced profound deafness with tormenting tinnitus, we have about as unhappy a person as can well be imagined. These patients are full of jealousies, often imagining that people are talking ill of them, or exciting a laugh at their expense. Besides this, tinnitus sometimes becomes almost unendurable, and cases are on record where its victims in moments of desperation have sought relief in suicide.

In marked contrast are the serene and kindly faces of the totally blind.

Causes.—From what has preceded it will be seen that whatever causes an acute inflammation of the drum cavities also is a pathogenetic factor in developing this affection. The often-repeated "colds" so prevalent in this climate, which usually act through the throat and Eustachian tubes, are a frequent cause.

Much has recently been written about the deafness of railway engineers, boiler-makers, and ship-calkers, which was formerly regarded rather as an affection of the labyrinth than of the middle ear. Later investigations, however, among which I may mention those of Dr. Holt, of Portland, Maine, point to the fact of middle-ear trouble as being the principal earlier lesion, followed subsequently by labyrinth trouble, the tuning-fork apparently proving the absence of nerve complications in a large number of the more recent cases. The explanation of boiler-makers' deafness seems to be twofold: first, the fact of exposure to draughts while heated and perspiring; and second, the violent vibrations of the middle-ear mechanism inducing a low form of inflammation of the tympanum, which results in great thickening of the parts. In ship-calkers, the observation was made that the ear next to the ship, and consequently receiving the more powerful concussions, became

deaf, while the other ear, for a time at least, was unaffected. Here the concussion had mainly to do with the
causation of the deafness. This also depended on middle-
ear trouble. In railway engineers, as in boiler-makers, a
twofold cause seems to operate.

It is frequently the case that an engineer on a train running 40 or 50 miles an hour is on the lookout without any
protection from the powerful current of air engendered by
the rapid motion of the train. The noise, of course, is of
the harshest variety.

A distinguished member of our profession in New York
has been in the habit of riding daily during the warm sea-
son to his country seat, in the steam cars. One day I noticed that he held his hand over one ear. He explained
that he had deafness of the opposite ear from exposure to
the rattling noises incident to railway travel, especially
when going through tunnels at a rapid rate of speed. The
affected ear was next to the open window. In this instance, again, we have the two factors—the draught of air
from the open window and the concussion from the loud
noises. It is no mystery that these patients subsequently
take on labyrinth trouble, as this has become sufficiently
well known to be a consequence of very chronic catarrhal
inflammations. Gestation sometimes conduces to chronic
aural catarrh. With each recurrence of gestation an accession to the deafness results, and the patient ultimately
becomes quite hard of hearing.

It is worthy of note that strumous and tuberculous subjects incline to chronic aural catarrh.

Constitutional syphilis, either acquired or hereditary,
predisposes to catarrhal otitis. I believe that the rheumatic diathesis, in many instances, has much to do with
the obstinate character of this affection; the rheumatic inflammation, according to its well-known predilection for
fibrous tissues, finding a lodgment in the muco-periosteal
lining of the drum. All possible unhygienic influences
enter into the causation of this affection, as well as depressed vital energy, from any cause whatever.

Subjective Symptoms.—The earliest and most important
symptoms are deafness and tinnitus. As the hearing is
normally much in excess, a portion may be lost without at-
tracting the patient's notice. As a rule, the disease commences in one ear, and this may become profoundly deaf
without the patient's being conscious of the loss, as the

fellow-ear may do duty for both. Subsequently, however, the opposite ear will become involved. The deafness may come on suddenly if the Eustachian tube is closed, but if not, it may develop very gradually. I have noticed elderly people, with a hearing distance of 2 or 3 inches for the watch, who were hardly aware that they had become hard of hearing, and never had noticed any trouble with their ears. It frequently is the case that the voice or musical tones are much better heard than the watch. The deafness may be very slight indeed at first, but subsequently the profoundest deafness with labyrinthine complications may result. If there is moist instead of dry catarrh, the hearing will vary with the weather, being better in cold dry weather than in cold moist weather or in very hot weather— that is, when the mucous lining of the Eustachian tube and tympanum become more congested, or secrete more freely. In the dry variety the hearing varies in this wise: it becomes a tiresome muscular effort for the patient to hear, as the contractions of the muscles of the tympanum in the accommodation of the ear are impeded by the thickening and rigidity of the tissues, and the consequent immobility of the middle-ear mechanism, and also by the diminished power of the tensor tympani and stapedius muscles themselves, owing to structural changes, such as atrophy of muscle fibres, fatty or calcific degenerations, etc. In this case the patient hears worse when fatigued, or after having lost sleep, or suffered any hardship. This class of patients will often present a variety of symptoms simulating those of brain lesions.

Tinnitus aurium may be the first symptom noticed, or it may be absent for some time, but it is nearly always present. A buzzing noise in the ear may first call attention to beginning trouble. I have seen tinnitus in a case where there were no middle ear symptoms, and the hearing was perfect as far as moderately careful tests went. This noise in the ear is the great opprobium of the aural surgeon. It may continue through a lifetime, being either constant or occasional. In some cases it seems, after a few years, when the ear is nearly or quite spoiled, to entirely disappear; but this is somewhat exceptional. Again, it may only give the patient trouble when he has a cold or during a damp period, when the lining of the Eustachian tubes and tympani become congested. Anything which congests the parts tends to produce it; a full meal, especially if stimulants are used, may develop it. It is more likely to come on in the after part of the day, when the patient is fatigued and the blood mounts up to the head. Any unusual exertion is likely to produce it, especially public speaking. The supine position in sleeping, by increasing the hyperæmia of the parts, is a factor in exciting it. The quietness of the night, however, probably may permit the patient to notice a sound in the ear of which he was not conscious amid the noises of the daytime.

The deafness may be nearly cured, but still the tinnitus may remain. A close or overheated room, or an insufficiently ventilated sleeping-room, may re-excite tinnitus.

Pain is not a very frequent symptom, and is not usually intense. Where the Eustachian tube is closed and the membrana tympani presses on the fluids of the labyrinth it may be quite intense, and may extend to the side of the head. Whenever the patient takes cold and has a relapse there may be pain. Some patients are so sensitive that they must wear cotton in the ears most of the time to prevent the wind or cold atmosphere from being painful to them. If you shout into the ear of a patient who is pretty deaf, he will sometimes start back much disturbed, as the concussion of the loud noise has been painful to him, there often being a certain amount of hyperæthesia of the nerve. I have noticed patients who were "run down" or depressed from any cause who would develop a pain in the ear as one of the symptoms of this condition.

It is a well-known fact that carious teeth will cause pain in the ear. The mental processes are often affected in this disease, especially when pressure is made on the fluids of the labyrinth or on the brain. In such a case the memory is weakened; the patient fails to grasp ideas and to maintain a consecutive train of thought. The mind seems easily fatigued, and the patient may be unable to do as much mental labor as formerly. He is often irritable and filled with forebodings, and is low-spirited. The intra-labyrinthine pressure will often seem to the patient as though there were actual pressure on the brain, and often give him the sensation of being in motion. Vertigo and vomiting sometimes also result from this condition. The pressure is more likely to be brought about where the ossiculæ are anchylosed, or the membrana tympani and the membranes of the round and oval windows are thickened so as to lose their normal elasticity, somewhat as the rigid and inelastic sclerotic assists in developing the symptom of increased tension in glaucoma.

Sensations in the tympanum dependent on the condition of the Eustachian tube and cavity of the ear.—The patient's voice may be heard with an unpleasant distinctness in the affected ear, and with an altered resonance, having a hollow, unmusical sound, for the same reason that the tuning-fork is heard longest in the affected ear.

Crackling sounds are often heard in the ear, due to the violent interchange of air between the tympanum and the throat, owing to the faulty permeability of the tube. During the act of deglutition air is frequently but not always interchanged between the tympanum and the throat, but in chronic catarrh this interchange may be almost constant, and sufficiently violent to cause movements in the membrana tympani. In other cases there may be no interchange at all in consequence of closure of the tube. The feeling of fulness in the ear may be due to hyperæmia of the tympanum, but it is more likely to be dependent on collapse of the membrana tympani. In some cases, where the ear has been too violently inflated and the air remains in the cavity, this sensation may be induced. The way to remove this surplus of air is to stop the nose and swallow repeatedly. In both of these conditions (hyperæmia and collapsed membrana) the ear may feel as though "stopped up" by something that ought to be removed. Sometimes the patient is annoyed by sounds proceeding apparently from the neighborhood of the Eustachian tube, which are produced possibly by spasmodic contraction of the tubal muscles, which force the walls of the tube tightly together.

Where there are important throat symptoms the patient may complain of much pain in deglutition, or during this process an unpleasant rasping sensation may be experienced, as though the walls of the pharynx were actually rough and rubbed together, producing a friction sound. In some cases there may be an aching pain in the throat, more likely to be found on the side of the most affected ear. In other cases there may be a burning pain in the inflamed spot, the parts feeling dry and hot. In pharyngitis sicca the patient may have no pain, but the dryness will be very uncomfortable, the patient frequently desiring to moisten the throat. Especially will this be the case if much mouth breathing is practiced. The source of the pains described as being in the throat may be in the ear, and sometimes it will be difficult to fix accurately the seat of the disease as evidenced by the pain. *The nostrils* will sometimes be dry, but generally moist, with points of special irritation or inflammation. These may excite frequent attacks of sneezing.

Objective Symptoms.—The meatus externus is usually preternaturally dry and more or less furfuraceous. In the earlier stages there may be impacted cerumen of greater consistency, and generally of a darker color than normal. Afterwards directly the opposite condition is found; the canal having little cerumen, and in some cases none at all. The sebaceous follicles likewise fail to secrete in normal amount. Near the membrana tympani the canal may be

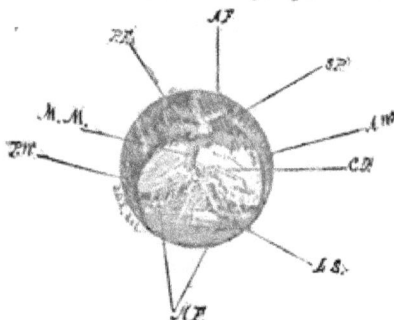

Fig. 46.

Fig. 46.—Collection in the tympanum: *H.F.*, height of fluid; *L.S.*, light spot broken and diffused; *C.D.*, calcareous deposit in ant. sup. part; *M.M.*, manubrium mal.; *P.F.*, post. fold; *A.F*, ant. fold; *S. P*, short process; *A.W*, ant. wall; *P.W*, post. wall.

somewhat reddened and tender. This symptom is sometimes found in very chronic cases. Occasionally there may be some widening of the canal from atrophy of its subcutaneous connective tissue, with absorption of fatty material.

The membrana tympani shows the greatest possible variety of changes. In more recent cases, where only the mucous layer is involved, the light spot will not only not be dimmed, but will be of unnatural brightness, from the increased reflection from the somewhat opaque mucous layer. In more advanced cases the dermoid layer will become roughened, furfuraceous, and no longer capable of reflecting light at the umbo. Inflammatory exudations into the fibrous or middle layer cause the membrane to become very opaque, giving it a dull, flat color, in strong contrast to its normal pearly-gray, translucent appearance. In earlier stages, where there is considerable hyperæmia of the mucous layer, the color may be of a reddish-brown tint.

If there are any collections in the tympanum, the membrana tympani opposite them will be darker than elsewhere, and if the color of the collection be decided, it may be apparent through the membrane (Fig. 46), but usually the membrane is too opaque to admit of this. By tilting the patient's head so as to change the level of the fluid, the different appearance of the membrane will give evidence of this fact. On inflation, bubbles of air may fill the tympanum, and show through the membrane somewhat conspicuously.

The position of the membrana tympani is of great diagnostic importance. Sooner or later the Eustachian tube will have been obstructed sufficiently to result in collapsed drum membrane. It is brought about in this manner: there not being the usual supply of air from the throat on account of closure of the Eustachian tube, the remaining air in the tympanum is absorbed and an imperfect vacuum is formed in the tympanum. Under these circumstances the pressure of the outer air on the membrana forces it inward, in some instances to the nearly complete obliteration of the tympanic cavity.

As the incus, head of the stapes, and short process of the malleus are only slightly movable, the membrana is pushed down upon them so tightly that prominent points of these bones show conspicuously behind it. The short process of the malleus will be very prominent, the descending shank of the incus will sometimes be apparent behind the malleus handle, passing in the same direction, but not extending so far; the region of the head of the stapes may be noticed, and the posterior ramus of the stapes itself sometimes shows beneath the membrane. If there is considerable anchylosis of the ossicula, the immovable malleus handle is often prominent and apparently divides the membrana into two unequal portions. A more usual sign of sunken drum-head, however, is a pushing inward or retraction of the malleus handle, giving it a foreshortened appearance.

Instead of the extending of the manubrium to near the centre of the membrana tympani, it will terminate at a point above the centre and usually somewhat posteriorly. This gives the manubrium a foreshortened appearance; it does not seem as long as the normal. In children this appearance is greatly exaggerated, the manubrium being almost in a horizontal position.

The tendon of the tensor tympanum emerges from the

anterior pyramid nearly opposite to its insertion near the
neck of the malleus.   There are, however, exceptions to this:
sometimes the pyramid is situated more posteriorly than
usual, and again it is located too far forward.   In this event,
if we admit that secondary retraction of the tendon of the
tensor has occurred, and acted as a factor in the indrawing
of the membrana tympani, then the malleus handle would
take the direction of the point at which the tendon emerges
from the anterior pyramid, and would be placed too vertical
in the event of the pyramid being too far forward, and more
horizontal if further backward.   Politzer and Gruber have
had much to say on this subject.   My own impression is
that the pressure of air on the outer surface of the membrana tympani is the cause of its collapes, for the most part.
It is a matter, however, of the most common observation, that
the inclination of the malleus handle does vary greatly, in
some cases being nearly vertical, and in others as nearly
horizontal.   Another argument against the theory of retraction of the tendon of the tensor is, that in marked cases of
collapse of the membrane of long standing, the muscles of the
tympanum become atrophic, their tendons undergoing fatty
and other degenerations, naturally causing them to lose
power.   I know it is stated that the cause of the drawing inward of the manubrium is retraction of the tendon of the
tensor, but if there is fatty degeneration very little retractile
power could be expected from it.

Indeed, the results of division of the tensor do not in the
least sustain this idea.   I have practiced the operation many
times, and in the main have met with failure.   Another observation I have made regarding the malposition of the
manubrium, which I also find mentioned by Tröltsch, is the
sabre-like shape of the whitish line in the membrana, placed
over the malleus handle.   This curve is convex behind and
concave in front.   For the upper half of its course it is reasonably clear that the manubrium is in apposition with the
membrana tympani, but for the remainder of the distance it
is evidently drawn away from the disconnected membrane,
as though the latter could not follow its extreme excursion
inward.   The anterior fold is nearly always observable in a
normal state, but in the collapsed membrane it becomes quite
prominent, extending from the short process forward and
downward in a curved direction, but sometimes nearly in a
straight line, running downward and forward.

Occasionally I have seen a rounded elevation at its commencement, in front of the short process, and much resembling the latter.

The posterior fold is not as well observed in the normal membrane, but in its collapsed state it is generally seen extending from a point behind and frequently above the short process, backward and upward, whence it passes downward in the direction of the posterior inferior periphery of the membrana tympani. It is not as sharply defined as the anterior fold, and sometimes has the appearance of merging with the wall of the meatus. If the malleus handle is drawn backward into a more horizontal position, the tendency is for the posterior part of the membrane to be less sunken and the anterior portion more so. The latter will be less sharply illuminated, and present a darker color, and the anterior fold will be much exaggerated.

The light spot is perhaps as important in a diagnostic point of view as any other landmark of the membrana. It is a truism that nothing can obliterate the light spot but loss of polish of the dermoid or outer layer. No conceivable malposition of the membrane is capable of removing the light reflex entirely. In the most variously sunken membrane some portion will still remain at right angles to the axis of the meatus, and will reflect light from this point to the observer's eye. Some have asserted that the position of the membrane may obliterate the light reflex, but I do not remember ever to have seen such a case, and on theoretical grounds it seems impossible.

The normal position of the reflex is, as has been already stated, in front of the malleus handle. It is of a somewhat triangular shape, with its apex near the extremity of the manubrium, whence it radiates towards the periphery, forming a somewhat obtuse angle with the anterior border of the manubrium. It may consist of one triangular spot, or several, or it may appear in the form of lines, its correct position and size being the indication for its normal appearance.

In the sunken condition it will form an acute angle with the membrane, or this will be too obtuse, it will be too large or too small, may be seen in almost any portion of the membrana, or may consist of several light spots; that is, wherever there is a depression or even an elevation there may be a light spot. I have seen them on the summit of the short process, on the rounded elevation at the commencement of the anterior fold, besides on other parts of the folds, and on every conceivable portion of the membrane.

In very great collapse a light reflex may often be seen at the periphery, more likely posteriorly, forming a long curved line, and running parallel to the periphery of the membrane.

Calcific and other spots in the membrana tympani seem to depend on a previous inflammation of the membrane, although it is not always possible to demonstrate this fact. Many varieties have been described. Wilde speaks of numerous small specks found on the membrane of a patient who had become deaf as a consequence of parturition. He makes observations on another class of cases in this wise; a membrane otherwise healthy, but with "a crescent-shaped opacity, about a line broad and three lines long, with a tolerably defined edge, and rather rough upon its surface, occupies the lower, and usually the back part, of the membrane." This spot is more insensible to the touch than the rest of the membrane, and has a well-defined edge.

It gradually spreads over the whole of the membrane and produces permanent deafness. He has no cure for it. He regards it as consequent on some form of inflammation of the tympanum (Wilde on the Ear: Phil., 1853, p. 274). According to the same author, on p. 276, atheromatous or calcareous deposits may be seen in the membranes of middle-aged females. These are usually situated in the anterior portion, having a yellowish color, with sharply defined irregular edges, which feel gritty on being touched with a probe.

They seem to be placed between the layers of the membrane, and resemble the atheromatous deposits in "the heart and arteries, and cornea." In each case there was great deafness; the sensibility of the part remained. Obviously, the spots of atheroma-like material, which Wilde describes as having a gritty feeling to the probe, must have invaded the outer layer of the membrana. Other spots resembling tendon, Politzer found to be placed between the fibres of the membrane, and were composed of fat corpuscles and granular matter.

The whitish lymph-like looking spots gradually shade off into the surrounding membrane, while the calcific spots are abrupt in outline. Many of these appearances are found in membranes where there has been suppurative otitis, or even a catarrhal inflammation of the tympanum.

Beyond question all these changes are the result of a previous state of hyperæmia or inflammation.

Politzer found in some cases opaque spots, where the exciting cause seems to have been the small violence resulting from brushing the membrana with a bit of cotton on a holder. Tröltsch (in Dr. Roosa's translation) speaks of whitish points on the anterior superior part of the membrane, located in the mucous layer, but he is at a loss in explaining their nature.

It is a fact worthy of note, that these changes in the membrana are not as often observed in this country as in Europe. The reason of this is not apparent. Politzer, in his book on the Membrana Tympani (translated by Mathewson and Newton: Wood & Co., New York, 1869), speaks of the periphery of the membrane as being somewhat opaque normally, in consequence of an excess of the circular fibres of the middle layer. The condition of opacity frequently seen, and resembling the arcus senilis corneæ, he explains in this wise: there is thickening of the fibrous layer at this point, and the development of fat granules; in some cases the mucous layer also becomes thickened. It has been stated that as a rule the opaque spots do not necessarily stand in the relation of cause to the existing deafness. Tröltsch is quoted by Politzer as saying, that when great deafness exists in conjunction with these changes, we may infer that similar appearances might be found in the oval and round windows.

These appearances seem, according to Politzer, to result in this wise: the fibrous layer of the membrana has few or no blood-vessels, and these changes first appear as exudations from the outer layers of the membrana, but more frequently from the mucous layer, which after a time undergo calcific degenerations. Ultimately a true bony formation may in a few cases result, the bone resembling that of the skull of a newly born infant. In chronic catarrh, Moos seems to have been the first to observe the formation of a calcareous deposit in the membrana. It occurred in a woman seventy years of age. Besides the fatty matter and amorphous material composed of carbonate of lime, dark-brown pigment molecules may be found interspersed between the fat globules. Politzer describes certain tendonous opacities which involve a large part of the membrane, the intervening normal portions appearing darker, and apparently but not really smaller.

In some instances Politzer found that the dermoid layer could be detached from the calcific formation, while in other cases it seemed to occupy all the layers of the membrana tympani. The mucous layer was more often incorporated with the calcific mass than the outer layer. In many of the old cases of inflammation the *membrana becomes atrophic and very thin, so as to bulge* considerably on inflation, or to be drawn out excessively by Séiglè's speculum, presenting the appearance of a thin cicatricial formation.

In adhesions between the membrana tympani and promontory the membrane will bulge on inflation, and may be seen to be fixed at the point of adhesion. A better plan is to use Séiglè's tympanic speculum and exhaust the air, when the point of adhesion may be easily seen.

Sometimes the malleus handle may be so adherent to the remaining ossicula or to the promontory, that no movement of the membrana is made on the most powerful rarefying or condensing of the air in the meatus by the pneumatic speculum. This must not be considered diagnostic, however, for many times the manubrium refuses to move noticeably, even when there are no adhesions. In the latter instance, however, we may suspect anchylosis of the ossicula.

The throat in chronic catarrhal otitis nearly always shows evidences of present or past inflammation.

The tonsils, more especially in children, may be enlarged and full of scars, showing the results of former inflammations. One tonsil is often more enlarged than its fellow, and probably will be found opposite the ear most affected. The uvula may be much swollen, but sometimes it is shrunken by atrophy, and may sometimes assume a spike-like shape, narrowing towards its termination. In many cases it will have a relaxed and somewhat œdematous appearance. I have seen what appeared to be a large drop of serum confined in its lower extremity. In phonation the velum is often seen to rise up unsymmetrically, perhaps less so on the side of the affected ear, the consequence of muscular relaxation or paresis, or it may be much limited in its general movements. The lateral half arches also contract in a similar manner. This also interferes to some extent with deglutition.

The voice is often affected, becoming unmusical. It may crack or break, and talking becomes very fatiguing, and singing is often impossible. The patient will talk in a high, shrill tone if there is considerable loss of hearing, which is so marked a symptom that the aural surgeon can sometimes make a diagnosis of incurable deafness from this sign alone.

These patients are very likely to need to use the pocket-handkerchief more than others; there may be a feeling of fulness in the frontal sinus, and headaches, and a dull feeling in the centre of the forehead due to the congestion of the mucous lining of the frontal sinus, or to obstruction to the flow of normal secretion from the sinus, through closure of the infundibulum, the result of catarrhal swelling. Again, the nostrils are unduly dry and scabby. On removing one of these scabs there may be a red or bleeding surface beneath, possibly granular. Sometimes the lining of the nostrils will be very pale and relaxed, giving rise to an excessive and thin discharge.

The septum is frequently bent to the opposite side, making it difficult or impossible to introduce a catheter in the obstructed nostril. The turbinated bones are often swollen, hypertrophic, and in some instances polypoid excrescences may depend from them. Mucous and fibrous polypi are found attached to the turbinated bones, the septum or upper part of the space. Granulomata are occasionally found.

The mucous lining of the nares resembles erectile tissue in many of its aspects. The slightest cause will often produce a swelling of the lining of one or both nostrils, so that the patient is unable to breath through them.

Often, while lying upon one side in bed, the upper nostril will be free and the lower one closed. This condition may be reversed on turning upon the opposite side. Again, if a patient becomes very much disturbed mentally, or is excessively fatigued, he may find, on retiring for the night, that his head is somewhat congested and the lining of the nares so swollen as to make nasal respiration difficult or impossible. This condition may disappear after the patient has become somewhat rested, and his condition of nervous irritation has passed off. I have no doubt but that the vasomotory nervous system is at fault in such instances. The upper pharyngeal space will often give the patient trouble from a feeling of irritation or actual pain, accompanied by collections of thin mucous easily removed by efforts at expectoration, strong nasal inspirations, etc., but dark or black or greenish masses of inspissated mucus may often be observed by the rhinoscopic mirror. These strongly adhere to the parts, and are only removed by syringing, and an act of screation on the part of the patient. Sometimes the violence necessary to their removal will cause bleeding. I have noticed these masses frequently on the posterior nasal septum. Occasionally they may be found on the vault of the pharynx, in the fossa of the Eustachian tube, or in that of Rosenmüller.

Fig. 47 will give a hint as to the normal appearances of these parts.

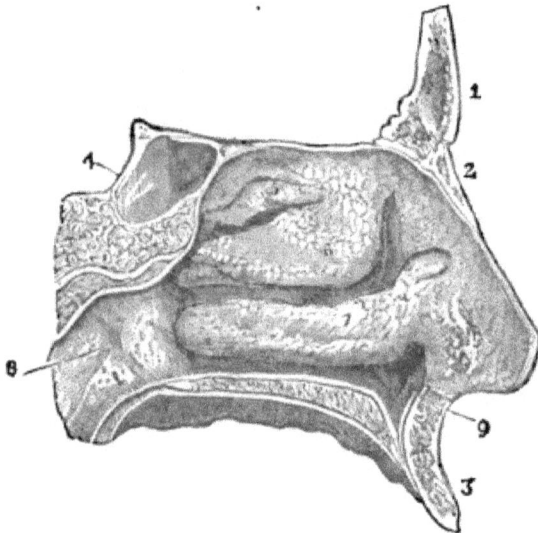

FIG. 47.

Outer wall of the left nasal cavity: 1, frontal bone; 2, nasal bone; 3, superior maxilla; 4, body of the sphenoid: 5, superior; 6, middle, and 7, inferior turbinated bones; 8, orifice of the Eustachian tube.

By rhinoscopic examination the Eustachian tube and its neighborhood reveal the following conditions: (Fig. 48), increased redness in or about the mouth of the tube, or an unnatural paleness like that described in the nostrils.

The mouth of the tube may be much swollen by hyperæmia. In other conditions œdema may produce a similar swelling, apparently obliterating the fossa of the tube, or so diminishing its size as to cause it to appear like a minute dimple; or the elevation separating the fossa of the tube from that of Rosenmüller may become greatly exaggerated, and sometimes surround the mouth of the tube like a collar. The posterior extremities of the middle and inferior turbinated bones are found enlarged by swellings, roughened, and the posterior nares generally much distorted in appearance. Granulations like those found in the pharynx are often met with here. Polypi in the nares, when situ-

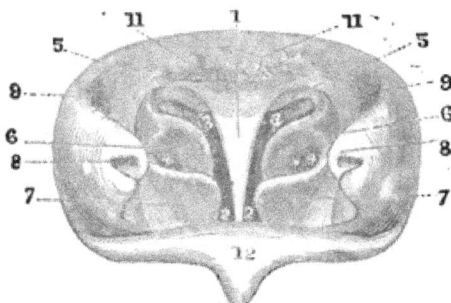

FIG. 48.

The rhinoscopic image in a normal pharynx: 1, nasal septum; 2, nasal passages; 3, superior meatus; 4, middle meatus; 5, superior turbinated bone; 6, middle turbinated bone; 7, inferior turbinated bone; 8, orifice of the Eustachian tube; 9, upper portion of the fossa of Rosenmüller; 11, glandular tissue at the vault of the pharynx; 12, posterior surface of the palate and uvula. (Cohen.)

ated somewhat posteriorly, are visible by means of the rhinoscope.

The mouth of the Eustachian tube is sometimes seen to move insufficiently or not at all during contraction of the tubal muscles, as in deglutition. This is better seen perhaps by Zaufal's nasal speculum (Fig. 49).

FIG. 49.—Zaufal's nasal speculum. Its three sizes are here indicated. Before introduction it should be anointed with vaseline.

Cicatricial striæ are often found in various parts of the upper pharynx, and sometimes the mouth of the Eustachian tube may be closed by bands of connective tissue passing across it, especially after the healing of a specific ulcer in the neighborhood. Quite as important as these changes are what W. Meyer of Copenhagen calls adenoid growths or vegetations, usually found in the vault or roof of the pharynx, and sometimes called pharyngeal tonsils. (Fig. 50). These are minute swellings or hypertrophies of the papillæ of the mucous membrane. They are often so numerous as to fill the upper pharynx, making a rhinoscopic examination impossible. By passing the finger behind the velum they may be felt, the sensation resembling that of a collection of worms. Their dimensions are one or two lines in breadth, with a somewhat greater length. They bleed easily, may be broken off and removed by rough manipulation with the finger, and behave in many respects like an ordinary mucous polyp.

There seems to be a large number of cases of adenoid vegetations on the continent of Europe, but here their oc-

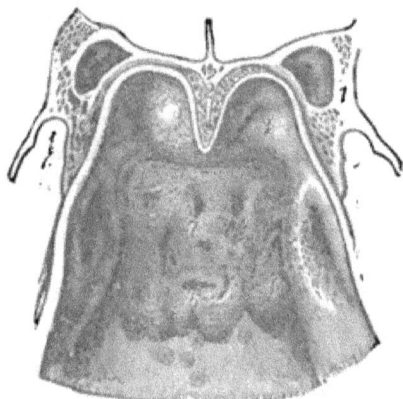

FIG. 50.—The normal appearances of the vault of the pharynx, showing its glandular structure: 1-1, pterygoid process; 2, vomer; 3-3, posterior portion of the nasal fossæ; 4, Eustachian tube; 5, bursa pharyngea; 6, fossa of Rosenmüller; 7, irregular surface of the glandular tissue. (Luschka).

currence is comparatively rare, according to the best observers.

Schwartze, in his "Pathological Anatomy of the Ear" (Dr. J. O. Green's translation: Houghton, Osgood & Co., Boston, 1878), states that on the tubal prominence and at the entrance of the pharyngeal orifice follicular ulcerations· may be frequently observed, and "in caries of the temporal bone with destruction of the osseous tube *ulcers from erosion* are seen on the ostium pharyngeum, if the foetid pus flows into the pharynx in large quantities." He mentions that an extensive tuberculous ulceration in a man 33 years old reached to the middle line of the fornix, and the posterior pharyngeal wall, involving the cavity of Rosenmüller, causing a deep excavation and destroying most of the tubal prominence. Syphilitic ulcerations on the tubal prominence and at the entrance of the pharyngeal orifice, with ulceration on other parts of the vaso-pharynx, as the septum narium, choanæ, fornix, and posterior wall of the uvula, can often be recognized by means of the rhinoscope. The Eustachian tube is rarely completely closed in chronic aural catarrh. During a number of years at the Manhattan Eye and Ear Hospital patients have been brought to me from another clinique, whose Eustachian tubes apparently could not be inflated. It was well known that I held the opinion that few tubes existed which I could not inflate with my faucial catheter. On actual trial I do not remember a single case for a number of years that I have not been able to inflate. Ordinarily, when air refuses to pass

FIG. 51.—TOYNBEE'S DIAGNOSTIC TUBE.

from the throat to the tympanum the cause may be looked for in a collection of mucus, generally at the faucial extremity, where so large a number of mucus-secreting glands are found, or the lining mucous membrane becomes swollen from hyperæmia, œdema, or infiltration into the sub-mucous connective tissue of the tube. The isthmus tubæ is the first to suffer by diminution of its calibre. In other cases the tubal muscles, according to Weber-Liel, fail to act, and the walls of the tube fall passively together, so that no interchange of air between the throat and tympanum is possible during deglutition or whenever contraction of the muscles ought to take place. In this condition Politzer's or Valvalva's inflation fails to open the tube, and only the catheter, possibly assisted by a bougie, succeeds in forcing air into the tympanum. It is in this condition that treatment to increase the power of the tubal muscles is indicated.

Objective Symptoms of Obstruction or **Closure of the Eustachian Tube.**—On inspecting the membrana during the act of deglutition, if there is no movement visible it points towards the fact of non-interchange of air between the throat and tympanum. As this may occur frequently in a normal state, it has only a moderate significance, and we cannot assert positively by this test that the tube is obstructed or narrowed. By means of the diagnostic tube of Toynbee (Fig. 51) we listen for the sounds resulting from the act of deglutition. We hear a certain distant and indistinct rumbling sound, which is caused by the contraction of the muscles concerned in deglutition, accompanied by a moist friction sound consequent on the rubbing together of the opposite surfaces of the mucous lining of the throat.

If air enters or leaves the tympanum freely, indicating a normal state of the tube, a tolerably distinct "click" or "crackle," caused by movement of the membrane, may be heard. If the tube is absolutely closed the first two sounds will be heard, but not the last. If the tube is pervious and the tympanum contains fluid, then a moist, mucous, or gurgling sound will be heard.

In making these tests the utmost quiet should prevail, and the diagnostic tube must be very carefully inserted into both the observer's and the patient's ear, and it should not be allowed to touch any intervening object, as a loud friction sound would result, greatly distracting the attention from the auscultation sounds. On inflating the ear by Valvalva's method, the only sound heard will be that produced by the entrance of air into the tympanum. This is accompanied by a rather sudden sound like a thud, and a rustling or crackling of the membrane, provided the latter moves. All of the sounds produced in the tympanum seem to be made almost *within* the observer's ear, so near do they appear to be. This is the best method of studying the sounds produced by inflation, for they are not modified by the blowing sound of the catheter or the deglutition sounds of Politzer's method. The moist crackling of fluid in the tympanum is quite easy to recognize. I do not believe that auscultation can differentiate between a tolerably dry tympanum and a normally moist one; the fluid must be in the tympanum in considerable quantities in order to be detected by this test. Where the tube is much narrowed, instead of obtaining a sudden and immediate thud-like sound, there will be a somewhat prolonged, faint, and possibly squeaking sound. If the tube is stopped with inflammatory products and the inflation removes them, then the sound will be slower in making its appearance, and will be more pronounced. If the patient makes a very violent effort at inflation by Valvalva's method, the muscular contraction concerned in the act may be faintly heard in the distance as an indistinct rumbling sound, but this is not a frequent occurrence.

In Politzer's operation the sounds of deglutition are the most distinct of any heard, and have been already described. The inflation is more prompt, the thud or crackle is of a sharper character and much more pronounced in quality than in any of the other methods of inflation. It is not as prolonged as that of either of the other methods, but the tube being widely opened a large volume of air fills the tympanum instantaneously, producing the inflation sound very promptly.

Naturally these remarks do not apply to the various modifications of Politzer's operation, where the patient blows air out of his mouth or utters the words Hick, hack, etc. A moist sound here, is distinguished in the same manner as in the other methods described.

In one case, on practising inflation I fancied I heard the rupture of an adhesion in the tympanum. It is safe to say that the imagination is sometimes called into play while studying these sounds.

On inflating with the catheter there is first the prolonged blowing sound incident to forcing air through a tube. As the proper insertion of the catheter includes the idea that it must fit into the tube so inexactly that most of the air forced into it escapes into the throat rather than the tympanum, there must be an addition to the blowing sound made by this escape of air into the pharynx. Whether air enters the tympanum or not, the blowing sound is heard well enough; but if it passes freely through the Eustachian tube into the tympanum the blowing sound is then heard as though it were produced in the observer's ear. This symptom is sometimes very startling. I do not remember to have made much out of the sounds produced by the movement of the membrana while using the catheter, though the moist tympanal sounds are easily heard. The auscultatory sounds while using the faucial catheter for inflating the tympanum are similar to those obtained by the use of the ordinary catheter. After making several efforts at inflation without eliciting any sounds, it will be well to direct the patient to swallow several times, while the nose is closed so as to empty a possibly overfilled tympanum, so that subsequently a considerable quantity of air may be forced in, which may elicit decided auscultation sounds.

A plan I have used very much to test the Eustachian tube, is easy to accomplish and quite striking in its results. While inflating, inspect the membrane carefully, and if even a small amount of air has entered, it may be determined by the appearance of the membrane—it has changed its position.

It may not be possible to see any movement; but if the position of the light reflex be carefully noted, together with the peculiar form of illumination of the membrana tympani, the slightest possible change of position of the membrane may be detected by the change in the size or position of the light reflex. Indeed, with the movement of the membrane the reflex may be obliterated, and perhaps one or more new ones caused to appear in other parts of it.

The change in the illumination of the membrane is also a conspicuous symptom. When the membrana was light it may become darker, and *vice versâ*. New features of *expression*, so to speak, may be brought out. Other methods still, of determining a pervious tube are more valuable than that by auscultation.

The effect of inflation on the hearing is one of the most valuable of tests, and is the easiest to apply of any.

If the hearing is improved by inflation, the tube is pervious; if the hearing is lowered this is still true; if tinnitus results, or if an existing tinnitus is relieved, the tube is pervious; if there is no change in the hearing, the tube *may* be pervious. If the patient feels something in the ear not before noticed, the tympanum has been inflated; if no sensation is produced the tympanum *may* have been inflated. Relief of pain in the ear, or the production of a pain from an attempt at inflation, proves a pervious state of the tube.

Another test of inflation of the tympanum, on which great stress was formerly laid, I believe to be of little moment— that is, the reddening of the membrane, or any part of it, as a consequence of the inflation, and therefore an evidence of it. In the *Transactions of the American Otological Society for* 1871, p. 49, I have recorded the results of 287 inflations of the tympanum. The cases were mostly of chronic aural catarrh, but a few were normal. Of the 287 inflations in 31 cases. 224 did not redden the membrane, 53 did, and 10 were doubtful. Thirty-two inflations were done by steam, in a very thorough manner, without reddening the membrane in the least. Forty-seven inflations were done in patients of normal hearing, with congestion of the membrane in 21 instances, no effect 25, and uncertain 1. Inflation of the old cases, with sclerosis and atrophic bloodvessels, as might have been expected, did not often result in reddening of the membrane, while the more recent ones much more frequently did. Occasionally I found a case where the slightest inflation produced a considerable reddening of the membrana. These observations seemed to point to the fact that inflation is capable of doing considerable violence to the tympanum, and it has been known to cause an acute otitis.

Pathological Considerations.—In chronic catarrh, the mucous lining of the tympanum, in the earlier stages, may be more or less swollen and reddened from increased vascularity, or preternaturally pale, secreting more than a normal amount of mucus or sero-mucus, the mucous glands being hypertrophic. The submucosa may be thickened by cell-proliferation, consequent on the inflammation. Occasionally the secretion may be sufficient to cause the membrane to bulge and to require evacuation. At a later stage the opposite condition may be found, when great dryness of the tympanum results, in which there is atrophy of the mucous glands, and a condition resembling that of pharyngitis sicca. In earlier stages "villous prolongations and slight elevations are seen on the surface of the membrane" (mucous lining of the tympanum); "the thickening may be confined to certain spots, the mucosa of the drum membrane, the maleo-incal articulation, labyrinthine fenestræ, or it may be equally distributed over all portions of the membrane, and may even completely obliterate the whole of the tympanic cavity.". . . "With the thickening of the mucous membrane the membrana tympani appears thickened, leathery, and but slightly yielding to the touch." . . . "Marked thickening from a new growth of connective tissue on the fenestræ of the labyrinth, and around the articulation of the malleus and incus, are specially injurious to the conduction of sound. The nich of the fenestra rotunda may be completely closed, and the ossicles may be wholly imbedded in the hypertrophied mucous membrane, so that careful preparation is necessary to render them visible."

Bands of connective tissue, the result of inflammatory proliferation, are frequently found in the tympanum. The membrana may be bound to the promontory by adhesions resembling the synechiæ of iritis. It may be brought about in this manner, according to Schwartze: When the membrana from any cause is pressed inward so as to touch the inner wall, the epithelial surfaces are destroyed by pressure, and the mucous membrane becomes changed into a vascular granulation tissue, which is subject to the usual cicatricial contraction." . . . "Still more common than this direct union are the so-called pseudo-membranous growths. They occur simultaneously in various forms in the same ear, and may be so numerous that the whole cavity appears to be filled with an irregular network. They are so common as to be found in about every fifth ear (Wendt); when recent they appear of a red or grayish-red color, soft and succulent from serous infiltration; when old, whitish-gray, or white and firm." . . . "They may unite the ossicula with each other, or with the wall of the tympanum; the drum membrane with the tympanic wall, the stapes, or the long process of the incus; the tendon of the tensor tympani muscle with the roof of the cavity or the ossicula; very frequently an arm of the stapes with the walls of the fenestræ ovalis. The fenestra rotunda and the ostium tympanicum tubæ may also be completely or partially covered by them, thus causing complete or partial closure of these openings." The tendon of the tensor tympani may in such cases also become involved. Sometimes the tympanum may be divided into several cavities by these bands. Where these adhesions are tense and rigid the sound conduction is interfered with, especially with synechiæ on the stapes, owing to its very limited normal vibration.

Many of these membranous bridges are the remains of the mucous tissue which fills the tympanum of the fœtus and new-born child; they are the result of incomplete retrogression of this tissue.

" The pathological connecting bands are produced". . "by the contact and union of portions of the mucous membrane when in a state of swelling and proliferation." If adhesions between surfaces are extensive, there will be wide bands or considerable membranous formations, but if only a narrow adhesion takes place the synechiæ will be narrow like strings or threads. Especially will this be the case if after an adhesion has formed it is stretched by movement of the drum membrane outward during inflation. This is analogous to the synechiæ of iritis.

It is not possible to distinguish between the adhesions resulting from the remains of the fœtal mucous cushion and those dependent on pathological processes. It is quite probable that these cicatricial bands and membranes act as exciting causes of recurring attacks of otitis, as iritis with adhesions provoke recurring attacks of iritis. These membranes and bands undergo the same retrogressive changes that other similar tissues do, as atrophy, fatty degenerations, sclerosis and cicatricial contraction, calcification, and ossification.

When the membrana has been sunken for a long time it may rest against the head of the stirrup, the incus being pushed aside in consequence of a subluxation dependent on a distended and relaxed capsule. Rigidity and immobility, with or without anchylosis in the annular ligament of the stapes and in the malleo-incus articulation, is very common.

"The capability of vibration in the ossicula is diminished by thickening or rigidity of the mucous membrane which covers these bones (sclerosis, calcification, or ossification of the periosteal connective tissue, with cellular and serous infiltration of the sub-epithelial layer), by synechiæ, and by the imbedding of the bones in hypertrophied connective tissue (membranous anchylosis)."

"If the whole annular ligament, or even the periosteal layer of the mucosa which covers it, is changed into a mass of lime, absolute immobility of the stapes results. Similar results may be obtained by osseous changes in this region, the latter being more likely to occur in those more advanced in years, although not necessarily so. If the stapes has been immovable for a long time the crura become atrophied so that they break at a slight touch.

"The malleus and incus may become ossified with the upper wall of the tympanum." . . "Exostoses on the ossicula" are common on the incus, but less so on the malleus and least so on the stapes.

"Enchondromata apparently are developed quite often on the sharply projecting processus brevis, such as is seen with a retracted drum-membrane." In synechiæ between the promontory and membrana resulting in arresting the functional activity of the tympanal muscles, the latter may undergo fatty or fibrous degeneration, or become atrophic."

. . . "Shortening of the tendon of the tensor tympani may result" . . . "from connective-tissue adhesions between the tendon and its sheath;" . . . "from retraction of the mucous covering of the tendon the result of thickening of the general tympanic mucous membrane," as described by Politzer, or "from membranous or thread-like synechiæ connecting the sheath of the tendon with the roof of the tympanum or with other parts of the cavity." . . . "Hinton found fibromata on the tendon of the tensor tympani." "In the Eustachian tube in the earlier stages there is congestion, swelling, and increased secretion of its mucous lining."

The mucous secretion, as in the throat, may become inspissated and block up the whole of the tube, forming a jelly-like mass, sometimes projecting from the faucial extremity of the tube. Similar masses have been found at the same time in the tympanum.

"There is also in chronic cases a marked projection and wrinkling of the mucous membrane perpendicularly to the axis of the canal (at the pharyngeal orifice), with hypertrophy of the glandular layer and thickening of the submucous connective tissue." Other changes occurring in the pharyngeal extremity of the tube are detailed under the head of Rhinoscopic Appearances.

Granular changes in the lining of the tube sometimes occur. "Small spots of ossification in the tubal cartilage have been described by Moos;" calcification of the same also sometimes occurs.

"*Contraction* or stenosis of the Eustachian tube even to complete closure takes place from swelling of the mucous membrane or thickening of the submucous connective tissue in catarrh, from hyperplasia" . . . "at the pharyngeal orifice."

"Hypertrophic thickening of the soft palate, by which the anterior lip of the tube may be pressed against the posterior lip," so as to close the tube, sometimes occurs. The pharyngeal tonsils may be so swollen from hyperplasia as to partly cover the mouth of the tube and compress it so as to render it a mere slit (generally the pharyngeal tonsil is above and behind the orifice of the tube, and does not encroach on it). "New growths in the naso-pharynx, such as naso-pharyngeal polypi, large cysts, cicatricial bands," . . . "great swelling of the lower nasal cartilages, or great hypertrophy of the palatine tonsils," result in closure of the tube.

"Real strictures, in the sense in which urethral strictures are formed, by thickening and atrophic shortening of the tissues, appear not to occur in the Eustachian tube."

"An osseous stricture of the tube" . . . "below the ostium tympanicum was seen and figured by Toynbee in the *Med. Times*, Feb., 1850. A bristle could scarcely be passed through it."

In *atrophy* of the mucous membrane of the naso-pharynx the ostium pharyngeum appears unusually wide open and deep. The tubal prominence then projects very much, and from the thinning of its mucous covering appears almost bare."

"Acquired enlargement of the canal" occurs in connection with sclerosis of the mucous membrane of the tympanum, when it may become three or four times its normal calibre." . . "The tube may also gap throughout its entire length, according to Rüdinger, in the *Monatschrift für Ohrenheilkunde*, 1868, No. 9, in consequence of atrophy of the musculus dilator tubæ. Fatty degeneration and atrophy of the tubal palatine muscles are the frequent results of chronic retro-nasal and tubal catarrh." . . . "A muscular hypertrophy of the musculus tensor veli palati vel dilator tubæ has been described by Moos as the result of chronic tubal catarrh."

In the above pathological observations the quotations refer to Herman Schwartze, M.D., on the Pathological Anatomy of the Ear, translated by J. Orne Greene, M.D.

The prognosis of chronic aural catarrh depends on many factors.

These are as follows: If the patient has a good constitution, without any strong catarrhal tendency, and the disease has produced no great change in the tympanic lining, he will make a good recovery, with hearing normal or sufficient for the ordinary avocations of life. The swelling of the lining of the tympanum and Eustachian tube altogether disappears, the exudations which may have taken place become wholly absorbed, the Eustachian tube returns to its normal functional activity, and the membrana assumes a perfectly correct position.

Again, if the patient be inclined to catarrh,—that is, if he fails to resist the irritating influences of our climate,—he may be cured of a first, second, or third attack, and so on, but at each relapse the hearing may fail to be perfectly restored, and after a time a high degree of deafness may result. From this it may be inferred that the hearing which fails to come up to the normal standard a few weeks after a relapse has occurred, is likely never to do so. For the most part this is true.

In this class of cases it is easy to infer that the important indication is to prevent the relapses, on which the unfavorable prognosis depends.

In the cases already referred to, an obstructed Eustachian tube as one of the factors on which the progsis turns, is presupposed.

In a few subjects, notably the tuberculous, who may rarely have a period even of complete recovery from catarrhal symptoms, the tube may never be normally pervious, in spite of our treatment; *this* symptom being the important one in determining the prognosis. They will often hear very well while the tympanum is inflated, but the tube fails to keep up a normal supply of air in the tympanum. Again, the Eustachian tube may have become pervious, but changes in the tympanum may have taken place, rendering recovery of the hearing impossible. A reference to the topic of Pathology in this article will render this point clear. Another class of cases will have a pervious tube and a collapsed membrane, with considerable deafness, dependent on the collapse. On inflation the hearing will be much improved, and if there has been tinnitus the latter may be relieved; but on the first effort at deglutition the tube opens, and the air in the tympanum rushes into the throat. The membrana, which had been kept in position only by the excess of intratympanic air-pressure, flies back to its old position, and the deafness and tinnitus return. These cases are not benefitted by treatment, except possibly by operations hereafter to be described. The membraneous bands in the tympanum are not removed by treatment, although occasionally they may be ruptured by inflation, much as synechiæ in the iris are broken by atropine, or they may even be divided by a tenotome. On the whole, however, comparatively little is accomplished. There is still another class of cases where the disease has gone on to atrophy of the mucous lining of the tympanum together with its glands, resulting perhaps from a previous condition of connective-tissue proliferation of its mucosa, where the membrana is much thinned, easily bulges on inflation, is too translucent, and the reflex from the promontory shows a bloodless condition of its lining, the Eustachian tube being not only pervious, but perhaps of increased calibre from atrophic changes, not only in its cartilaginous but its osseous portion. Naturally under such conditions no improvement could take place by inflation, and nothing remains to be done by way of effecting absorption of inflammatory products, for they have disappeared in the slowly retrogressive changes until a condition of sclerosis is the result. Naturally this class must occur in persons advanced in years, for nothing but a considerable lapse of time could eventuate in such ultimate changes. As a matter of fact, no amount of treatment can benefit these cases.

It has not been quite fully admitted in treatises on the ear, but it is nevertheless a fact, that most if not all the cases of profound deafness from this disease result, in part at least, from labyrinthine complications. Under the head of Diagnosis will be found the mode of determining what class of cases have these complications. As a rule, and I hardly remember an exception, the deafness in these cases which depends on the labyrinth trouble is incurable under all circumstances. Another rule in prognosis is invaluable; in a case of chronic catarrh, where no further diminution of hyperæmia may be hoped for from treatment, and where there are plainly no intra-tympanal adhesions, the amount of hearing to be expected from treatment is that which may be gained by inflation. Once in a long time I have increased the hearing beyond the highest point obtained at first. It is difficult to explain this: it may require several visits before the exact amount of air is injected into the ear which will ensure the highest possible amount of hearing, and the failure to carefully settle this point may have been my source of error.

Tinnitus aurium in more recent cases may often be relieved, but if the tinnitus has continued several years or perhaps six months, it is rarely cured. Many patients are solicitous as to whether their hearing will be still further impaired. The answer to that appears in what has already been stated: if inflammatory processes are still in operation a greater amount of deafness may be expected, but in others just described, where the fire of inflammation is as it were burned out,—that is, where sclerosis exists,—no further impairment of the hearing may be expected. The same rule sometimes seems to apply to tinnitus aurium.

In certain advanced cases it seems to disappear without

treatment, but as numerous labyrinthine conditions, undiscovered during life, act as causes of tinnitus, it will appear that an accurate statement of prognosis in a given case is impossible.

Whee the patient's voice is considerably altered so as to produce the loud, harsh, and discordant tones previously described, the prognosis of any considerable improvement to the hearing cannot be made. The same is true if the patient has a peculiarly observant manner, so to speak, eying one at every turn, showing that he depends on his eyes rather than his ears to determine what is said. The younger the patient the more favorable is the prognosis.

The average aural surgeon is able to say to an old or elderly subject who has been hard of hearing for several years, that there is no hope of relief. If there are periods of comparatively good hearing, or if the patient hears better in fine weather, the case is more hopeful. If the patient hears much better in a noise, it is of ill omen for a favorable prognosis. If there are occasional deep-seated aching pains in the ears, or if they are sensitive to cold air, there is likely to be progressiveness of the hardness of hearing. Tröltsch, in his Treatise on Diseases of the Ear, translated by Roosa, quotes Politzer to the effect that if a patient has a persistent tinnitus the prognosis is bad, even though decided improvement to the hearing may suddenly occur. The latter will probably not be permanent.

When the first sign of trouble is tinnitus, continuing some time without perceptible loss of hearing, the prognosis is unfavorable. Large calcareous spots on the membrana with great deafness, according to Tröltsch, "are, as a rule, very unfavorable for treatment." "These degenerations are generally connected with morbid processes on the fenestra rotunda and ovalis."

Treatment of Chronic Aural Catarrh.—General considerations.—Inasmuch as this disease is a consequence of inflammation of a mucous membrane, which usually commences in the throat and nares, and travels up the Eustachian tubes to the middle ears, causing the mischief already detailed under the head of Pathology, our endeavors should be directed to the subduing of this inflammation and remedying as far as possible its consequences. The throat and nares should be restored as nearly as possible to a normal state and maintained in that condition. The nares especially should be looked after, as a free passage of air through them is essential to the well-being of the Eustachian tubes and drum cavities. This has been discussed under the heading of Acute Catarrh. The Eustachian tube must be rendered pervious, and any atony or paresis of its muscles or a collapsed condition of its walls should receive appropriate treatment. The membrana needs to be forced outward by inflation or other means; any hyperæmia of the drum cavity must be disposed of; bands of adhesion in the tympanum should be broken up or divided if possible; rigid or sclerosed conditions of the membrana require attention, and when the Eustachian tube is impervious to air, the membrane may be punctured or incised to maintain the proper amount of air in the tympanum. Although this is a chronic disease, the relapses of inflammation that are frequently met with are more of the nature of an acute affection, and on the management of these will largely depend our success in treatment. It should be clearly understood that when a catarrh has reached a certain stage of development very little is gained by treatment. In earlier stages, if the patient has some pain, especially if it be of a throbbing character, it will be proper to use one or two leeches on the posterior surface of the tragus, every two or three days, until all pain or feeling of fulness has disappeared from the ear. Care needs to be taken, however. Do not mistake the pain and feeling of fulness due to a collapsed membrane for the pain of hyperæmia. I purposely avoid speaking of inflation at this juncture, for one would infer, from reading some of the books on the ear written since Politzer devised his method of inflation, that this was almost the principal treatment of chronic catarrhal otitis. I well remember the mistake a pupil of mine once made. He had heard me say much about inflating the ear and rendering an obstructed tube pervious, so he tried the treatment on a patient with chronic aural catarrh, but the hearing failed to improve as he had expected; he then brought the patient for advice. I found the tube moderately pervious and inflation did not improve the hearing.

It was obvious that the cause of the deafness was in the congested state of the tympanum and not in the obstructed tube. I applied one leech to the posterior face of the tragus every three or four days; each time the hearing improved, and at last it was so perfect that the patient found no difficulty in practising an avocation quite trying to the hearing —that of a stenographic law reporter. Other means are serviceable in diminishing the hyperæmia of the tympanum. Warmth, moist or dry, preferably the latter, is very effective. The most elegant method of applying dry heat is by means of a rubber bag containing warm water and laid on the ear. The details of this method are found under the heading of acute inflammation of the middle ear. I have often noticed in old cases that occasional aching pains would appear from time to time in the ear, with reddening of the inner end of the meatus, and perhaps slight reddening of the membrana. Such cases have been relieved by painting tr. of iodine on the inner extremity of the canal, and sometimes lightly on the membrane itself. Care must be taken to make the application lightly, as it may be painful when too much is used. Besides the warm applications to the ear, the latter should be kept warm, if need be, by a bit of cotton in the meatus, or an ear muff may be worn. If the ear is becoming congested from the cold, there will be a painful sensation, or at least a feeling of insufficient protection from the cold when the temperature is too much lowered. The prevention and cure of hyperæmia of the tympanum are at all times very important steps in the treatment of this disease.

Where in the earlier stages of inflammation there still remains hyperæmia, mucous or sero-mucous collections are likely to be found in the tympanum. These should always be removed by paracentesis or otherwise. Inflation of the tympanum, if the collection is not of too great consistency, may be sufficient to remove it through the Eustachian tube. A physician under my care had serous collections in his drum cavities. I removed them repeatedly by paracentesis, but he, at length evidently becoming wearied with the punctures, caused its removal in the following manner: he inflated his tympanum by Valsalva's method, turned his head forward and somewhat to the opposite side, when the excessive air-pressure in the tympanum, aided by the elastic membrana, forced the collection into the throat. Large doses of iodide of potass. will sometimes put an end to these intra-tympanal collections, or full doses of quinine will sometimes accomplish this result. A strong solution (20 to 40 grs.) of arg. nit. applied to the mouth of the Eustachian tube, or even painted on the membrane, will often accomplish the same result. When the collection depends on a membrane collapsed from insufficient air-pressure on the walls of the vessels of the tympanum, the obvious procedure is to inflate and restore the normal air-pressure upon the tympanal vessels. Where puncture becomes necessary, the rule already laid down in the chapter on acute inflammation of the tympanum may be followed. It is not a very painful operation, and in some cases causes no pain whatever. There will sometimes be no bulging of the membrane, so the puncture may be made almost at any point not high enough to touch the chorda tympani, which passes across the tympanum above the short process of the malleus, nor in the vicinity of the extremity of the long process of the incus, for by so doing we might disarticulate it from the stapes, or wound the oval window. If it is done rather high up and posteriorly, by which means a deeper portion of the tympanum will be reached, permitting more thorough penetration of the instrument, and consequently a wider incision in the membrane will result; the paracentesis needle should be held so lightly in the fingers that by the sense of touch we may determine just when the membrane is punctured. After the opening is made the tympanum is to be inflated while the head is turned towards the affected ear. If the fluid is removed with difficulty in this manner, apply a rubber tube or Séiglè's speculum to the meatus, and by exhausting the air (and if need be repeat the inflation at the same time) the tympanum may be thoroughly emptied. Do not, however, exhaust too strongly, or blood may flow and the ear become hyperæmic.

This process may require to be repeated many times. The following case from Dr. Burnett, in his Treatise on the Ear (Henry C. Lea, Phila., 1877, p. 428), illustrates this point: A gentleman 80 years of age had a brownish transparent fluid in the tympanum, visible only through a thin depressed cicatrix. Eustachian tube impervious. Hearing for the watch was $\frac{1}{80}$. Twenty or thirty drops of fluid were removed by incision; hearing arose to nearly the normal degree. In one week the tympanum filled again: another puncture was made, with results similar to the first; in another week a slight return of a "muffled feeling" was relieved by a third puncture, with the escape of a small amount of fluid, when the hearing became normal and remained so.

Another case from the same author: July 1, 1874; a gentleman, aged 55, for the last year has noticed a gradual diminution of hearing in the left side. Treatment by inflation was employed by other physicians, which did some temporary good. Hearing distance, watch, $\frac{4}{40}$. Inflated the ear several times a week for over a month, which temporarily improved the hearing. On September 12th the patient stated that he felt something like a drop of fluid moving in the ear whenever he turned his head about; when he lay down he heard better. As this seemed to indicate the presence of fluid in the tympanum, puncture was resorted to, resulting in the evacuation of about twenty drops of a brownish, transparent, serous fluid, with streaks of opaque mucus. Hearing distance increased from one inch to five feet. Complete relief continued until March, 1875, when, after a cold, he had a feeling of fulness in the ear, but there was little sensation of movable fluid. A paracentesis again gave relief. By the 23d the ear filled again, and was punctured with similar results. On April 15th he had the muffled sensation in the ear, and again the membrane was punctured with relief. On May 8th and on June 8th the same procedure was followed by similar results, and the patient had no further trouble until September 8th, when the symptoms returned, and a puncture was again made; also one on October 26th, November 24th, and January 3d, 1876, February 19th and March 28th. On February 19th air-bubbles were visible in the tympanum through the membrane, which could be seen to move under Valsalva's inflation. Complete relief resulted from the last operation. It is astonishing how small a collection in the tympanum will suffice to lower the hearing. Some time since I had a case of chronic otitis, in which the patient grew suddenly deaf without apparent cause. I observed that the light reflexes on both the drum membranes were extraordinarily bright. The membranes were both much sunken, and at first it did not seem possible that fluid could be lodged in the tympanic cavities. The excessively bright reflexes on the membranes, however, led me to suspect the presence of fluid. I punctured, and by means of inflation removed not more than two drops of fluid, with the result of at once rendering the hearing nearly normal. The collection returned several times, and after each puncture results similar to the first were obtained. The patient noticed subsequently, besides the sudden deafness, a fulness as though the ear were stopped up or had something in it. There was no sign of the fluid by inspection, the membranes being too opaque for the fluid to be seen through them.

I am of the opinion that in view of the fact that it is often impossible to diagnosticate fluid in the tympanum by inspection, that an exploratory puncture is justifiable, it doing no harm, causing little or no pain, and only leaving a little soreness behind which is of little consequence. Some years since I read a paper before the New York Medical Journal and Library Association on operations on the membrana tympani, in which between two and three hundred punctures of the membrane were recorded. Only a few resulted in inflammation and pain, and in none was the hearing lowered.

Counter-irritation behind the ear may do something towards diminishing an hyperæmic tympanum. This is at least an unfashionable measure at the present time, and there is danger that its merits will not be properly recognized. Cantharidal collodion may be used for this purpose, being applied with a brush over the mastoid region every four hours until vesication results. Care needs to be taken not to touch the sulcus behind the auricle, for a very disagreeable ulceration may result. Dress the blister in the usual manner. It may be kept open by occasional applications of the cantharidal liniment, or the collodion may be reapplied when the first application no longer causes sufficient irritation.

A plan I like very much is to paint the part with tr. of iodine every day, until considerable irritation is produced. This is less painful, and does not cause the rather offensive moistening of the parts characteristic of the blister. It may be well to keep up irritation for weeks, and sometimes for a month or two.

When the throat comes under consideration much may be said about the effect of counter-irritation applied to that part as favorably affecting the tympanum, possibly by reflex action.

Any means whatever which lessens the flow of blood to the head is likely to diminish hyperæmia of the tympanum. When the cerebral vessels are full, and the tympanic injection seems to be in any way connected with it, it will be proper to administer bromide of potass. or ergot, with a mustard plaster, perhaps, at the nape of the neck. A warm foot-bath will often give relief.

Under the head of Acute Catarrhal Inflammation of the

Ear, it was recorded that change of climate often relieved a hyperæmic condition of the tympani in a few hours or days. It is well to remember that if the patient is fatigued or harassed, or has been insufficiently nourished, the tympani will become congested if the patient has been affected with chronic aural catarrh. Then the obvious treatment is rest and food, and sometimes stimulation will have the effect of overcoming any local congestions.

Means for the correction of the faulty position of the drum membrane may be used at any time, even when relapses occur. For the most part this will consist in blowing air through the Eustachian tube, to produce a normal supply in the tympanum, and by excess of air-pressure to force the membrane outward. On the whole, Politzer's mode of inflation is the best. By it there is no forcing of the tube. During the act of deglutition the latter is opened, and air naturally passes into the tympanum in consequence of the inflation. The amount of inflation necessary is to be determined as follows: as long as the hearing improves, continue to inflate. If after a time the hearing is lowered, then an excess of air has entered the tympanum, and it will be proper to desist. If pain results, inflation has been overdone. If tinnitus is caused by the inflation, or if the patient is rendered giddy or faint, there has been too much air forced into the tympanum. The inflation ordinarily may be repeated as often as the good effect of the first has passed away. I would not, however, give the patient a bag for self-inflation. In many instances this practice has resulted in destroying the elasticity of the membrane, so that when it is forced outward it remains in an unnaturally flat or bulged condition, or it has become so limp as to flap to and fro with every change in the intra-tympanal air-pressure.

Enough has already been said of the mischievous effects of excessive inflation. By my own experiments previously recorded, it has been seen that hyperæmia and even inflammation of the tympanum may result from injudicious inflation.

Where the tube walls are collapsed, the Eustachian catheter will become necessary to inflate the tympanum. It has been argued by many, and notably by Dr. Roosa, of New York, that the contact of the end of the Eustachian catheter with the faucial extremity of the tube, by its irritation, produces a remedial stimulation on the lining of the tube, and unloads blood-vessels by the increased discharge caused by the contact of the catheter. It certainly is true that the tube becomes more pervious to Politzer's method of inflation after the catheter has been used.

On the other hand, it has been argued that whatever causes irritation to the mucous membrane aggravates the catarrhal condition. This seems certainly to be true in the treatment of the lachrymal sac and nasal duct. A few years since, bougies were introduced into the latter and allowed to remain from a few hours to a day or two. The practice was found to aggravate the previous catarrhal condition, and was abandoned. Now, many only leave the probe in from a minute or two to not more than ten minutes. This is my own practice. I certainly feel that the Eustachian tube should not be unduly irritated.

Another objection to the use of the catheter is its painfulness. Even if not actually painful, it is extremely disagreeable. Introducing it in one's own nostril will sufficiently settle this point.

Again, it is frequently impossible to introduce it in consequence of malformation of the nostril, or changes in its calibre incident to the catarrhal condition. It is true that Dr. Noyes' double-curved catheter (see Index) may be introduced into the tube through the opposite nostril, if that happens to be permeable.

Again, it is often desirable to inflate only one ear instead of both, which the catheter does much better than can be done by other methods. I am aware that Dr. Knapp of New York and Gruber of Vienna have added a long tapering tip to the Politzer nose-piece, so that the ear on the same side as the nostril into which the tip is introduced is more likely to be inflated. My faucial catheter will also inflate one ear only at a time, and is therefore desirable for the same reason that the Eustachian catheter is. Dr. C. J. Blake, of Boston, in the Progress of Otology in the Report of the First International Otological Congress, New York, 1877, quotes Josef Gruber in the *M. f. O.*, No. 10, 1875, as suggesting a new method of practising Politzer's inflation: While the air is forced into the nostrils, the patient is directed to pronounce the words hack, heck, hick, hock, huck, by which means the base of the tongue rises up, presses against the soft palate, and pushes it upward and backward against the posterior pharyngeal wall, and shuts off the upper from the lower pharyngeal space. If the patient's head is inclined to one side during inflation, in this manner, the air passes into the ear which is uppermost. Dr. E. E. Holt, of Portland, Me., suggests that by filling the cheeks with air during inflation of the tympanum, that the act of deglutition may be dispensed with.

Dr. J. O. Tansley, of New York, directs the patient to purse up the lips and blow air out of the mouth strongly, as though a lamp-flame were to be blown out. I have latterly practised Dr. Tansley's blowing method, and find it effective. Repeating the letter K successively is also a good manœuvre.

The other methods are more complicated, and the patient gives trouble in understanding what is wanted of him. I am sure that Politzer's method, with a hospital patient of the usual intelligence, is sometimes quite troublesome to apply.

With children, none of these details are necessary. Their tubes are of large calibre and easily forced. The more frightened they become the more likely is the soft palate to fly back upon the posterior pharyngeal wall, and shut off the upper from the lower cavity. Valsalva's method is objectionable, as it often imperfectly inflates the tympanum, and frequently does not inflate it at all. Congestion of the cerebral vessels sometimes results from the straining effort in inflating by this method. It is claimed by many that the pressure of the column of air in the tympanum acts as a stimulant to the mucous lining, acts curatively on the inflamed membrane, and excites absorption of inflammatory material. I am by no means certain that the effect is more than mechanical, dislodging obstructions from the tube, dilating it, and forcing the membrana into a better position. Where there is reason to believe that adhesions between the membrana and the ossicles, or inner wall of the tympanum, exist, it will be proper to inflate with considerable violence, using at the same time Siéglè's otoscope to assist in the outward movement of the membrane. I believe I have ruptured synechiæ by these means. Ordinarily I would not use traction by Siéglè's instrument, nor any other method of suction, to aid in the outward movement of the membrane, as practised by Dr. Howard Pinckney, of New York, and others. Dr. P. has, however, reported some good results from this practice. In my experience it has congested the drum cavity, increased the tinnitus, caused giddiness and faintness, and has shown a tendency to lower the hearing. It seems to cause too much violence. When speaking of the management of the Eustachian tube it was stated that other methods were more to be depended upon for restoring a normal supply of air to the tympanum than these. Where the Eustachian tube is impervious, or of such small calibre as to admit of insufficient ventilation of the tympanum, it will be proper to puncture the membrane for the purpose of increasing the supply of air.

A large incision is not necessary, as in any event the aperture closes in a day or at most a few days. This being temporary in its effects, repetition is called for.

I have a chronic catarrhal patient who comes to the hospital occasionally to have his tympanic membrane punctured. The symptoms of insufficient air-supply in the tympanum are annoying enough to cause him to apply for the relief which he has so often experienced. In his case it causes very little pain or subsequent soreness, and he always expresses himself as relieved by the operation. I do not always take the trouble to use a speculum, but depend on the sense of feeling in making the puncture.

The main cause of failure in this is the wonderful power of the membrana to repair itself, so that it matters not how large a portion of the membrane has been removed, nor by what means, repair is likely to take place sufficient to close the tympanic cavity, and defeat the object for which the operation was performed. F. E. Weber first recommended the division of the tensor tympani muscle near its insertion on the manubrium, together with any adhesive bands which might prevent the membrana from assuming its normal position. His theory was to this effect: The tendon of the tensor, during a long-continued collapse of the membrane, had undergone a secondary retraction, so that the membrane could no longer assume a normal position, even though the Eustachian tube was pervious and sufficient air-supply could be obtained for the tympanic cavity. Besides the pressure upon the contents of the tympanum, perpetuated by this enforced collapse of the membrane, a constant exciting cause of catarrh of the tympanum existed. Weber-Liel divided the tendon by entering the tympanum through the membrane, in front of the malleus handle, and catching it at that point. His instrument consisted of a handle of sufficient size and length to be conveniently grasped. To this was attached the tenotome, which consisted of a hooked knife attached to a narrow shaft, the end of which was fixed to the extremity of the handle at a somewhat obtuse angle. At the latter point there was a cog attached to a movable button or slide in the handle, enabling the thumb, when the in-

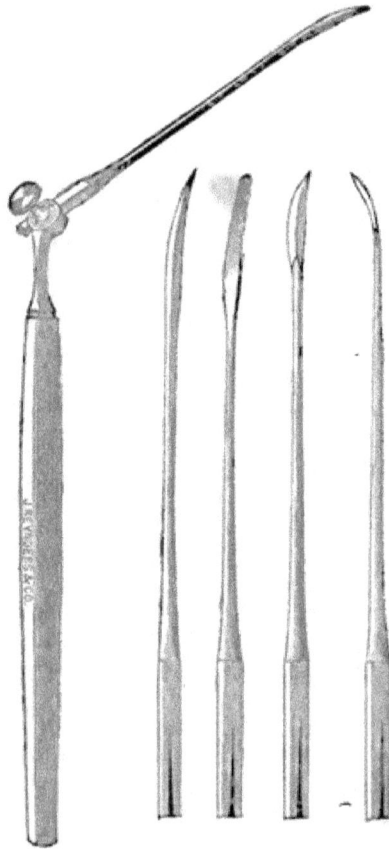

FIG. 51.

1, Gruber's tenotome and adjustible handle; 2, Gruber's tenotome; 3, J. O. Greene's tenotome; 4, 5, two views of Hartman's tenotome.

strument was grasped, to push this slide or button downward, causing the knife to make a revolution of about one fourth the circle.

After passing the knife into the tympanum, and hooking it on to the tendon of the tensor, the latter is divided by causing the revolution of the knife as before described. In removing the instrument from the tympanum the hook should be revolved in an opposite direction to that which accomplished the division of the tendon.

I much prefer Gruber's instrument. The knife is quite different from Weber's, as seen by the accompanying cut (Fig. 51). The blade of the knife is one cm. long, slightly curved at the end. It is fastened by a screw in a handle, which admits of its being adjusted for either ear. Following the suggestions of Voltolini and J. Orne Greene, I have always done the operation posteriorly to the malleus handle, instead of in front, as Gruber suggests, and have used Greene's modification of the Gruber knife oftener than otherwise. The latter, represented by Fig. 3 in the cut, is a blunt, spatula-shaped knife, curved flatwise, as is Gruber's. Being of this shape, it is less liable to wound any important parts in the tympanum; moreover, it admits of being somewhat wider near its extremity, and consequently may be made to cut better. Greene penetrates the membrana with a pointed instrument preparatory to using the tenotome, but I have found that this tenotome may be pushed through the membrane without undue violence, thus simplifying the operation. My mode of operating is to enter the membrane near the extremity of the manubrium, extend the incision upward until the tendon is reached; divide this by an up and down motion of the knife. The numbers 4 and 5 in the figure refer to Hartman's tenotome. It has a double curve—one upward on its cutting edge, and the other forward on its flat surface, so as to prevent injury to the stapes and to the chorda tympani (Burnett). I have purposely written briefly on this subject, because I believe it to be comparatively unimportant. In Germany some years since the operation was frequently resorted to by many of the more eminent aural surgeons, and comparatively good results were reported. In this country the operation has not been done extensively, as the indifferent results of the operations done by the few who have attempted it have deterred others from doing it, Bertolet, of Philadelphia, and myself being among those who were comparatively unsuccessful in this operation. I occasionally succeeded in relieving tinnitus aurium by the operation (about one in twelve or fifteen), and usually not improving the hearing more than temporarily, which indeed a simple paracentesis may do.

As far as I am at present informed, the operation is infrequently done in this country, and much less frequently in Europe than formerly. It is sometimes justifiable where everything else has failed to relieve tinnitus aurium, for this does sometimes succeed.

The operation of puncturing or opening the membrana to add air-supply to the tympanum, where the calibre of the Eustachian tube is insufficient for the purpose, has been performed in a great variety of modes; and as it is often desirable to maintain an opening for a considerable time, or permanently, a variety of expe-

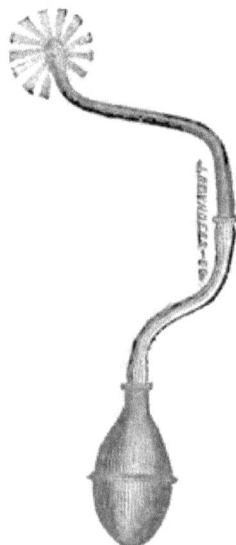

FIG 52.—WARNER'S POST NASAL SYRINGE.

dients have been resorted to for the purpose of accomplishing that end, such as making an opening by the galvanocautery, the trephine, dissecting out a piece of membrane, burning holes in the membrana with sulphuric acid, the insertion of an eyelet, etc., etc. It is safe to say that some of these procedures have been successful.

On the Management of the Eustachian Tube.—The first point is to free the tube of inflammatory products. Very often the faucial end of the tube will be filled with mucus, either inspissated or of a tenacious consistency. It is well to inject a little tepid salt-and-water, of the strength of a drachm to the pint, behind the velum, by any properly constructed syringe having a bend in the tip, so as easily to pass behind the velum. It will be seen that I have a fear of passing fluids into the middle ear by these injections, and in all the cleansing processes of the upper pharyngeal space I act with reference to that fear.

It will be well to first throw up only a few drachms of the warm salt-and-water; it may be done in this manner: draw up the syringe piston to the whole distance, even though there is only a drachm or two of water to be used, then forcibly send the piston home, by which means a somewhat coarse spray will be produced. If a soft rubber bag is used, like Warner's instrument (Fig. 52), the same idea may be carried out. Any spray-producer may be sufficient for our purpose, or a small quantity of water may be thrown into the nostrils by an atropine dropper, the head being thrown somewhat backward during the operation.

If the tube then inflates easily, enough has been done to free the pharyngeal extremity from mucus. Frequently, however, firm adherent masses of mucus may be seen by the rhinoscope to cling to the fossa of the tube, completely closing it; then more effective means may be necessary for its removal. The syringe may be entirely filled with the salt water, and thrown in with some force. It will be well to pause now and then, and request the patient to make an effort to remove these masses in the natural manner. Sometimes I make the remedial application at once, and the suddenly increased secretion of the part may throw off the mass, but in any event it assists the patient's efforts to remove it. If the treatment progresses favorably, this forcible removal of the masses of mucus will soon become unnecessary. In all these efforts to cleanse the tubal orifice the patient should not cough, nor sneeze, nor swallow, nor violently catch his breath, nor for a little time even blow his nose, for fear of forcing fluid into the middle ear. If on attempting inflation by Politzer's method, or its modifications, difficulty is experienced in rendering the tube pervious, the catarrhal application may at once be made, when, from the increased discharge resulting, the tube may at once become pervious to the renewed effort at inflation. I believe the explanation of this to be the same as that recently given for a similar action of the catheter. The same end may be accomplished by the use of the faucial or the Eustachian catheter, introduced previous to the Politzerization. I lay great stress upon using Politzer's method, be-

lieving, as I have previously stated, that it is the most
thorough and satisfactory of all methods of inflation.

If there seems to be few catarrhal symptoms, and the
patient does not have recurring attacks of lowered hearing,
it may not be necessary to make any application to the
mouth of the tube. In increasing the calibre of the tube
by catarrhal applications to its mouth, is it necessary to go
further and treat the mucous lining of the nares and
pharynx, wherever diseased, if not absolutely necessary,
it certainly is desirable; and most American authorities, as
Roosa, Buck, Burnett, and others, express the belief that
in treating the Eustachian tube the pharyngeal and nasal
mucous membrane requires attention. Von Tröltsch is
decidedly of the same opinion. I shall, however, first consider the topical applications suitable for the mouth of the
tube, and the best method of applying them.

The remedy I have the greatest confidence in, and which
I have used, almost since I first commenced the practice
of otology, is the nitrate of silver. The selection of the
proper strength to use requires great skill on the part of
the surgeon. If used in small quantities, as a drop or two
or less, a very strong solution may be used. If used in
larger quantities, as one or two drachms thrown into the
pharynx, a weak solution should be employed—from one to
ten grains to the ounce. When used in the form of spray,
the remedy being diluted, so to speak, with air, a strong
solution may be used, and if the spray is a·fine one, a still
stronger solution is proper. If the application has been
excessive, the catarrhal symptoms may be aggravated for
months afterwards; at least I have seen patients who insisted that this was so. Nitrate of silver seems poisonous
to some throats. There should not be an unpleasant irritation of the·part, lasting more than an hour or two. I fear
we are often at fault in making our applications unnecessarily strong. Some time since I had a medical man under
my charge with chronic aural catarrh. As he was skilled
in auscultation, his hearing could be appreciated with most
acute intelligence. After a few weeks of treatment the
hearing would come up to the normal. He had several relapses, but at present his condition is perfect. He made
this observation about the treatment: while actually under
treatment his condition was not as satisfactory as some
days or a week or two after the discontinuance of the treatment. He was conscious that the good effects continued

some time after its discontinuance. I inferred from this that I had been too heroic in my applications.

The methods of applying nitrate of silver to the Eustachian tube which I have the most experience with are as follows: A drachm or two of a weak solution may be thrown behind the velum, the patient being directed to expectorate as soon as the application is made. Send it up as directed for the salt-water injection. Again, a stronger solution, perhaps one of twenty grains, may be used by the same method; that is, draw into the syringe six to ten drops, the valve being extended as though it were to be filled, then send it home sharply. The spray instrument is deservedly popular. I use the hard rubber instrument which is sold in the shops in New York as Pomeroy's (Fig. 53). It

FIG. 53.—POMEROY'S HARD RUBBER SPRAY INSTRUMENT.

is not my practice to use with this a weaker than a ten-grain solution, and often I use as high as a sixty or eighty grain solution. The spray may be directed to the mouths of the tubes or to the whole pharyngeal space. My faucial catheter (see Index), according to my experience, affords one of the most exquisite methods of applying arg. nit. to the mouth of the tube. It may be used as follows: With a dropping tube, deposit a part of one drop, or two or three drops, upon any convenient surface, such as a piece of writing-paper or a palette; then compress the air-bag, having the thumb on the aperture, until the fingers on the opposite side of the bag touch the thumb, the compressed bag being between them; then apply the tip of the instru-

ment at its perforated extremity accurately on the drop of
solution; then allow the air-bag to fill sufficiently to draw
the fluid within the catheter, the thumb being still on the
aperture in the bag. Allow the bag now to fill with air,
mostly by relaxing the close application of the thumb to
the aperture; then introduce behind the velum, to a point
as near the mouth of Eustachian tube as possible; then
compress the bag strongly. When a small quantity of the
fluid is used (and it is perfectly easy to deposit on the
palette one tenth of a drop) the solution may be made
very strong. I have thus used a saturated solution. I use
this instrument in private practice, for the above purpose,
much more frequently than any other. Some years since,
when the instrument was new, and I needed to test its
capacities fully, I found on rhinoscopic examination that it
was quite easy to throw the solution into the fossa of the
Eustachian tube in nearly every instance. The fluid is not
intended to pass into the tympanum, nor even more than a
little way up the tube. In a few instances I have injected
the tympanum while using it, but without doing harm.
Some years since, Dr. Robert F. Weir, of New York, used
my instrument for making an application to the Eustachian
tube, and a suppurative otitis resulted. I have never heard
of another instance of the kind. If the bag is compressed
strongly, the fluid passes out of the catheter in the form of
a coarse spray.

The Eustachian catheter is an admirable instrument for
making applications to the pharyngeal entrance of the
tube. Some years since I appropriated a method of using
this instrument which Dr. Agnew, of New York, suggested
to me, and which he had been using for a long time. The
catheter for this method has a moderate curvature, so as
readily to pass into almost any nostril; it was attached to
a light rubber tube, which again fitted the tip of an ordinary Politzer bag having a perforation. The fluid was
drawn up into the catheter by means similar to those used
in charging the faucial catheter, although I believe Dr.
Agnew was in the habit of inserting the catheter into a
bottle of nitrate of silver, finding that he could draw up in
that manner a sufficiently small quantity; he then introduced it as near to the faucial extremity of the tube as
possible, and compressed the bag with some force. By
this means the tube was treated, and the tympanum injected with air at the same time. This method admits of

rapid work, although the catheterization of the tube is likely to be done somewhat imperfectly.   Another method, which is more exact, is to introduce the catheter, having a maximum bend, carefully into the mouth of the tube, hold it in position by the hand, or a catheter holder (my own— see Index—is a convenient one), or if it is perfectly inserted it will remain in position without aid; then, the patient's head being thrown somewhat backward, the nitrate of silver

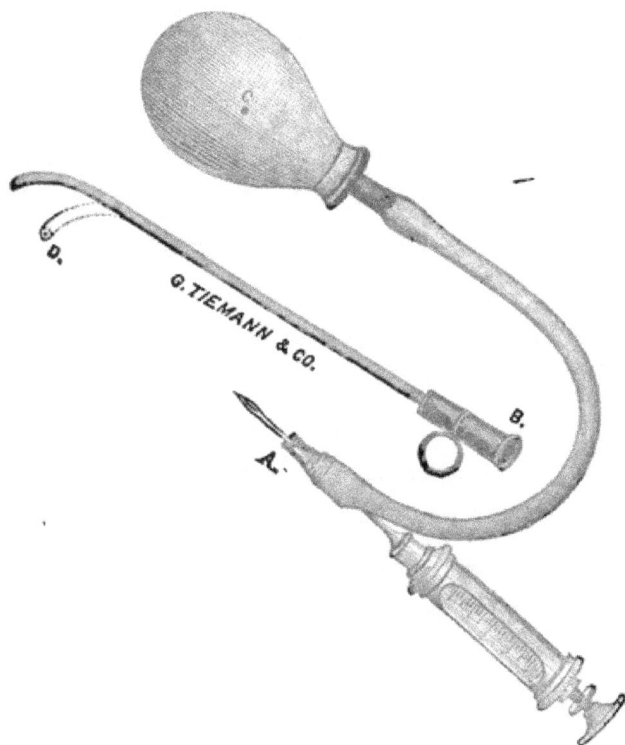

Fig. 54.—Hackley's Instrument for Spraying the Mouth of the Eustachian Tube.

may be passed into the opening in the catheter by a medicine-dropper.   The air-bag, provided with a tip accurately fitting into the catheter, which for obvious reasons may be attached to the bag by a short rubber tube, is then ready for use.   Several compressions of the bag may be necessary to sufficiently blow the medicament out of the catheter.   Two or three extra drops may be allowed, as that amount is likely to adhere to the walls of the catheter.

Many are in the habit of causing the patient to swallow during compression of the bag, but the objection to this is the greater tendency of the fluid to pass into the middle ear.   Fig. 54 represents Dr. Hackley's method of spraying the mouth of the Eustachian tube by means of the Eustachian catheter.   A Politzer bag, with tube and attachment for a catheter, is penetrated near its extremity by a hypodermic syringe charged with a fluid.   The rubber tube by its tip is pushed into the catheter, either before or after it is applied to the mouth of the tube.   Then, by pushing down the piston of the hypodermic syringe, a proper amount of fluid is injected into the catheter.   By compressing the Politzer bag, the fluid is forced into the Eustachian tube. Another method of making applications to the Eustachian tube as well as any part in its vicinity, is by the use of Zaufal's speculum (see Index).   Select the largest one which will pass into the nostril, and make the application by means of a cotton holder, armed with a small, tightly rolled bit of cotton immersed in a strong solution of arg. nit.   If any granulations are found, a saturated solution would be of a proper strength.   Dr. Prout, of Brooklyn, who introduced the instrument into this country, recommends anointing the speculum with vaseline previous to its introduction.

I have little to say about probangs, brushes, or the cotton on a holder, as means of making applications behind the velum, although many men of the highest authority use and recommend them.   It is true that an application thus made has a certain quality of thoroughness, as the rubbing of the part incident to the application removes secretions which often prevent the remedy from reaching the surface of the membrane.   In making applications to the posterior surface of the pharynx, velum, etc., it is preferable to any other method.   When passed behind the velum, considerable violence is likely to result in the endeavor to reach the proper point.

If this is not positively injurious, it is at least very annoying and often painful to the patient.   Even if he is directed to phonate, breathe through the nose, etc., to cause the velum to hang loosely down, there certainly is likely to be spasmodic contraction upon the probang, making it difficult to remove, and thus causing pain and irritation.   It certainly is proper to use these methods, although I do not personally like them.   In using Zaufal's speculum, a most elegant plan, when a strong effect is desired, is to fuse a bead of arg. nit. on a probe and make the application to the mouth of the tube. For an account of the indications for the use of this speculum (see Index). Many other applications may be made to the Eustachian tube. Tincture of iodine may be cautiously applied by means of the catheter, syringe, or spray instrument. If undiluted, only a small quantity should be used—a drop or two by the catheter or two to four with the syringe. In using the spray, a momentary application will suffice. This may also be diluted with alcohol and used as described for the weak solutions of arg. nit. It is often painful. In the older and dryer catarrhal conditions it is more applicable, being much more stimulating than astringent. Carbolic acid is a mild, not very effective remedy, but its action is agreeable to the patient, and I have used it in many mild cases with satisfactory results. It should always be used rather weak—from two to five grains to the ounce of water. Dobell's solution is a pleasant preparation of carbolic acid. ℞. Acid carbol., gr. i.; sod. biborat., sod. bicarb., āā. gr. ij.; glycerinæ, ʒ i.; aquæ, ℥ i. M. The spray instrument is an admirable means of using these solutions, or the posterior nasal syringe; with the latter, one to three drachms may be thrown behind the uvula and removed by expectoration. It should not be allowed to touch the borders of the nostrils, as considerable soreness may result. It may also be used, poured into the nostrils with a pretty large atropine dropper, or by a little earthenware instrument called a sick feeder, having a nozzle, which adapts itself to the nostril. In using this, the patient throws the head slightly backward, and the solution should pass into the pharynx, to be removed by expectoration. I sometimes add to this wash, with good effect, forty to sixty grains of tannin to the pint. Ferric alum latterly has become somewhat popular in this connection, used in from one to four grain solutions in water. It has a somewhat unpleasant metallic astringent taste, and is likely to soil instruments and clothing. Its predominant quality is astringency. I believe it to be somewhat more effective than common alum. The latter may also be used in the same manner and same strength. Chloride of zinc in from one to four grain solutions, used in the same manner as the last, may be recommended. This remedy is more stimulating than those just mentioned, somewhat resembling the nitrate of silver in its action. Occasionally it is rather painful. Sul-

phate of zinc has a well-established reputation in this affection, and may be used in a two to four grain solution. I
have no rule by which remedies may be certainly selected.
If the remedy acts well, continue it; if not, use some other.
No one remedy should be continued more than two or three
weeks at a time. Sulphate of copper sometimes acts well,
but it will occasionally cause considerable irritation, and it
has an excess of astringent action not always easy to manage. Use it in solutions of two to four grains to the ounce
of water, or it may sometimes be applied in substance, but
in the latter instance the part should be touched very lightly, and only momentarily. Applications to the mouth of
the Eustachian tube may ordinarily be made every second
day. Some of the severer remedies may cause so much reaction as not to allow of a repetition under three or four
days. If at any time the Eustachian tube closes up for a
few days, the application has probably been of excessive
strength, and a weaker effect must be subsequently sought
for. On the other hand, if the tube still continues to supply insufficient air to the tympanum, known by the evanes-
cent effect of inflation on the hearing, it is well to increase
the strength of our remedies. I have repeatedly made applications to the tube of a thirty or forty grain solution of
arg. nit. without success, when on changing to an eighty-
grain solution—using only a drop or two in the faucial catheter—the good effect at once became apparent. A too
weak solution is just as much to be avoided as the opposite.
The toleration of a given patient always requires to be
very carefully studied. I have discussed this point more
fully while on the subject of acute catarrh of the tympanum.
Many of the milder remedies suggested may be used daily.
I sometimes give the patient the carbolic-acid solution to
throw into the nostrils once or twice a day, using for this
purpose the atropine dropper, the sick feeder, or in some
of the more skilled, a Warner's nasal syringe (see Index),
the nozzle of which is passed behind the velum. If at any
time a coryza is produced by the applications, they have
been too strong, or have been injected too far upward, approaching or entering the frontal sinus by means of the infundibulum. All applications should be warmed, except
those used in very small quantities. If the discharge becomes thicker after an application, too much irritation has
been produced. A thick discharge, under proper treatment, should become thinner and diminished in quantity;

an unnaturally dry membrane should secrete more freely, and become moist. If the patient should sneeze after a certain remedy, and show other symptoms of continued irritation of the naso-pharynx, the treatment is not properly borne. If the patient is doing well, the parts are likely to feel better and the hearing should cease to be variable, and the inflations should not be required after a few weeks, to maintain the hearing at its maximum standard.

Of the Management of the Naso-Pharyngeal Space in Chronic Aural Catarrh.—The first indication is to thoroughly cleanse the part. As the nasal douche, first devised by E. H. Weber, and in England known as Thudicum's, has become popular among the profession, I must needs speak of it, although I do not often employ it, for the reason that water is sometimes sent into the ear by its use, causing occasionally violent inflammation of the tympanum, with rupture of the drum membrane and even extension to the mastoid cells. This douche is based on the principle that if water is forced into one nostril it will pass out of the other without entering the lower pharyngeal space. The contact of the water with the hanging palate produces a reflex excitation, which causes it to fall back upon the posterior wall of the pharynx, shutting off the upper from the lower pharyngeal space. This movement is also assisted by somewhat rapid and continuous mouth respiration, or if the patient holds his breath a similar result follows. In order to avoid accidents to the ear, the vessel holding the water should be only high enough to allow of a stream to pass through the nostrils. It is not advisable to have it higher than the forehead. The fluid used may be salt, one drachm, and water, a pint, always warm—about the temperature of the water used in syringing the ear. Be careful not to force fluid into the frontal sinuses, by not allowing the patient to incline the head forward more than a very little, if at all. During the operation the patient should not be in the least disturbed or excited, lest he cough, make an effort to sneeze or swallow, when the fluid would very likely be forced into the middle ear. Before commencing this irrigation, observe whether both nostrils are free; in case one only is free, the douche, if used at all, must be made to enter through the partially occluded nostril. After using the douche, unless it be very warm weather, the patient should remain a few minutes in a warm room. I would not approve of passing a large quantity of water through the nostrils at a sitting

—just enough to fairly cleanse them; perhaps twelve to twenty ounces, although I have heard enthusiastic laryngologists speak of the perfect safety of passing unlimited quantities through them. For instance, one of those gentlemen said he had sent *barrels* of salt water through the nostrils, and he believed that a large quantity acted curatively, and cleansed with greater thoroughness. There is no doubt but that the irrigation acts pleasantly in removing secretion, especially when the nostrils are somewhat dry and "gummed up" with thickened mucus, as one of my patients felicitously described it, but whether it acts as a corrector of abnormal mucous secretion is a question in my mind. I am aware that *any* mode of injecting the nares, whether in front or in the rear, is liable to cause fluid to enter the tympanic cavities. In the cases Dr. Roosa tabulated in his work on the ear, occurs one of my own, where a suppurative otitis resulted from my use of the douche. In this case I think the douche was properly used, the patient, however, was not highly intelligent, and it is possible enough that he himself did something which interfered with the proper action of the douche. These conditions, however, are certainly too exacting to permit of patients using it themselves. I recently had a patient from the country who, without any interrogations from me, volunteered the information that the doctors there had recommended the nasal douche, and many people had become deaf in consequence.   •

In douching a patient I would throw in only enough of the fluid to dislodge the accumulation; stopping from time to time to see if the patient himself could not dislodge any remaining collection. Other means for cleansing the nostrils may be used, and I decidedly prefer the catarrh syringe, throwing up a larger or smaller quantity according to the amount of collection to be removed. The practice of snuffing up salt water into the nostrils I do not like, having observed it to pass into the middle ear, but so far without doing harm. Pouring the fluid into the anterior nostrils by any convenient vessel is a good method, although I have known fluid thus to pass into the tympanum.

A medical gentleman of this town was some time since injecting his nostrils by means of a Davidson syringe, with warm salt-and-water, in a gentle, cautious manner, when he suddenly dropped to the floor in a faint, in consequence

of the fluid passing into his ears; he, however, soon recovered. Forcing even warm salt water into the tympanic
cavities is sometimes a serious matter.

What I recently said under the heading of Management
of the Eustachian Tubes applies with equal force here: Remove collections from the throat by the aid of the patient,
as far as possible, by stimulating injections, or by the
smallest amount of salt-water injections that will suffice to
remove the masses. I suppose spraying the nostrils both
anteriorly and posteriorly with salt-and-water is the mildest possible mode of removing secretion from the naso-
pharynx, aided also when practicable by the patient's own
efforts. The general condition of redness of the naso-

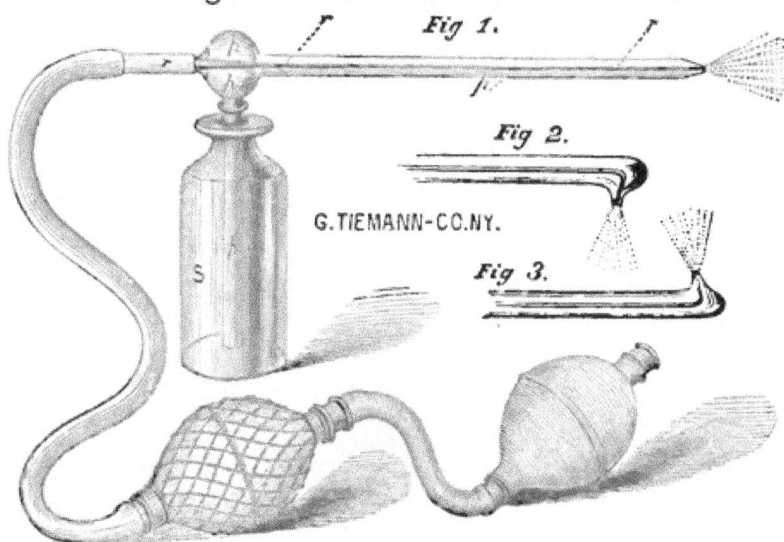

FIG. 55.—NEWMAN'S SPRAY PRODUCER.

pharynx is met by the astringent applications mentioned
when speaking of the Eustachian tube. The spray may be
used both anteriorly and posteriorly. In spraying the nostrils a minute funnel may be inserted, having an expanded
portion, which may catch any excess of spray. Fig. 55
shows Newman's spray-producer, and Fig. 56 represents
Sass's spray instrument as it is propelled by compressed
air. No applications to the nostrils should excite symptoms of coryza or any prolonged irritation. Whenever reddish or granular patches appear in the nostrils, they should
be touched with a forty to eighty gr. solution of nitrate of sil-

ver. I have also sometimes found the saturated solution of carbolic acid to do well. These reddish spots are often covered by an incrustation of catarrhal secretion, which clings tenaciously to the part, and often causes laceration and bleeding on being forcibly removed. If the nostril is anointed with vaseline several times a day, or, in case that irritates, with a little freshly prepared mutton suet, the crusts will form much more slowly, and often not at all. These patients occasionally have minute patches of herpetic vesicles form here and there on the lining of the nostrils, near the outer orifice. Both of these conditions are best treated by brushing on a rather strong solution of nitrate of silver, one of thirty to sixty grs. to the ounce, or tr. of

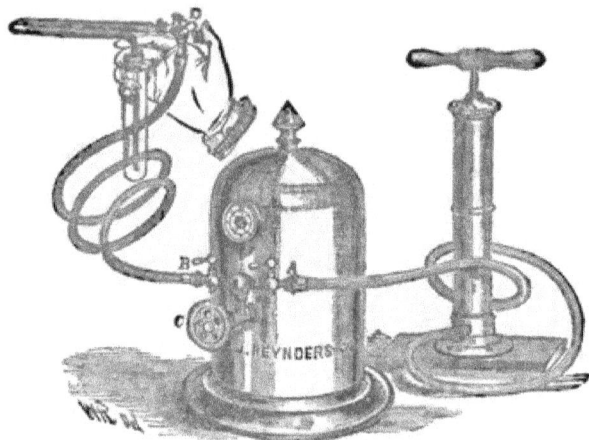

FIG. 56.—SASS'S SPRAY INSTRUMENT WITH COMPRESSED AIR APPARATUS.

iodine often acts well. For the general catarrhal condition of the naso-pharynx, powders properly medicated are often blown in, both posteriorly and in front, by an instrument here figured, and called a powder-blower. This was de-vised by Dr. A. H. Smith, of New York. (See Fig. 57.) It is well known that injecting the nostrils with a fluid is often very irritating, exciting coryza, sometimes causing severe headaches, with a feeling of fulness in the frontal sinus. The advocates of the powder method of treatment claim that the latter produces no unpleasant symptoms of this kind. I have used the powder only a little, and have a general feeling that it somewhat lacked in effectiveness. The active remedy in an ounce of the powder may be of somewhat greater proportion than where water is used. For instance, the nitrate of silver powder, frequently used, is made up thus: ℞. Arg. nit. gr. xx.; bismuth subnit., q.s. ft. ℥ i; potass. sulph., ℥ ij.   This would correspond to about a twelve or fifteen gr. sol. of arg. nit. in water.   The sulph. potass. is for the purpose of facilitating the reduction of the mass to a sufficiently fine powder.   The tubes used are made of hard rubber, but any person may extemporize an instrument using glass tubing instead of rubber. The tip for turning in any direction cannot be made of glass.

It is of the highest importance to keep the nostrils free, so that mouth respiration may be avoided.   It is not always easy to ensure a free passage for air through both nostrils simultaneously.   I have known many patients who usually had only one nostril at a time open, the trouble constantly changing from one to the other and back again.   Lying in bed on one side will render the lower nostril obstructed

FIG. 57.—A. H. SMITH'S POWDER-BLOWER.

and the upper one free, the condition being reversed on turning to the opposite side.   The tendency of this tissue (the mucous lining and the submucosa) to suddenly become engorged with blood, and swell considerably, presenting many of the characteristics of erectile tissue, has already been alluded to.   This symptom is not easy to manage.   Some nervous patients are likely to suffer from it when anything disturbs them greatly, or if excessively fatigued.   To such, a nervous stimulant would be beneficial in restoring a patulous condition of the nostrils. Whatever growths obstruct these passages must be removed.   Usually a mucous or fibrous polypus will be the variety found in this region.   These may be removed by forceps, snare, écraseur, galvano-cautery, or electrolysis. About the same rule obtains here as in the removal of aural polpyi. What remains of the pedicle of the polyp should be cauterized. Other tumors of this region, malignant or benign, may be removed by a somewhat formidable surgical operation, as removing a portion of the superior maxilla, a part of the palate bone, etc. These, however, come more in the domain of general surgery. Often, in old cases of naso-pharyngeal catarrh, the mucous lining of the nares, with the submucous connective tissue, become so hypertrophic as to occlude the nostrils. This is more likely to be the case with the parts covering the turbinated bones, especially of the middle and inferior. No amount of stimulation and cauterization will in certain cases cause this condition to disappear, and a variety of operations have been resorted to for their removal.

Dr. A. H. Smith, of New York, has devised an instrument for this purpose, which he calls a canula scissors (Fig. 58), which in a very effective and not violent manner removes these hypertrophic swellings. Herewith is presented

FIG. 58.—SMITH'S CANULA SCISSORS.

the doctor's description of the instrument from *The Planet:*

" The instrument consists essentially of two canulæ with closed rounded ends, one canula revolving within the other. One half of the circumference of each canula is cut away for the distance of one and a half inches from the distal extremity, on a line somewhat slanting. The cut edges of the outer canula are brought to a knife-edge, while those of the inner canula are finely serrated. This part of the inner and outer canulæ, respectively, represents the two blades of the scissors. When the inner canula is so placed that it corresponds with the outer one the scissors are open; when it is turned half round the scissors are closed. The slant given to the cutting edges insures a scissor action."

" At the opposite end, the inner tube projects beyond the outer one, and is furnished with a milled head, by which it is revolved."

" The instrument is introduced into the nose open, that is to say, with the inner blade lying in the concavity of the outer one, and the open side is pressed firmly against the part to be removed. Whatever tissue is contained in the hollow of the inner blade is cut off when a half turn is given to the milled head."

" The removal of hypertrophied tissue from the turbinated bone by this method is followed by hemorrhage, which may be profuse, but is usually easily controlled. The results in cases of obstinate catarrh are often extremely satisfactory."

The denuded surface subsequently becomes covered by a fine cicatricial membrane doing duty in the place of normal mucous membrane quite satisfactorily.

Also, for the purpose of reducing the swelling, so that the nostrils may become pervious, tubes have been inserted into them and allowed to remain from a few hours to several days, if not provocative of undue irritation. Bougies of laminaria are occasionally used, but as they expand rather rapidly, and sometimes unevenly, much irritation is caused on removing them, especially as the laminaria is not very smooth, and may lacerate the delicate mucous lining. Any ulcerated places in the nostrils may be touched with a caustic, as arg. nit., carbolic acid in crystals, nitric acid, or acid nitrate of mercury. The probability of some of these ulcerations being specific will suggest appropriate constitutional measures.

A variety of ostotic processes greatly narrow or altogether close the nares. These require removal. This is now frequently done by the dental engine and drills. The ostosis may be perforated, and a sufficient aperture made for purposes of nasal respiration, and cleansing the nostrils by blowing the nose, etc. A case of my own, published in the Transactions of the New York State Medical Society for 1881, illustrates this point. The patient, a man 44 years of age, could not force a particle of air through his right nostril. The catarrhal collection could not be blown out, and was a source of great discomfort to him, as he could cleanse the nostril only by using the syringe. There was considerable deafness on the side of the affected nostril. Anteriorly I could see nothing, but a probe struck an obstruction deep in the nostril. Rhinoscopy revealed nothing. I then passed my finger behind the velum, and by a strong effort pushed it forward into the posterior nares of the right side for the distance of half an inch, when it touched a solid wall of bone. By measuring the length of each nostril, and estimating the distance of this obstruction from the entrance to the posterior nares, I could determine accurately its depth in the canal, and that it was of no great thickness. I attempted its removal in the line of the inferior meatus, knowing it to be too difficult to remove the whole of it, neither was it necessary. I at first used a number of drills of different shapes, which enabled me to perforate the bone, but did not permit of enlargement of the aperture. I then had constructed what is known among dentists as a cross-cut burr drill, the head having an almond shape, and as large as could be conveniently introduced into the nostril—about $2\frac{1}{2}$ lines in diameter. It required to be covered with vaseline, as the sharp edges of the burr would catch the mucous membrane of the nostril, and prevent its introduction without laceration. In three sittings I enlarged the opening sufficiently

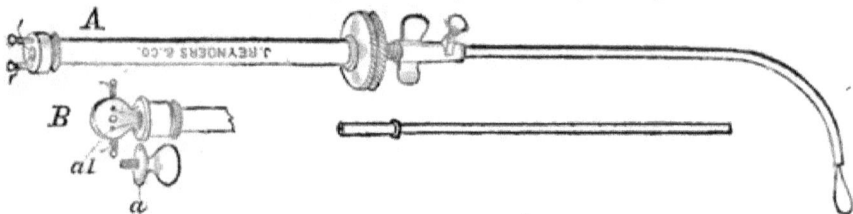

FIG. 59.—JARVIS' WIRE SNARE ÉCRASEUR.

for moderately difficult respiration through that nostril. After each operation the opening at first would be quite free, and by the next day it would be nearly closed, but would open again in a day or two more.

A more usual method, however, for removing these obstructions is to use a rather fine drill and make successive borings in the periphery of the mass until a central portion is cut off, when this may easily be extracted by forceps (Goodwillie). The objection to a large-sized drill is that the power is insufficient to effectively propel it. The Morrison engine is usually used, although a modification has been made by Dr. White. A somewhat more powerful apparatus, called the Elliott suspension dental engine, is, according to Dr. Arthur Mathewson, of Brooklyn, superior to the one ordinarily in use. *In the removal of adenoid growths from the vault of the pharynx* the same procedures are employed as in the management of aural polypi. Politzer thinks that cauterization is sufficient to dispose of these, but the weight of authority is against him. Forcible removal by some

mechanical appliance seems strongly indicated. Where a
delicate snare can be passed through the nostril and looped
upon the growths by the aid of the fingers passed into the
upper pharynx, the extraction becomes an easy matter.
For this purpose Jarvis' écraseur snare (Fig. 59) is valuable. Sometimes by passing the finger behind the velum
and scraping or tearing the growths, if not too fibrous, and
if they are sufficiently pedunculated, they may be removed.

FIG. 60.—FORCEPS FOR THE REMOVAL OF ADENOID GROWTHS.

An instrument resembling a lithotrite (Burnett) may destroy them by crushing, when they shrivel up and disappear, or are removed by sloughing. This is not, however, a very reliable method. Sometimes an instrument
resembling Buck's curette for the removal of aural polypi
may act serviceably here.

Forceps passed through the nostrils, if of sufficient deli-

FIG. 61.—BELLOCQ'S CANULA.

cacy and strength, act very admirably. Dr. Noyes' polyp
forceps, elsewhere figured, are very strong in the jaws, and
if a magic catch be attached to the handle so as to allow of
twisting the growths, they may thus be removed. The forceps may be introduced at once, and in the possibly large
multitude of growths one may be easily caught by the
sense of feeling to guide. Afterwards the finger passed
behind the velum, will cause the growths to be pushed into

the bite of the forceps. By using Zaufal's speculum, a single growth may be brought under inspection, and its peduncle sought out and destroyed by nitric acid or acid nitrate of mercury, or the nitrate of silver. When employing the more powerful caustics, the smallest quantity must be used, and no other points touched than the growths. After the latter have been removed, cauterization of the bases may be necessary. Occasionally it will be possible to catch the growths from behind the velum by means of a pair of forceps having a short curve similar to Mackenzie's laryngeal forceps (Fig. 60), or a wire loop may be passed from the throat through the nostrils by the aid of Bellocq's canula (Fig. 61), the wire first being attached to a small cord passed from the nostril to the throat by the aid of the canula, and this to be attached to the Jarvis instrument. By this manœuvre the loop is readily passed around the growths. These growths often obstruct the Eustachian tube; but if they do not, there is a catarrhal condition perpetuated by their presence which may aggravate any ear symptoms. It is well to attack only one or two growths at a sitting, as considerable disturbance is likely to result from the operations. The same rule may be applied in removing granulations or polypi from the same region. The galvano-cautery may be used. (See Fig. 62, representing Shroetter's instrument with all the necessary attachments.) The objection to it is that it is an expensive mode, and the instrument gives considerable trouble in keeping it in repair, and to operate effectively the wire loop must generally be used, which involves the same difficulty as in the ordinary snare. Sometimes the porcelain burner or the knife may be used. There is this gained over the snare, however: it cuts through the growth without violence, and leaves a surface already cauterized, from which there is little or no trouble with hemorrhage, which in other methods is likely to occasion embarrassment. *The uvula* often requires attention. When it is so elongated as to impinge upon the base of the tongue, it excites coughing and aggravates the catarrhal condition. Before attempting its removal, an astringent gargle may be employed; if this fails it may be pencilled every day or two with a strong solution of nitrate of silver, sixty or eighty grains to the ounce of water, or nitric acid may be cautiously used, or even the acid nitrate of mercury. It is very tolerant of strong applications.

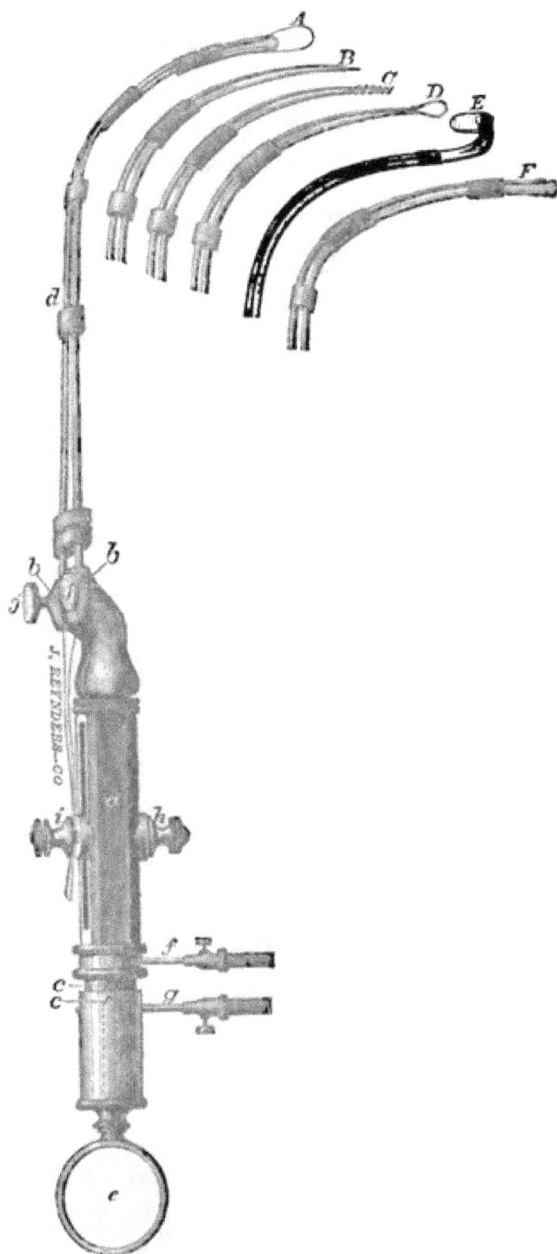

FIG. 62.—*A*, snare; *B*, knife; *C*, porcelain burner; *D*, knife; *E*, Voltolini's knife; *F.* ligature carrier. The universal handle is attached at *A*; The galvanic connections are made at *g* and *f*.

If this fails, then ablation must be practised. Catch the tip of the velum in the bite of the forceps, and divide it by a single stroke of a stout pair of scissors. The piece removed should be about one fourth of an inch in length. Frequently too much is removed, unnecessarily mutilating the organ; besides, the uvula becomes, after a little time, still shorter from natural shrinkage.

For a few days after removal, a mildly astringent and emollient gargle may be used, as biborate of soda, chlorate of potass., etc. The velum sometimes suddenly swells to such a size as to fill the cleft in the pharynx and seriously interrupt respiration. This accident once occurred to me while making an application behind the velum of nitrate of silver by means of the pharyngeal syringe. In an instant the velum swelled to the size of one's finger, and the patient was horrified to find that he could not breathe. To gain a few seconds time, I compelled him to breathe through the nose, what little was possible, when I immediately relieved him by making some dozen minute punctures in the velum with a Graefe's cataract knife. The uvula returned to nearly the normal size in from four to six minutes. This accident happened to me once besides this. I had made an application to the Eustachian tube of nitrate of silver, when I lacerated perhaps, the mucous membrane covering the velum. After this I performed Politzer's inflation, when suddenly the velum swelled as before described, and the patient fell to the floor, perhaps as much from fright as anything, and for a few seconds could not breathe at all. Relief was obtained as in the first instance and as suddenly. Dr. J. Solis Cohen, in his work on the throat, recommends cutting off the end of the velum in similar conditions.

The lateral half arches, and adjacent parts are often of a more intense red color than other parts, which require special attention. I am in the habit of pencilling on a rather strong solution of arg. nit. or tr. iodine to meet this indication. When this congestion presents the appearance of a distinct red line at the border of the arches, a syphilitic infection may be suspected.

The tonsils are often so swollen and hypertrophic as to require treatment. Very strong cauterization with arg. nit., nitric acid, or acid nitrate of mercury or chromic acid crystals, may be used to diminish the size. In order to accomplish anything, a large part of the surface of the tonsil must be cauterized, which will be rather difficult to accomplish. If incisions are made in the tonsil previous to cauterization the effect will be greater.

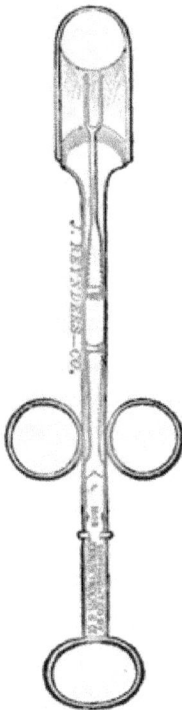

FIG. 63.—McKENZIE'S MODIFICATION OF PHYSICK'S TONSILLOTOME.

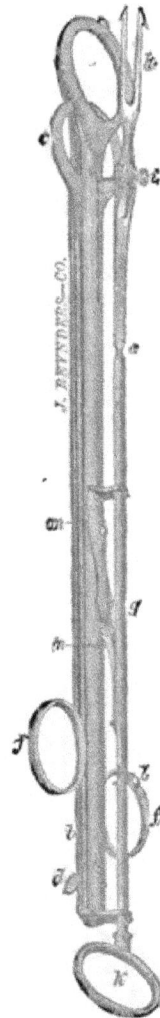

FIG. 64.—MATHIEU'S TONSILLOTOME.

Dr. J. Solis Cohen, in his book on Diseases of the Throat, quotes Morrill Mackenzie to this effect: He uses on enlarged tonsils London paste, which is composed of equal parts of caustic soda and hydrated lime, a portion of which is moistened with water at the time of its employment. The caustic is applied with a rod of aluminium wire. The application must be made many times to sufficiently destroy the excess of development of the tonsil.

It is much less painful than caustic potash or Vienna paste, and causes less reaction. The galvano-cautery, by means of the loop, answers well for removing a considerable piece of the tonsil. It is passed over the latter, and as it becomes heated it is tightened, when it readily cuts its way through with little or no bleeding. Electrolysis is a good method for diminishing the size of the tumor, but it takes many sittings for its accomplishment. Before doing an important operation such as ablation, it is well to remember that enlarged tonsils are likely to diminish in size at the age of puberty. If an operation for removal is decided upon, McKenzie's modification of Physick's tonsillotome is perhaps the best instrument to use (Fig. 63), or Mathieu's (Fig. 64) is a most elegant instrument. It is advisable, if an operation be done at all, to remove a good-sized piece, sufficient to extend to the pillars of the fauces. In the removal considerable hemorrhage may occur, even in some rare cases to a fatal result, but ordinarily nothing is to be apprehended from the bleeding. The main points to be observed are, to avoid a fold of mucous membrane lying somewhat in front of the tonsil, called the anterior faucial fold, or the palato-glossal fold, which by being incised in the removal of the tonsil may bleed very freely, besides causing an unpleasant sore throat subsequently, and the avoidance of removing the tonsil too deeply so as to reach the branches of the internal carotid, or ascending pharyngeal arteries. This may also be obviated by drawing the tonsil away from its position, and avoiding in the incision anything except the tonsil itself. In children use the tonsillotome, but in adults a good method is to catch the tonsil with a properly constructed forceps, and remove it by cutting with a probe-pointed bistoury, commencing above and cutting downward, backward, and terminating the incision by coming out forwards. The resulting hemorrhage may be controlled by pieces of ice held in the mouth, or a piece of ice held in forceps pressed against the part, or by simple pressure with the fingers or a sponge on a holder, against the bleeding surface. The persulphate of iron is a good styptic, but it coagulates the blood and makes a dark-looking mass, which obstructs the throat and is troublesome to remove. The best plan is to carefully examine the wound and if any artery is found to be divided, or if the blood oozes considerably from any particular point, pass in a pair of forceps and twist the tissue in the expectation of including the cut end of the artery, when it (the bleeding) will be easily arrested. Injecting water as hot as may be borne is a good expedient. It is rarely that the common carotid or the internal carotid require ligation to arrest the hemorrhage. After the removal of a tonsil, bland and fluid food should be given for a few days, and gargles of salt-and-water, biborate of soda or chlorate of potassa, may be used. It is well to remember that hemorrhages may occur after the first or second day and to be on guard for such an accident.

Applications to the upper pharyngeal space may be made in a variety of ways. I have for years discarded the probang, brush, or cotton on a holder, for reasons sufficiently discussed heretofore. The spray instrument, with a tip for throwing the spray upward, is indicated. The glass instruments I do not use in this connection, for the following reasons: the tip is liable to be broken off, but if in some cases this objection does not exist, there is another reason —the tip will not hook behind the velum, and if the latter chances to be thrown backward against the posterior wall, as often happens, no spray can enter the posterior pharyngeal space. In using the hard rubber instrument the tip is sufficiently long to pass behind the uvula, and the spray is readily applied. If the uvula is strongly drawn against the posterior wall it may catch the tip of the spray instrument, and close it so that no spray can be produced; this condition is met by moving the instrument from side to side, or by a forward and backward movement, which disengages it from the velum and a spray is easily formed. In rapid dispensary work I never hold the instrument perfectly still, but keep it moving constantly. It is easily known when a proper spray is made by the soft peculiar sound accompanying its production. Next to the spray instrument I prefer the posterior-nasal syringe, as has been before indicated. Use the same remedies and in the same manner as described under the head of Applications to the Eustachian Tube. The posterior wall of the pharynx will often need somewhat strong applications of the nitrate of silver. The brush, cotton-holder, or porte-caustique, with the mitigated or clear stick, may be used in making applications to this part. It may be well to cauterize only a small spot at a sitting; be guided, however, by the patient's toleration of the remedies. I do not forget that sufficient cauterization to cause an eschar may destroy mucous membrane, and lead to the development of cicatricial tissue, having no glands to moisten the part. Tr. iodine is a good application when cicatricial changes have occurred, but it may sometimes be painful. The whole naso-pharynx, when swollen, perhaps pale, succulent, and relaxed, will often be much benefited by tannin ʒi., glycerine, ʒi., brushed on the posterior wall, and injected behind the velum. Hypertrophic glands may be cauterized by a saturated solution of arg. nit., or the stick. Occasionally the part may be brushed with a solution of carbolic acid, one drachm to the ounce of water. The pain from the application is quite momentary. Occasionally the applications of nitrate of silver to the hypertrophic glands may produce no effect; in that event the latter may be cut across and the caustic in substance may be insinuated into the incision. In the advanced stage of chronic pharyngitis, called pharyngitis sicca, where the mucous membrane is thin, dry, shiny, and devoid of glands from atrophy, the treatment is palliative. The glands cannot be restored and the moistening of the part must be done by applications. For this purpose almost any of the stimulating lozenges or so-called troches in the market are useful, such as salicylic acid, carbolized or chlorate of potassa lozenges. Applications to the posterior wall of glycerine or vaseline will diminish the dryness and sense of heat. In speaking, such patients will find that moistening the throat with a draft of water frequently has a pleasant effect. Internal remedies calculated to act on the mucous secretion may often do good and benefit the catarrh in a general way. These are cubebs, copaiba, muriate of ammonia, or iodide of potassium. Inhalation of steam often has a pleasant temporary effect.

Gargles for chronic pharyngitis are beneficial in a variety of ways. It is true that very little of the gargle passes behind the velum, but it is a well recognized fact, that if the parts which are in sight are improved in condition, the good effects almost certainly extend to those adjacent. A very important factor in the results of gargling is the effect on the muscles of the region, notably those of the soft palate and Eustachian tubes. In chronic catarrh of the pharynx, the mucous membrane is often thickened, and it may be accompanied by infiltration of the underlying connective tissue. Where this changed mucous membrane lies over muscles, it interferes with their action, and atrophic degeneration may result as a consequence. Again, the mucous membrane covering the muscles is peculiarly rich in glands, which are located upon and between the muscular fibres, the latter in part surrounding them, according to Von Tröltsch (Roosa's translation), so that, whenever a muscle contracts, the glands are more or less pressed upon, and when obstructions to the latter exist, their contents are often squeezed out by this muscular contraction. Thus, the condition of muscles and mucous membrane react on one another. One of the best means of improving the faulty muscular condition, according to Von Tröltsch (loc. cit.), is by a species of gymnastic training. This is accomplished by gargling. It should not be done in the usual manner ; the patient should throw the head very far backward, almost as though he were lying down ; make a partial attempt at swallowing, but do not complete the act ; expire very forcibly and produce a loud rumbling phonation; endeavor to bring into action all the muscles of the throat during the gargling. This may be done several times in a day.

The *kind of gargle to use* may be selected as follows : At first simple water, or plain carbonic acid water, may be used ; the rinsing of the part and the gymnastic effect being the main object gained in using these remedies. Chlorate of potash in saturated solution, is an old and valuable gargle. Carbolic acid, one drachm to the pint of water, with or without tannin, acts satisfactorily. Tr. of Iodine and Iodide of Potass., of each one drachm to the pint of water, with a half-ounce of Spts. Vin. Gal., is a gargle strongly recommended by Von Tröltsch. Alum, one drachm to the pint of water, with a little alcoholic addition, may be recommended. A very old remedy is alum in an infusion of sage tea. The latter disguises the taste of the alum. We may go on adding to our gargles, by selecting any mild remedy suitable for the treatment of naso-pharyngeal catarrh.

In the constitutional treatment, I simply repeat what has already been said, when treating of acute catarrh. Whatever lowers the tone of the general system is damaging to a

catarrhal patient, so every possible means must be resorted
to which may accrue to the patient's general well-being.
The functions of the skin need special attention, to compensate in part for incomplete function of mucous membranes.
Bathing frequently, with friction of the skin, by means of
a crash towel, hair mittens, flesh brush, etc., will naturally
suggest itself as appropriate management for this indication.　Means for resisting the pernicious atmospheric and
climatic influences need to be resorted to and are detailed
under the head of Prophylaxis, in the article on Acute
Catarrh.

Where the muscles of the Eustachian tube are in a relaxed condition, or the innervation is insufficient, electricity may be employed for the restoration, as far as may be, of these
muscles to a normal energy of action.　Many methods
may be practised.　On the whole, the constant current will
be preferable to Faradism.　The positive pole may have an
extension resembling an Eustachian catheter, to be applied
to the orifice of the tube, while the negative pole is applied
by means of a moist sponge to the meatus externus.

ACUTE PURULENT INFLAMMATION OF THE TYMPANUM.

The symptoms are as follows : The patient is attacked
with a pain in the ear, which may have been preceded by inflammation of the naso-pharynx, or the disease may have invaded the ear from the direction of the membrana.　In either
case, *pain* is the most prominent symptom.　It is characterized by a feeling of pressure or fulness in the tympanum,
which in the severer forms goes on increasing until the
suffering is well-nigh unendurable.　It is accompanied by
the rhythmical pounding of the heart beats, which still further aggravates the suffering of the patient.　This goes on
from one to several hours until relief is experienced,
usually by the rupture of the membrane and the discharge
from the ear.　Previous to this, however, the patient experiences certain prolonged crackling or hissing sounds
accompanied by a sharp pain which seems to dart through
the depths of the ear.　This is repeated a number of times
until the discharge is well established and the pain relieved.　The whole side of the head may be tender on pressure and painful.　If the disease commences in the throat,
the pain may be easily followed up the Eustachian tube, the
cavity of the ear will begin to feel oppressed and full, and

ere long the disease will have wholly invaded the tympanum. This pain is aggravated by opening the mouth, by coughing and sneezing, and by the act of deglutition. Any movement of the auricle causes pain. It is worse while lying down than when sitting upright for hydrostatic reasons undoubtedly. The pain oftener comes on in the after part of the day, when the patient is fatigued, or more likely, perhaps, in the night, possibly towards morning. After a first attack, relapse may occur a number of times, if the symptoms are not promptly met at the outset. Sometimes the pain may be of the dull, heavy kind, with less fulness than usual. This is occasionally the case when leeches fail to effect relief, or where the tympanum is full of fluid. There are apparent exceptions to the general proposition that pain precedes the discharge. In a few instances, if the patient is somewhat unobservant, a discharge may make its appearance without any pain having been noticed. This is more likely to be the case in tuberculous subjects. If, however, such patients had carefully observed their symptoms, undoubtedly a feeling of fulness or a slight throbbing in the ear would have been noticed. In every severe attack of acute otitis media, we are likely to have the canal much swollen, tender and often completely closed. Together with the pain, we have decided *constitutional symptoms*—fever, possibly delirium, and in children sometimes convulsions. The patient is very likely to be ill in bed, and if the pain continues for a considerable time, the face, by its distressed and haggard expression, reveals the severe nature of the affection.

The discharge does not make its appearance at first, usually not until the membrane is perforated. This may be a mixture of serum, mucus, pus, and blood. Its most constant characteristic, however, is pus, which gives to the disease its name. Other subdivisions of otitis, determined more exactly by the character of the discharge, might be given, as otitis media, serosa, hæmorrhagica, etc. The severer the attack the more is the discharge likely to be purulent or sanguinolent, and to be excessive in quantity. After a few days, if the meatus becomes macerated and loses its epidermis in consequence of the maceration and excoriation incident to the discharge, the latter may become somewhat flaky from the epidermic admixture. As the patient convalesces the discharge grows thinner. *The hearing* may be at first unnaturally acute, and loud sounds will become painfully distinct, but subsequently deafness may become profound and even bone conduction may be nearly obliterated. Tinnitus aurium adds to the patient's discomfort, and it may be accompanied by giddiness, nausea, and symptoms resembling Meniére's disease. The patient hears his own voice with unpleasant distinctness (autophony), provided the bone conduction be good, causing a very unpleasant hollow reverberation within the skull. *The appearances on inspection* are, at first, as follows: In a severe case nothing may be seen but a canal so swollen as to be obliterated. If, however, the drum membrane be visible *before there is a discharge*, there will be more or less redness. The light spot, the malleus handle, the short process, the folds, and all the landmarks may be recognized, and the redness of the membrane may be seen along the handle of the malleus and about the short process, and possibly the periphery may be injected, or the whole of the membrane may be of an uniform red color, resembling raw beef. Naturally in this state none of the "landmarks" of the membrane are visible. If the tympanum is full of fluid, as pus, blood and mucus, the membrane will be seen bulging to a greater or less extent. The umbo may perhaps still be seen, as the malleus handle has a strong disinclination to be pushed outward, and the membrane may be seen bulging in front and behind it, but more frequently the prominence will be above and behind, and so near to the canal wall as to seem to be a part of the meatus. The apex of the prominence may show a faint light reflex, if the polish of the membrane still remains. *When the perforation has taken place* it may be known by several signs: the perforation may be *seen*, but frequently not, a jagged spot or line only indicating its site. Again, after carefully cleaning the ear, if there appears a light spot of small size and of great brightness, it undoubtedly is a reflex from fluid which has been caught by capillary attraction in a fissure-like perforation. If there is seen a *pulsation in the membrane*, there is very little doubt of a perforation. This may be explained on the theory that a considerable sized blood-vessel has been partly uncovered by the break in the membrane, and its pulsations become visible. If, on inspection, during inflation, fluid or bubbles of air are seen issuing from the membrane, there is no doubt of perforation. When the fluid is all blown out of the tympanum, the *perforation whistle* is heard. If the ear has been carefully cleaned, and a few minutes subsequently there is again a quantity of fluid on the membrane, there probably is a perforation, as the external car is very infrequently capable of secreting so rapidly. Often, by using a diagnostic tube during inflation, a very faint squeaking or hissing sound may indicate the fact of perforation. We will admit, however, the possibility of eliciting a sound, produced by air passing through the Rivinian fissure, although we have never satisfactorily demonstrated the fact.

The bulging of the membrane cannot always be determined by inspection. If it be touched by a probe, the peculiar sensation elicited, of added depth to the tympanum, the result of the collection, will probably indicate the fact.

By alternately drawing out and pushing in the air of the meatus by Siégle's otoscope (Fig. 65) the mobility of the mem-

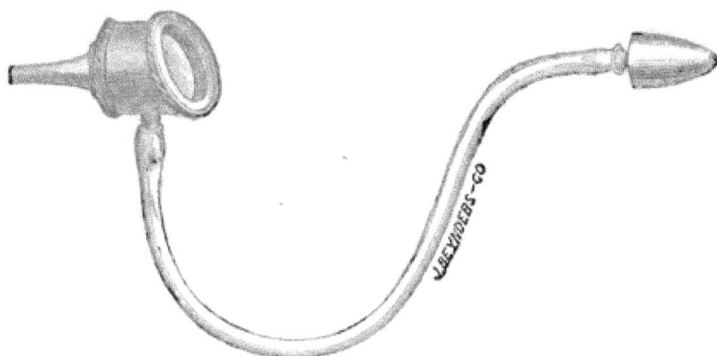

Fig. 65.—Siégle's Otoscope.

brane will give us a hint as to whether the cavity contains fluid or is empty. As soon as the membrane has been perforated and the discharge has moistened the parts, the polish of the membrane will be destroyed and its translucency lost. Even in high degrees of congestion the red color will give place to the whitish gray appearance of a macerated epidermis. The latter is easily thrown off, and indeed often may be wiped away by the cotton used in cleaning, when the red color of the congested fibrous layer of the membrane may be seen.

Throat symptoms are usually present, and may be of any variety of acute inflammation from that of diphtheria and fevers, to the simplest form of naso-pharyngeal catarrh. Where the disease involves the tympanum from the direction of the meatus, there will not necessarily be any throat complications. It will readily occur to the reader that the angina from scarlet fever is peculiarly liable to cause this form of disease, and that of the greatest intensity. Many of the old cases of chronic suppurative otitis which are so intractable to treatment are caused by scarlet fever. The otitis peculiar to typhus and typhoidal fevers will be treated of in another place.

The duration of this disease is extremely variable. In a case of moderate severity, and properly treated, the pain is usually relieved at once or in the course of two or three days, while the discharge is likely to continue for from two to six weeks. With the cessation of the discharge the perforation usually heals promptly.

In cases, perhaps improperly treated, or of unusual severity, the disease may continue for a much longer time. We have observed patients who have had attacks of pain every night or two for five or six weeks. Other cases, where the membrane fails to heal, continue on under the form of chronic suppurative otitis, and may last indefinitely. One point must not be lost sight of in determining the prognosis, to estimate how frequently relapses are likely to occur. If the patient is debilitated and also is imprudent in exposure to inclement weather . there will almost certainly be relapses. If the mastoid becomes involved it may be a long time before recovery takes place. If necrosis occurs it will render the case tedious.

The prognosis is, on the whole, favorable. In a healthy adult, with proper treatment, the pain should be promptly relieved and the inflammation arrested, and the membrane healed with good, if not perfect, hearing. In children, after scarlet fever, it is not as likely to result in resolution and repair of membrane, and is frequently found involving the mastoid, and causing bone destruction. Brain complications are met with occasionally. Cerebral abscesses occur from the destruction of the bony septum between the brain and tympanum. Septicæmia and pyæmia may result, as well as metastatic abscesses. Thrombosis of the sinuses in the vicinity of the ear sometimes is observed.

Etiology.—The first fact we are impressed with in discussing the causes of this disease is that the patient has "caught cold" and has a sore throat, the throat affection antedating the ear symptoms, and, as has before been hinted, the pain and unpleasant sensations are often found on only one side of the throat which soon invades the ear of the same side. Exposure of the ear directly to cold will excite an otitis without there necessarily being any throat symptoms. Listening at a keyhole will give a patient otitis consequent on the draft of air. Cold water introduced into the canal may cause an otitis, and in sea bathing it often does. When the patient goes under water a certain amount may get into the naso-pharynx, and in the disturbed respiration, coughing, sneezing, swallowing, and the like, water is forced up the Eustachian tubes into the tympanum. It is true that in salt-water bathing the water is salt, but it is *too salt*, and it is *cold*. Besides this, the breakers may sometimes strike the side of the head and produce such a concussion upon the ear as to rupture the membrana tympani ; but in any event, inflammation of the tympanum may follow the introduction of cold water into the meatus.

Many skin affections about the head and face may creep into the meatus and involve the tympanum. The tuberculous condition, with its throat complications, often induces suppurative otitis. Tuberculous matter, in the tympanum or what is analogous to it, acts as a foreign body and excites inflammation. In Bright's disease of the kidneys, the diseased bloodvessels in the tympanum sometimes rupture and excite inflammation. *Traumatism* sometimes causes suppurative otitis. It is a well-known fact that the concussion upon the membrane of the tympanum, resulting from standing near a cannon when fired, will rupture the membrane, and sometimes excite a violent otitis, although it is more likely to injure the labyrinth. The directions given by gunners to meet this difficulty, is to stand with the mouth wide open. The explanation of this is probably that the concussion falls equally upon both sides of the membrane. This phenomenon would cause us to suspect that the Eustachian tube is either habitually open, or is very easily forced. A box on the ear may, by suddenly compressing the air in the meatus, rupture the membrane and excite inflammation. Recently a boy came to me with an otitis which plainly pointed to the traumatism inflicted by a snowball which had been thrown by another boy, and had struck him full in the meatus, and had filled the latter with snow. In this case there was added to the traumatism the intense cold of the snow as an excitant to inflammation. Dr. Burnett, of Philadelphia, in his Treatise on the Ear, refers to a case where the disease was excited by a blow from a playing-ball thrown by a comrade. He explains the mode of the injury by the sudden compression of the air in the meatus, which, striking on the membrane, inflicted the injury. Another patient under my observation accidentally thrust a lead pencil into the meatus, its pointed extremity penetrated the membrane, and caused a violent purulent otitis, with loss of a portion of the membrane. He, however, recovered with a perfectly healed drum membrane and good hearing. The nasal douché has, I have not the least doubt, caused many attacks of otitis purulenta.

The action of scarlet fever in developing otitis, is mainly through the throat, though occasionally the external meatus is first involved. This is usually a severe form of otitis, with a strong tendency to destruction of tissue. The labyrinth may also become implicated. It does not, however, in my opinion, differ in kind from other purulent inflammations of the middle ear. Nearly the same may be said in regard to the otitis of measles, but it is not as severe as the scarlatinous variety. Other fevers also cause otitis by throat complications. Pneumonia sometimes causes otitis, especially in children. In them the patulous condition of the Eustachian tubes allows air to pass into the tympanum during the rapid respiration incident to this disease. The air being taken in rapidly and through the mouth, is not sufficiently warmed, and this alone may cause the otitis, besides the violence inflicted by the passage of air into the tympanum. It is not denied, however, that an angina in pneumonia may produce an otitis as in the same affection elsewhere.

The diagnosis of this disease is mainly from acute catarrhal otitis. From this it may be distinguished by the presence of pus in the discharge; by the constancy of perforation of the membrane and the tendency to its destruction; its greater severity, its symptoms of constitutional disturbance, and its tendency to become chronic unless cured by treatment. We regard this disease as the same in kind as the catarrhal, but differing principally in its greater intensity. Otitis externa somewhat resembles this disease, and the differentiation has been given under the heading of the former. In small children it is very difficult to determine whether signs of distress manifested in a given case are dependent on ear trouble or on some other affection. There is violent crying; the head may roll about as though there was brain trouble; there may be intolerance of light, and even convulsions; and whether the trouble be of the brain, the bowels, or the ear, is difficult to tell. It is true that if a good view of the membrana be obtained, it will probably appear intensely red and perhaps bulging; but the parents do not usually think of having an examination made. After a time the discharge appears and the mystery is solved. ·

In the treatment, the first indication is to relieve the pain. Fortunately, most of the means commonly used for this purpose tend also to subdue the inflammation. Abstraction of blood by leeches is perhaps on the whole the most potent remedy. From one to three leeches may be applied to the outer opening of the meatus, but preferably on the posterior face of

the tragus.   After the falling off of the leeches the bleeding
may be allowed to continue for an hour unless the patient is
completely relieved from pain, and there are special symptoms
of depressed energy.   We have usually found one or two
leeches sufficient for our purpose.   It is well to abstract only
enough blood to relieve the pain, as the patient tends, in consequence of the grave nature of the disease, to rapidly become
depressed.   It is a good plan to repeat a single leech once in
three or four hours, if the first gives insufficient relief.   It may
be necessary to repeat the leeching every one, two or three
days to thoroughly deplete the engorged tympanum.   As long
as there is a feeling of fulness in the ear plainly pointing to
hyperæmia of the tympanum, it is well to repeat the leeching.
The point at which the leech is applied seems to me important.
It has been my experience, that a leech in the position previously mentioned, accomplishes much more than when placed
elsewhere.   It is well not to allow the leech to bite on the
cartilage of the concha or helix, for quite serious erysipelatous
inflammation has resulted from such a leech bite.   The management of the leech requires skill and judgment.   If some
violence is used in the efforts to arrest the hemorrhage, the ear
may again begin to throb and grow painful.   Rholand's styptic cotton has been highly recommended for this purpose, but
I have often seen it fail.   It is well to touch the bites with a
bit of cotton wound on a cotton holder previously immersed in
a solution of *liq. ferri persulph.*   This application should be
made to the leech bite in the most thorough manner.   Pressure
on the leech bite by the finger will also often arrest the hæmorrhage.   If both of these fail, use a solution of *arg. nit.* grs.
xx. to xl. to the ounce of water, in the same manner as the
styptic iron was applied.   When the leech fails to relieve the
pain we may resort to large doses of morphine administered
hypodermically.   Under the circumstances the remedy is likely
to act satisfactorily, and indeed morphine has nearly as much
power in relieving inflammation of the ear as it has in peritonitis, for which it has such a reputation.   Next in value to the
leech is the application of heat or warmth to the ear.   As a
rule dry warmth may be selected instead of moist, as the latter
produces a macerating effect on the canal, which encourages
otorrhœa and the formation of polypi.   There is this exception,
however: if the moist warmth is only continued for a short
time and the part well dried, afterwards there is little danger,
and the moisture of the warmth *does* facilitate the unloading of
blood-vessels.   My experience, however, is on the whole favor-

able to dry warmth. A rubber bag filled with hot water is the most elegant and convenient method of applying warmth. The temperature of the water may be carefully adjusted to that degree which contributes most to the comfort of the patient. This detail is quite an art. I remember a patient who had a severe pain in the whole side of the head from an otitis, which was aggravated by either very hot or moderately warm water; by adjusting the temperature to a middle point the pain was relieved. Very rarely cold applications do good, but, as a rule, cold is banished from the therapeusis of aural surgery.

Common salt heated and placed in a bag and laid on the ear is a good remedy; a bottle of hot water may be used for the same purpose, but instead of laying *it* on the ear, the latter should be laid against the bottle, as its weight would be painful to the sensitive ear. No objection need really be made to a flaxseed poultice, especially as laudanum or some other anodyne may be poured upon it, if it is not kept on too long, and allowed to unduly soften the parts. Excessive poulticing will however cause a dull, aching pain in the ear. I remember a patient who, for a considerable time, poulticed an ear which notwithstanding continued to be painful; the poultice was ordered to be discontinued, and the pain subsided without other treatment. A roasted onion with the outer part stripped off and applied by its conical portion to the meatus is a good domestic remedy. Corn-meal, heated and placed in a bag as directed for the salt, is also useful. Escaping steam falling on the meatus as hot as can be borne is a valuable remedy. The ear needs to be stopped with cotton after these applications, to prevent taking cold. Remedies dropped into the ear are often worthy of recommendation. Magendie's Solution of Morphine may be dropped into the ear; repeat every half-hour to an hour until relieved. If the drum cavity is exposed, opium poisoning might result from using this too frequently. Dr. Theobald of Baltimore (Amer. Jour. Otol. 1879, No. 3) recommends the instillation of a solution of Atrop. Sulph. gr. iv. to the ounce of water. He directs that ten drops be placed in the ear and remain ten minutes, then allowed to run out. This to be repeated every half-hour until relief is obtained. In one case, that of a child, with a ruptured membrane, the pupils became dilated after the atropine instillation, but he considers the remedy safe. We have very little experience with it, and do not, as a rule, use anodynes in the meatus. It certainly would be proper to carefully observe the effects of the belladonna upon the general system. In a severe case, all these remedies are ineffective. I remember having myself a severe otitis where the pain was relieved for three or four minutes only by inhalation of chloroform, when after the application of a single leech to the tragus, the pain was completely relieved in less than half an hour. Another remedy that may be classed as warmth is douching the ear with Clark's Ear Douche (see Index). The stream of quite warm water (somewhat warmer than is agreeable to the hand) may be allowed to fall very gently on the membrane for from ten to twenty minutes at a time. It is most highly thought of by many aural surgeons. I do not use it to any great extent, as it occupies considerable time and seems less effective than many other remedies. It is, however, convenient in children.

Another good remedy in the earache of young children or infants is a piece of cotton with black pepper in its interior. There should be only a small amount of pepper, for fear of

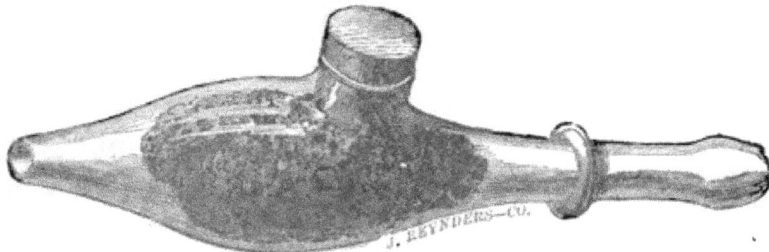

FIG. 66.—POMEROY'S MODIFICATION OF THE BUTTLE'S INHALER.

causing intense burning pain. If the remedy is first tested in the ear of an adult and only a warm sensation results, it will be quite proper to use it on the young child or infant. Chloroform vapor, blown into the ear by the Buttle's inhaler which has been adapted for this purpose by Dr. Roosa, of N. Y., and also by myself in a glass instrument (Fig. 66), will sometimes prove very soothing to the painful ear. *When the drum cavity is filled with products of inflammation, the indications undoubtedly are to puncture or incise the membrane, and evacuate its contents.*

The indications of the presence of fluid in the tympanum have already been given under the head of "appearances on inspection." My own practice is to use what ophthalmic surgeons call a broad needle with a tolerably long shank and *puncture* rather than make an *incision*. It will hardly be credited, that in passing a needle into the membrane, there is some risk of pricking the external meatus, but it is quite possible, as the canal is one and one-fourth inches long, is narrow and not quite straight.   To obviate this I am in the habit of passing the needle down as near to the membrane as possible, meanwhile using the forehead mirror and carefully inspecting every step, and when the apex of the bulging membrane is carefully brought to view, give a quick light thrust until the point seems to touch a firm bottom, the inner bony wall of the tympanum.   In this step of the operation, the sense of touch is more important than that of sight.   It is quite necessary for the surgeon to *feel* that the instrument has gone through a thin elastic membrane and touched a hard substance beyond. If the surgeon is confident of his power to manipulate skilfully, he may incise instead of puncture, making an opening in the membrane of about one eighth of an inch in length. This generally will be in the posterior and superior quadrant of the membrane.   The operation may be done with a narrow curved bistoury, with either a straight handle or one placed at an angle with the blade.   Gruber's curved knife (see Index) is a good instrument for this purpose.   It is not impertinent to remark here, that the operation done with an instrument in which the handle is placed at nearly a right angle to the blade is very difficult of performance.   If the surgeon ever feels awkwardly it is when operating with one of these instruments.   I know of no stop needle which does well in this operation; it is impossible to make it sufficiently delicate to penetrate with the requisite facility; neither is it necessary, as too little force is used in the operation to do harm even if the inner wall of the tympanum is gently pricked.   After the membrane is punctured, it will generally be necessary to inflate the tympanum by Valsalva's or Politzer's method in order to blow the fluid out of the tympanum.   If the Eustachian tube is impervious, then use a rubber tube to the meatus, exhausting the air by the application of a syringe or rubber air bag. If Siéglè's otoscope be used, the process may be inspected. Where the pus is removed by inflation this should be repeated until the perforation whistle is elicited, the patient's head being so placed that the perforation will be the most dependent point.   Consequently all the fluid is removed before air makes its appearance.   This operation may be repeated as often as the incision heals, provided there is fluid to evacuate.   If the ear is frequently inflated, however, this is less likely to become necessary.

After evacuating the tympanum it will be interesting to note the improved hearing and also the bettered bone conduction. Paracentesis or incision of the membrane may be done usually *without ether*, even in children; but as the latter are etherized in from one to five minutes, the anæsthesia becomes a small matter. If an adult is peculiarly sensitive, the operation may be performed with a Graefe's cataract knife, which, when done with gentleness, hardly causes pain. The patient had better not be placed against any unyielding object during the operation, for the head may be jerked towards the operator and mischief might result. It is well to hold the needle so gently in the grasp, that if its point unexpectedly strikes against an obstacle it will slip in the grasp before being pushed violently forward to the infliction of possible injury.

The febrile symptoms require treatment at times, though not always. If the fever runs high it may be well to give the patient a brisk purge—a large dose of calomel is excellent—and possibly a few small doses of Tr. Aconit. Rad. to diminish temperature and pulse; occasionally where there are head symptoms, the bromide of potass. may be given in from ten to thirty grain doses every two hours until relief is obtained. Occasionally the potass. iodid. in large doses is indicated, especially in the latter stages of the disease, when it may be suspected that the brain may be involved. With the same view of diminishing cerebral hyperæmia the fluid ext. of ergot may be given: 20 drops four or five times daily in a little water.

After the acute symptoms have somewhat subsided it would be well to use an astringent to diminish the discharge if it is not inclined to cease spontaneously. The plumbi acet. gr. ij. to v. to the ounce of water may be instilled twice a day, the ear previously having been syringed with warm water and carefully cleansed with cotton. The acid. carbol. ℨi. to the pint of water makes a harmless and cleansing wash which assists in keeping the ear free from any unpleasant odors.

Before using too many astringents in the ear, observe carefully whether it does not incline to recover without much treatment, as there is always a natural tendency for the membrane to heal and the discharge to cease of itself. Indeed, we shall find hereafter, that it is nearly impossible to prevent the membrane from healing, when we sometimes desire to do so.

A great variety of astringents may be used in this connection if it be found necessary; for arresting the discharge, arg. nit. perhaps stands at the head of these remedies. It may be used in strength of from 2 grs. to 80 grs. to the ounce of water, dropped into the ear and allowed to remain a few minutes.

Sulph. zinc. gr. ij. to the ounce of water is useful. Tannin with glycerine ʒi. to the ʒi. is valuable. In this connection remedies may be selected somewhat as in conjunctivitis. We should retain the one we have the best results from, but not continue a single agent too long at a time. Boracic acid sometimes does wonderfully well after the more active symptoms have subsided; the canal is to be packed full of the finely powdered acid. This should be renewed as soon as the discharge moistens it and finds its way out of the ear. Previous to each application to the ear, the patient should be Politzerized, the discharge blown out of the Eustachian tube and tympanum, and then removed by syringing with warm water, followed by gentle wiping out with the cotton. If the membrane shows a reluctance to heal, or the edges of the perforation begin to grow granular, the part may be touched with a saturated sol. of sulph. cup. or with a xl. or lx. gr. solution of argent. nit. A slough should not be produced. The absorbent cotton may be kept in the meatus until the patient is convalescent; remove it whenever it becomes in the least moistened. A very small bit loosely applied is sufficient and does not perceptibly diminish the hearing power. All *mastoid* or *cerebral* complications will be considered under their headings as a separate matter.

Constitutional Management.—As a rule the patient inclines to become much depressed by this disease; demoralized would perhaps be a better word. The pain often undergone is sufficient to profoundly interfere with nutrition, and the patient will have the appearance of having been seriously ill for a long time; the face will have a haggard look, the features will be shrunken, there may be no appetite, and digestion will be interfered with. The bowels may be sluggish and the liver inactive. One important matter is to see that the patient has made up what sleep may have been lost by his severe sufferings. General attention should be given to his secretions, so that he may again be prepared to digest food. Tonics and stimulants almost always will be indicated, and his diet must be very carefully attended to. The most nourishing food that can be digested should be selected. As soon as possible he should resume as far as may be his usual avocations. Going out and breathing fresh air, with change of scene, will often do more for the appetite and digestion than medicines. His courage and hope may also spring up anew. As far as possible endeavor to prevent the mischievous effects of cold. Relapses frequently result in consequence of injudicious exposure. The dress requires to be regulated to the production of just sufficient warmth, but perspiration should be avoided.

CHRONIC PURULENT INFLAMMATION OF THE TYMPANUM.

A certain proportion of cases of acute purulent inflammation of the tympanum fail to recover at the termination of the period of active hyperæmia, and the inflammation degenerates into a state of more or less passive engorgement of the vessels, and we then have the disease under consideration, viz., chronic purulent inflammation of the tympanum. This results from the acute disease not being always self-limiting, or very likely treatment has been neglected or unskilfully applied. The reparative action may still be going on, but so tardily as not to accomplish a cure within the limits assigned to the acute disease, but may, however, subsequently terminate favorably.

Some constitutional conditions may interrupt the progress of the disease towards recovery, as struma, tuberculosis, syphilis, Bright's disease, or even a state of general debility. In a few instances the disease seems to commence as a chronic affection. Many tuberculous subjects may be found with a discharge from the tympanum without apparently having had any previous acute symptoms. I have seen many cases of chronic otorrhœa, where the patient would stoutly maintain that there had been no pain nor even uncomfortable sensation in the ear previous to the advent of the discharge. The membrana must have been ruptured, except in rare instances, and the perforation been of considerable size, or it would have healed before this period. There is *always* a discharge, of greatly varying products, but usually it is of the characteristic purulent quality. As a rule, the discharge continues until the membrane is healed. The mucous lining of the tympanum is always swollen, being much thicker than normal. Frequently it is pale and has lost most of its original characteristics. The general tendency of the disease is towards the destruction of tissue. The membrane may slowly be wasted away by ulcerative processes until not a vestige of it remains. The ossicles frequently have their ligaments destroyed by the erosive action of ulceration and become detached from each other, and may even fall out of the ear. The stapes, however, is much less liable to be separated from its position in the oval window.

Associated with the faulty repair attendant on the inflammatory process we have the so well-known and dreaded aural polypi or granulations, which may recur frequently after having been removed, especially if there be uncovered or dead bone in their vicinity. It is from this disease that chronic mastoid affections result. Destruction of the bony wall of the tympanum and adjacent osseous tissues depend upon suppurative otitis, and when the dura mater is in contact with the diseased bone, meningitis, cerebral abscess, thrombosis and obliteration of sinuses adjacent to the ear result; septicæmia and pyæmia also occur. Fatal hæmorrhages have been observed from the extension of these destructive processes to large arteries, notably that of the carotid in its canal, passing so near to the tympanum. Paralysis frequently is seen, conspicuously that of the facialis, as it is often protected in its Fallopian canal by a very thin septum of bone, or sometimes even by a membrane only, from the tympanum. The hearing may be impaired from a very slight degree to that of absolute deafness. Tinnitus aurium is much less likely to occur than in many other ear affections. Chondritis and perichondritis of the meatus and auricle may result from this disease by extension of inflammation from the drum cavity. The external canal often is ulcerated in consequence of the *ichorous* discharge from the tympanum, and deep grooves are excoriated along its inferior wall. The discharge is a prolific source of eczematous affections of the auricle and meatus. It has recently been found surcharged with cocco bacteriæ carrying mischief in its wake. So much is this disease dreaded, that life insurance companies refuse to insure a patient having a chronic otorrhœa.

It is scarcely possible to anticipate all of the results accruing from suppurative otitis. In the Tr. of the Amer. Otol. Soc. for 1874, p. 539, Dr. Arthur Mathewson, of Brooklyn, reports a case where the destructive processes resulted in an opening by a narrow sinus extending from the tympanum to the outer side of the tonsil through the anterior palatine arch. On pressing the finger against the mouth of the sinus, pus was forced in large quantities into the meatus, through an opening apparently situated in the lower wall of the meatus close to the membrana tympani. This could not be followed by a probe. On widely opening the mouth, pus was forced into the meatus; pressure on the neck below the mastoid also caused a discharge into the meatus. He concluded that there was a burrowing abscess deep among the tissues of the neck; no fluctuation. On July 1st, an incision was carried downwards, from the posterior part of the mastoid, as deeply as was deemed prudent; no pus was found; this was kept open; care was taken with the drainage and medication. These means being unsuccessful, trephining was done; only a few cells were found and no pus; some hypertrophy. This gave only partial relief to pain; subsequently this became excessive, requiring morphine. The discharge from the ear and from the sinus in the throat again became abundant. Pain extending downward along the spine was apparently relieved by counter-irritation. In January, the discharge from the ear and throat suddenly ceased, and the patient rapidly recovered. Another characteristic of this affection is that acute exacerbations are liable to occur which may inflict great injury; mastoid affections as well as fatal brain complications are often traceable to these acute exacerbations.

The discharge is the characteristic symptom. It varies in quantity from the smallest possible amount, just smearing over the membrane or tympanic cavity, to an amount so excessive that the ear may need to be cleaned every hour or two. In the latter instance the external canal may contribute to the augmentation of the discharge. It is greater in recent cases than in those of longer standing. It will usually be greater during an acute exacerbation, although occasionally it may be less.

Children usually present a greater amount of discharge than adults.

Its character is predominantly *purulent*, and of a yellowish or greenish color, and it readily mixes with water. In some cases *mucus* may be observed; this does not dissolve in water, but appears as grayish ropy streaks. A serous character is often noticed when the discharge may be transparent or somewhat chocolate colored. When the discharge is mixed with flakes of epidermis or epithelium, it assumes a turbid, whitish or rice-water appearance. Frequently there is a tinge of blood in the discharge. This ordinarily indicates the presence of granulations or polypi. There is sometimes a thick cheesy-looking material, which may be an admixture of tuberculous matter with epidermis and other débris.

The offensiveness of the discharge is due mainly to decomposition. Pus, mucus, serum and blood have very little smell of themselves, and if the discharge is removed often, there is an absence of any special offensiveness; but let decomposition take place, and a great variety of most offensive odors ensue. The discharge from carious bone is an example of the effects of decomposition upon the odor. As a further consequence of the decomposition of the discharge, B. Löewenberg, in the *Archives of Otology* for September, 1881, speaks of the production of micrococci in very great numbers. Few, however, were present when the proper cleanliness had been practised. The microphytes formed concentrical gelatinous envelopes about every minute particle of decomposed matter; these were composed of the spherical bacteria, together with a few of the rod bacteria.

Appearances on inspection.—The meatus presents a great variety of appearances. If the discharge moistens the canal, there will be signs of macerated and detached epidermis. This will present a whitish or grayish color, caused by the presence of numerous loosened epidermic scales, which prevent exact inspection of the parts. The macerated epidermis often has a wrinkled appearance. When it is stripped off, the meatus, especially near its tympanal extremity, may be considerably reddened, and present somewhat the appearance of mucous membrane. In some instances the canal will be much wider than normal, as though there had been atrophy of its soft parts. In others, and especially in children, the canal will become so much infiltrated, that it may be entirely closed. At the New York Foundling Asylum this is a very common occurrence. Associated with the macerated and excoriated condition of the canal, will be found, in some instances, an auricle swollen, red and possibly moist from eczema, the consequence of this condition. Grooves are often channelled in the lower wall of the meatus, the consequence of the ichorous discharge, which often presents a red and bleeding appearance. When the discharge is less than usual, it forms, by evaporation, scabs and crusts which obstruct the canal. Impacted cerumen is sometimes seen. Perfectly hard swellings of exostoses are often observed. Aural polypi are infrequently seen in the canal, their attachment usually being in the drum cavity.

The membrana tympani is either perforated or has been perforated at some previous time. If mostly intact, and covered with discharge, the signs of maceration described as characteristic in the meatus will be seen in this instance, gray where the dermoid layer is still adherent, and reddish where it has been removed. There is almost always a higher degree of redness about the short process and malleus handle than elsewhere. The folds of the membrane are not important as diagnostics in this as in other conditions. The malleus handle is almost always retracted and foreshortened, and does not present the normal inclination, it being in some instances too vertical, and in others too horizontal. When the membrane has become much thickened and opaque, the site of the manubrium may be obliterated, but its location may often be determined by a minute red streak showing the manubrial plexus of blood-vessels which are upon and on either side of it. *The perforation in the membrane* is more frequently seen in its inferior portion, and may be from the size of a pin-hole to the destruction of its entire area. As a rule the manu-

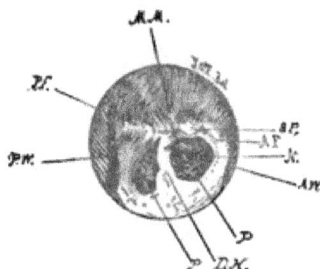

FIG. 67.—A double perforation, one in front of the malleus handle and the other behind, a cicatricial new formation being interposed between the perforations. M. M. Manubrium; S. P., short process; A. F., anterior fold; N. A., necrotic point on inner wall of tympanum; A. W. anterior wall; P., perforation; D. H., cicatricial band attached to the end of manubrium; P., perforation; P. W., posterior wall; P. F., posterior fold.

brium is in position, and it seems to protect the membrane in its vicinity from destruction. A large sized perforation may leave a sickle shaped rim in the lower part of the membrane, extending upward on either side of the manubrium, forming a kidney shaped opening, having the malleus handle projecting into its hilus. Again, there may be two smaller perforations, one in front of the handle and the other behind, or there may be a third perforation in Shrapnell's membrane, or in the region of Tröltsch's pocket. I have with considerable frequency seen a bridge of membrane extending downward, somewhat in the direction of the malleus handle, fairly dividing a large perforation into two unequal parts (Fig. 67). If the perforation is of large size, and the edges of the membrane are much sunken, so as nearly or quite to touch the promontory, the exact coloring of the lining of the tympanum may appear. Sometimes it is very pale, and again it may be deep red. This may give the usual moist reflex, or in some cases where the mucous lining of the promontory has become nearly or quite dry from the conversion of its epithelium into a surface very analogous to that of the skin, it will be red, but may have a dry shining appearance. Where the perforation is small, and the membrane is not much sunken, the tympanum will be badly illuminated, and the perforation will appear like a minute dark spot. Sometimes the perforation will be only a fissure which may not be distinctly or even at all visible. *To diagnosticate a perforation* is sometimes difficult. Where the membrane is entirely swept away, it is not always easy to determine this fact. By touching the part with a probe, the absence of the elasticity felt when the membrane is touched will usually settle the doubt, but sometimes the membrane may be red, and moist, and in contact with the inner wall, when by using atmospheric traction with Siéglè's otoscope, any remnant of membrane may be seen to move. The perforation whistle, elicited by inflating the drum cavity, often settles a doubt. If the tympanum contains fluid and air, an air bubble may appear opposite the aperture during inflation. If the cavity is well filled with secretion it may be seen slowly to ooze out of the aperture during inflation, and not until it is emptied will the perforation whistle be elicited. If, after cleaning the membrane with cotton, discharge makes its appearance in a few minutes, it probably comes from the tympanum, through an aperture in the membrane; or if, after cleaning the ear properly, a small very bright light reflex makes its appearance, it will depend upon a reflection from a fluid which is oozing through a minute perforation, and which the cotton could not thoroughly remove. It is extremely unsafe to give an opinion as to the appearance of an ear before thoroughly cleaning. Numerous light reflexes may depend on the presence of a fluid. The color of muco-pus when rather thick and resting on the membrane, may give the impression that the latter is the object inspected, and not the muco-pus covering it. I have seen many grave mistakes in diagnosis made by not first cleansing the ear carefully. Where there is a point of pulsation in the membrane

there is usually a perforation. A blood-vessel of considerable size is probably well-nigh uncovered by the destructive
process, and exhibits its normal pulsations. Other explanations are given. Its chief interest, however, centres
in its diagnostic value. - The ossicula may often be visible
where the membrana has been partially or wholly destroyed. Sometimes the malleus handle is so drawn inward and possibly imbedded in the swollen lining of the
promontory, as to be invisible, but the short process is
almost always to be seen. The long shank of the incus
may frequently be observed passing behind the manubrium, and parallel with it, terminating a little more than
half-way towards the end of the malleus handle. The
posterior shank of the incus, or processus brevis, passes
backward and somewhat upward and is rarely visible,

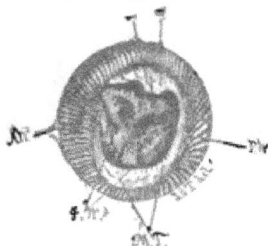

Fig. 68.—Results of suppurative inflammation of the drum cavity. A cicatricial membrane replaces the normal membrane, which is adherent to the promontory. The
remains of the old membrane appear sickle-shaped. Ossicles are in situ except the
manubrium. *I.*, incus; *M.*, malleus; *P. W.*, post wall; *M. T.*, remains of old membrane; *I. W.*, internal wall; *A. W.*, anterior wall. Tansley's case.

being placed above the field of vision. The ramus of the
stapes may sometimes be seen opposite the end of the long
shank of the incus, but somewhat higher up and much
foreshortened. The round window is not visible, being
too far towards the posterior inferior quadrant of the tympanum. Its anterior edge may sometimes be seen, however.
In the absence of the malleus and incus, the head of the
stapes may occasionally be seen in the superior and posterior part of the tympanum, as a minute nodule, usually
reddish, and covered by swollen membrane. When the
perforation heals, the membrana may assume a tolerably
normal position and appearance, the substitution membrane (Valk) may be possessed of all the characteristics of
the normal membrane, but it is more likely to be composed
of cicatricial tissue, without any division into distinct layers; thinner and more yielding than the normal, as is shown

during inflation or by the use of Siégle's speculum. If the perforation be large, the repair is accomplished in this manner: the edges of the perforation become attached to the promontory, reparative material is thrown in, filling the aperture and causing still further adhesions of the membrane to the inner wall of the drum cavity (Fig. 68). If the perfora-

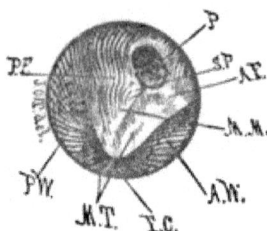

Fig. 69.—Represents an old case of necrosis of the superior wall of the osseous meatus near the membrana. The latter is perforated. This necrotic point extends upward and backward, so as to involve the tympanum, being separated from the meninges by a very thin lamella of bone. A probe could not be satisfactorily passed into the tympanum, and no perforation whistle could be elicited. No operation for the removal of dead bone was thought justifiable. There was a slight purulent discharge. The canal was so hypertrophied that only a small portion of the membrana was visible. P., perforation; S. P., short process; A. F., anterior fold; A. W., anterior wall; F. C., floor of the canal; M. T., membrana tympani; P. W., posterior wall. The membrana was much thickened. Tansley's case.

tion has extended so as to involve most of the membrane in a given meridian, the cavity of the tympanum may be divided into two chambers; and if only one of these communicates with the portion of the tympanum in communi-

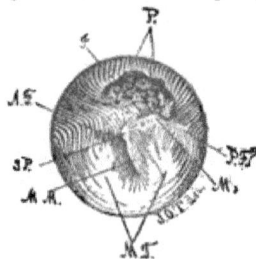

Fig. 70.—Shows the results of suppurative otitis. P., perforation; M. M., manubrium; S. P., short process; M. T., membrana tympani; A. F., anterior fold; I, incus; P. F., posterior fold. Tansley's case.

cation with the Eustachian tube, that only can be inflated (Fig. 69). Occasionally the edges of the perforation come in contact with the ossicula, and form adhesions with them, instead of with the promontory (Fig. 70). In the repair of the membrane, the process may be observed step by step. I have seen fibres shoot across a perforation, gradually

closing it in, and, evidently, from their radiate appearance
entering into the formation of the fibrous layer of the
membrane.   Great care should be practised not to break
up these incipient formations, for at first the material is exceedingly soft, and is easily ruptured by rough manipulation.   These reparative processes have also been repeatedly
observed by others.   It is well to be always on the lookout
for any redness, swelling or fistulous openings, on the posterior wall of the meatus, near the membrana, for here is
frequently found a communication with the mastoid cells,
which lie directly in contact with this portion, and which
may be the source of the persistent discharge.   In any
prolonged and unexplained discharge, it is well to look
over every part of the middle ear as far as possible, to find
any minute polyp, or granulation, which might give rise to
the trouble.   About the entrance to the mastoid antrum,
as Dr. Buck has carefully pointed out, are often found
polypi or granulations which keep up irritation and discharge.

In the treatment of purulent otitis, the first consideration
is absolute cleanliness.   The discharge must all be removed
in a gentle but effective manner, and as often as it accumulates in any considerable quantities.   Syringing with tepid
water containing a little salt or carbolic acid or castile
soap, is the proper procedure to begin with.   I am aware
that latterly many aural surgeons of undoubted ability have
objected to the syringe.   It is true that some patients are
rendered giddy and faint by having the ear syringed, even
gently.   I once had a patient fall in a faint while I was
syringing his ear.   In a case of suppuration with moderate
discharge, excessive syringing may irritate so much as to
increase the discharge.   In syringing, all violence should
be avoided, the patient must not even be rendered uncomfortable, and as a rule no harm will result.   I use my
one flange syringe, which is figured elsewhere.   For the
patients' use I order the ordinary one-ounce hard-rubber
syringe sold in the shops.   After syringing, there will be
some discharge hidden away in various nooks and cor-
ners, and possibly in the Eustachian tube, which the
stream of water fails to reach.   It will be well to fill the ear
with warm water, hold the head towards the opposite side,
then perform Politzer's inflation (Fig. 71).   Any matter collected in the Eustachian tube or its vicinity will be blown
into the tympanum, and will mix with the water, when it

is easily removed. Some use the Eustachian catheter to
inject water through the Eustachian tube into the tympanum. Ordinarily this is unnecessary, and may sometimes be too violent. The same objection obtains to Hinton's method of fitting the nozzle of the syringe to the
meatus water tight, and forcing a stream through the
middle ear and Eustachian tube into the throat. I believe
this method to be more objectionable than the other, but it
may sometimes be employed, especially if there is a large
perforation in the membrane and the Eustachian tube is
sufficiently pervious. The principal objection to the
method is, that the column of water may press too
strongly on the round and oval windows, causing labyrinthine pressure. It is well to always select the simplest
and least violent methods for cleansing the parts from the

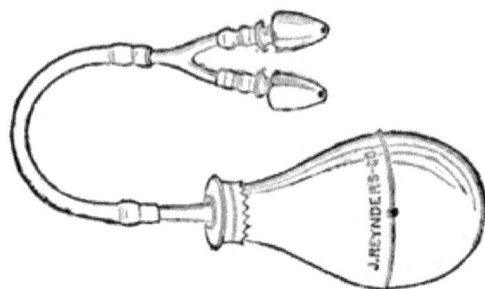

Fig. 71.—Pomeroy's Apparatus for Performing Politzer's Inflation.

discharge. An ordinary glass or earthenware bowl may be
used in syringing. If very great cleanliness needs to be
observed, use a second receptacle for the water. The
practice of Wilde, of using a basin with a strainer in the
centre, is well enough perhaps for hospital work, but it is
not altogether cleanly. After all that is possible has been
done by irrigation, use the absorbent cotton on a holder to
perfectly dry and cleanse the part. The dentist's cotton
holder may be used for this purpose. What is really better
is an ordinary *hair-pin* straightened. The ends may be
roughened by rubbing against a file, a piece of stone or a
brick, so as to enable it to readily "catch" the cotton.
Before winding on the cotton, dip it into water. The hair-
pin is *superior* to other cotton holders: it may be found of
any convenient size, and being very light, the tactile sensibility is not at all dulled by its weight. It naturally
requires a little mechanical dexterity to properly straighten

it, for no cotton holder works properly if in the least degree bent. A piece of iridium wire, being very light, makes a good cotton holder. The wiping out of the ear should be done with the forehead mirror in position, and every step of the operation should be carefully observed. Where granulations or polypi are present, do not use the holder violently enough to cause them to bleed, for this will embarrass our operations; or if the membrane is in process of healing, care needs to be taken not to break or lacerate the newly formed, soft and fragile tissue. Frequent inflations during this process may be necessary, as muco-pus may be blown into the tympanum or through a perforated membrane into the meatus and require to be removed. Any detached scales or crusts of epidermis, and dried discharge or minute particles of cerumen, may be gently wiped away by a twirling motion of the cotton holder. Avoid causing the parts to bleed if possible. Where great gentleness is essential, Clark's aural douche may be used, and the cotton swab may possibly be dispensed with. This careful cleansing of the ear should be done by the surgeon himself three to six times a week. In the meantime the patient must carry out the instructions of the surgeon as carefully as possible. *Never syringe the ear unless it is necessary.*

In the selection of *astringents* for the treatment of the lining of the tympanum and its contents, and for the arrest of the discharge, great care and judgment are necessary. We cannot definitely assert which remedy will be the best in a given case; this can only be ascertained by thorough trial. In the selection of remedies, however, we are to be guided by a few cardinal principles, and by the results of experience. I will here enumerate the remedies I have found most serviceable, and also will give preference for those that have stood the test of extended experience in the practice of others. I know this has in it an element of empiricism. I may premise, however, that whatever remedy is selected, should be discontinued as soon as it seems to have lost its effectiveness.

On the whole, nitrate of silver in solution is one of the most effective remedies in the treatment of suppurating ears. It has the advantage that it is of service where there are granulations or polypi, a point I had noticed at the time Schwartzer was making his studies on the action of the nitrate of silver in suppurating ears, but he did not regard the remedy as useful in granulations. There is no way of determining the proper strength of the solution except by trial. It may be effective in solutions varying from two grains to the ounce of water, to a saturated solution (more than 480 grains to the ounce). An average strength for children would be from 2 to 20 grains to the ounce, and for an adult from 20 to 80 grains. In the *N. Y. Medical Journal* for December, 1872, p. 631, I reported a case of chronic suppuration of the tympanum in an elderly subject, with symptoms of tuberculosis, in which there was a moderate sized perforation of the membrana with a few small granulations. I had treated the patient by dropping in solutions of arg. nit., of about the strength of 100 grains to the ounce. It never was painful, and it diminished the discharge, but did not arrest it. I felt that the remedy was a good one, but that its power required to be increased; so I dropped in a saturated solution, and watched carefully its effects. The ear was syringed out with water in a minute or two. Beyond a slight feeling of fulness in the ear, and a little warmth, there was no pain. From that time on there was no more discharge, and six months afterwards the membrane was found to be healed. (The patient had not returned after the application, until that time.) I have used the same solution many times since. It has sometimes caused pain, but never has there been any considerable reaction. I remember a case of suppuration where the aperture in the membrane was so small that intratympanic injection was impossible, and the discharge persisted. Two applications of the saturated argent nit. sol. to the membrane arrested it, but its use was accompanied by a little pain. I never have seen sloughing of the membrane result. *The mode of application* is very important. My practice is to use a medicine dropper and inject a sufficient amount to fairly inundate the parts, the head being turned so that the meatus is placed pointing upward. While in this position inflation may be done to still further insinuate the remedy into every part of the tympanum. I also frequently, with the pipette inserted into the canal, alternately draw out and reinject the remedy again and again. A still further effect is obtained by passing the cotton wool over the parts, performing a sort of "rubbing in process." If there is pain, syringe with warm water. The effect is greater, however, if the solution is allowed to remain in the ear, a bit. of cotton moistened with salt and

water being applied to the meatus.   This may be allowed
to remain for some hours.   Obviously there will be no
staining of the meatus by the nitrate of silver.   There may
be applications made every second or third day; as long
as the effect of the first is still operating, do not repeat.
Over-treatment is always to be avoided.   I am not satisfied to
recommend nitrate of silver indiscriminately, for it sometimes does harm.   If there is not prompt reaction from its
effects all the symptoms may be aggravated; the indications then will be to weaken the solution or abandon the
remedy altogether.   I have seen granulations grow in a
very vigorous manner, under the irritation of the nitrate,
and also have seen them rapidly disappear under its influence.   No remedy requires more careful observation.
While it is one of the most effective of agents, it is also a
hazardous one.   *Boracic acid*, in very fine powder, as recommended by Bezold, is a most valuable remedy.   It should
· be pulverized by a machine made for the purpose, which is
used by the wholesale druggist.   The ear having been
properly cleansed and dried, the meatus may be filled with
the powder, and even packed, by means of the cotton
holder or any of the approved powder blowers.   Dr. J. B.
Johnson, of Paterson, N. J., uses a quill of about the size of
the meatus attached to a rubber tube.   By plunging it
into a mass of the acid it is readily filled, when it is inserted into the meatus, and by the pressure of a column of air,
propelled by the breath, it is readily pushed deeply into the
meatus.   This process may require repetition once or twice
to completely fill the meatus.   The powder does not fly by
this manœuvre.   Another method is to place in the meatus
a long, rather narrow speculum, and fil¹ the ear by means
of an ear spoon, packing it by a bit of cotton on a holder.
If the discharge is moderate the boracic acid may remain
in several days; if there is considerable discharge the acid
will be moistened and dissolved sufficiently to allow free
exit to the discharge, when it may be syringed out and renewed.   Recently I was much gratified to witness the effect of this remedy at the Manhattan Eye and Ear Hospital.
A considerable number of cases of suppuration were not
doing well under nitrate of silver and other astringents, when
boracic acid was used.   To my great delight, in most of
them the discharge ceased in a few days.   It is serviceable
even in minute granulations, and is a valuable disinfectant to the decomposing matters always found in a dis-

charging ear. Dr. Löewenberg, of Paris, adds alcohol to the boracic acid, thinking it more effective in destroying the microphytes found in the decomposing discharge. He uses the alcohol diluted at first until the patient can bear it in full strength. I have not on the whole been favorably impressed with the efficacy of the alcohol treatment. Neither do I believe that any remedy needs to be used to destroy organisms incident to decomposition, provided proper cleanliness be practiced. The above comes under the head of "dry treatment" so much spoken of at this time (1883). I am sure that after what have said about the ill effects of the maceration and soaking of the parts by excessive syringing and douching of the ear and the avoidance of undue moisture in treatment, that no one will accuse me of going too far in the direction of moist treatment. Dr. H. N. Spencer, of St Louis, has written in the *Amer. Jour. of Otology*, vol. ii. p. 184, on the dry treatment of suppurative otitis. He quotes Becker as authority for the dry mode of *cleansing* the ear, and gives cases illustrative of his mode of "dry dressing" for the ear, as follows :—Iodoform is blown into the ear so as thoroughly to come in contact with the parts; a bit of absorbent cotton is then pressed down quite upon the membrana tympani and allowed to remain. This dressing with the iodoform in one case was renewed twice a day for four days, when the discharge had ceased. After that the cotton alone was used. He thinks that the cotton by itself will be sufficient treatment in some cases; it protects from the air, absorbs moisture, and gently stimulates the parts. He is in the habit of applying the cotton, then inflating, and the moisture forced through the perforation sticks the cotton to the membrane and affords an admirable dressing. In an article published by me in the *N. Y. Med. Jour.* some years since I advocated packing the tympanum with cotton, but at that time there was no absorbent cotton, the latter being so superior to that which has not been prepared. Dr. S. thinks that the cotton exerts traction on the edges of the membrana tympani, bordering the perforation and prevents sinking. I have had excellent results with iodoform. Its unpleasant smell may be diminished by mixing a little balsam of Peru with it, (Mittendorf). Powdered alum, burned or otherwise, is a valuable remedy blown into the ear. I have known it to shrivel up and disperse small polypi in a few days, while in other cases it seems quite ineffective. Some years since

Dr. Agnew of New York was in the habit of using what he denominated "alum mud." A supersaturated solution of alum was made, and the resulting sediment was pushed into the ear instead of the powder. One important contra-indication to the use of alum is the resulting coagulum, which often acts as a foreign body in the ear and is difficult to remove. It is quite proper to give the patient a saturated solution to pour into the ear daily, after syringing. Gruber objects to using powders in the ear on account of their collecting into masses and acting as foreign bodies. He is also in favor of seeking out diseased places in the tympanum or membrana, and making a stronger application to these parts only. I have often done this with arg. nit.

Some years since the permanganate of potassa was quite fashionable, especially as it was a good disinfectant. I used it for a time, but it has latterly somewhat fallen into dis-use. Recently Dr. Lucien Howe (*Tr. Am. Otol. Soc.* 1879, p. 360) has revived its use, and is very favorably impressed with it. He uses a solution of from two to eight grains to the ounce of water, instilled once or twice daily; allowing it to remain five or ten minutes, or syringing it out sooner if it causes any smarting or burning. Whatever agent is used may be placed on a cotton pellet and pushed down upon the part. Dr. Gruber, of Vienna, passes a small quantity of a mild astringent into the posterior pharynx, inclines the head to the side of the affected ear, and the agent is blown into the tympanum by Valsalva's method of inflation. Sulphate of zinc in solutions of from one to six grains to the ounce of water is a valuable astringent. It may be used in the ear two or three times daily. The chloride of zinc, in weaker solutions, may also be used; it resembles somewhat the arg. nit., but it is apt to cause pain at times. Tröltsch recommends solutions of cup. sulph. in from one to five grains to the ounce of water where there is carious bone. This is one of the most valuable of all the remedies for con-junctival diseases, but in the ear it is sometimes rather pain-ful, and its intense astringent action seems occasionally to produce unpleasant irritation. Plumb. acet., in from two to ten grains to the ounce of water, has long borne an ex-cellent reputation in otorrhœa. In eye practice it has been abandoned on account of its forming insoluble crusts of lead upon the cornea, and the same objection to its use has been made in ear practice, but I never have observed any-thing in its action to cause me to abandon its use. It sometimes does admirably in diminishing the gelatinous softness of a mucous polyp previous to its removal. Carbolic acid, although a few years since, as a new remedy, was somewhat too highly extolled, is really a good agent, even to diminish a discharge from the ear. It still holds its place as a good antiseptic, and deserves to be used. From two to five grains to the ounce of water is a proper strength. Any pain it causes quickly subsides. Other remedies to diminish the offensive odor from a suppurating ear are the liq. sod. chloriuat. about ʒj to an ounce of water. The permanganate of potassa is valuable also for this purpose, in strength of, from one to three grains to the ounce of water. At the New York Foundling Asylum we have had many cases of suppurative otitis, in which the meatus was much narrowed and in some cases even entirely closed by the swelling of the canal, induced by the acrid discharge. Many of these cases had swelling also of the auricle from the eczema accompanying this affection. I tried a variety of expedients for the relief of this condition. The canal was incised at first, and afterwards an extensive incision was made behind the auricle opposite the meatus, so as to communicate with the deeper portion of the latter, which was not usually narrowed; this it was hoped would allow free exit to the discharge and permit the ear to be cleaned and astringents applied. These measures failed. Dr. Chadbourne, of New York, then the House Physician of the Asylum, suggested and successfully carried out an idea of his own—to use rubber drainage tubes. His practice was as follows: If the canal was not too much narrowed, a piece of black or red rubber was forcibly pushed into the bottom of the meatus, and then cut off on a level with the concha, so the child could not pull the tube out. In other cases, where only an ordinary sized probe (say a No. 8 Bowman) could be passed in, a rubber tube was drawn over the probe, the end of the latter was filed so as to produce four short prongs or teeth, on which the opposite end of the tube was caught; then by drawing upon the tube it was so attenuated as to be scarcely larger than the probe within; in this condition it was passed into the canal. When pushed in sufficiently, the end next to the surgeon was allowed to spring back into its natural shape, when it drew itself into the canal, and most if not all of its former calibre was restored. It was then cut off and allowed to remain in, from one to several days. With even a very small calibre to the rubber tube, the ear could be properly cleaned by syringing, and even cotton on a very small holder could be pushed in for purposes of cleansing. After a few days the canal, by the constant pressure of the elastic tubing, increased in size so that a larger tube could be inserted, and after a considerable size was reached, it was not necessary to do more than simply push the piece of tubing in without the aid of the probe. Rarely has any irritation been produced, the patients wearing the tubes for weeks together. One little patient was sent into the country with a tube in the ear. By some oversight it was not removed for several months, when, on his return, the ear was found dry and healthy, and the tube seemed to have *grown into the canal*. It was easily removed by forceps without injury. Besides restoring the calibre of the canal, the result of the constant pressure of the dilating rubber tubing, the parts are protected from the irritation of the discharge, the source of the trouble. Any granulations in the canal are likely to be absorbed by this elastic pressure, and altogether its action seems to be satisfactory. The eczema of the auricle usually disappears quickly without any other treatment. Granulations in the tympanum may require treatment, when the tube is removed for any operative procedure, Dr. C. often found that a little carbolized absorbent cotton resting at the end of the tube, within the concha, collected the discharge by absorption and prevented its mischievous effects on the auricle. This was renewed often enough to prevent maceration and irritation of the parts. The softer variety of tubing only is proper to be used. Some varieties of white, stiff and hard American tubing should never be used. They act as irritating foreign bodies. Sometimes the patient recovers from an otorrhœa without treatment. I was called some years since to see the wife of a physician who had been suffering for some weeks with a suppurating ear, which persisted in discharging; the membrane would not heal, and there was much of the time a dull heavy pain in the ear. It had been syringed, douched, and seemed to be constantly growing worse. The epidermis of the meatus was soddened, detached in places, and looked thoroughly soaked. I advised the stopping of all treatment for a few days, to see how far the symptoms depended on the moist applications. The discharge ceased in about a week, there was no more pain, and the membrana healed promptly. The ear was cleaned with absorbent cotton several times daily to keep the discharge from irritating. Politzer's operation was frequently performed, simply to more effectually cleanse the ear. I do not infer that a *very* chronic case would have recovered so promptly. This would come emphatically under the head of "dry treatment." The acute exacerbations which are likely to occur must be treated antiphlogistically, as recommended under the head of acute suppuration; astringents and stimulants should not be used until the exacerbation has somewhat passed off. The Eustachian tube is often obstructed or narrowed by inflammatory swelling of its lining membrane, and calls for treatment; this is sufficiently detailed under the heading of chronic catarrh of the Eustachian tube. A very important object of treatment is to heal the membrana, as the discharge is not likely to be arrested permanently, with an open tympanum, the atmospheric vicissitudes being likely to re-excite the disease into activity. Naturally, when the tympanum is open, we endeavor to protect it. This may be done by wearing cotton wool in the ear, loosely applied so as not to diminish the hearing; change as often as it becomes in the least degree moist. We may also use the Yearsley pledget of cotton pressed upon the aperture in the membrana tympani, which has again become somewhat popular. The subject will be referred to under the heading of artificial drum membranes. Exposure to cold must always be avoided. The ear is never in a proper state when insufficiently warm. Even when the membrane has healed there may only be cicatricial material in the new formation, which readily breaks down on the first onset of inflammation. In general terms, I would not recommend any specific application to the membrane to cause it to heal. If the granulations and polypi are carefully removed, and astringents have been judiciously used, and the ear kept sedulously clean, there is a *strong and natural tendency of the membrana tympani to repair.* In fact a healthy membrane will heal in spite of being punctured, incised, and even where portions have been removed. There is a class of cases in which there is no tendency to the formation of granulations and polypi, where the membrana tympani looks pale, sodden, and shows little evidence of vascular activity. The mucous membrane of the promontory is pale, and if much swollen will be so from passive serous infiltration, and possibly *no treatment whatever will induce reparative activity.* The patient may or may not be in a good constitutional condition. If the edges of the

perforate membrane are considerably reddened, and especially if there is a little nodular roughness of outline, pointing to the possible production of granulations at some
future period, it may be admissible to touch them with a
saturated solution of cup. sulph. or a 40 grain sol. of arg.
nit. The latter should not be strong enough to destroy any
tissue, the stimulating effect only being required *In tuberculous subjects*, there is a form of otitis which seems to be
chronic from the commencement; there is little or no pain;
the tendency to destruction of the membrana tympani is so
great that it is often described as rapidly *melting away*, in
spite of all efforts to prevent the destruction. As a rule
treatment does very little good indeed. It is true that otitis
in phthisis sometimes does not show these characteristics;
the case I reported as cured by the saturated arg. nit. was
not of this kind, but when the peculiar appearances of a
typical tuberculous case shows itself, there is a dubious out-
look for the patient. I have seen a few cases only, where
repair was at all active. Dr. J. O. Greene, in the *Amer. Jour.
of Otology* for 1881, No. 2, p. 137, details two cases where an
exception to this rule is recorded. The first patient was a
man 50 years of age, with empyema and an undoubted
tuberculous condition of the lungs. In a short space of
time, as a consequence of purulent otitis media, two thirds
of the membrana tympani melted away; the mucous covering
of the promontory was destroyed, and the bone beneath
was exposed and roughened. Reparation soon commenced,
the promontory was covered with mucous membrane, the
discharge ceased and did not return for the number of
weeks he remained in the hospital. On leaving, the perforation in the membrana tympani showed a decided tendency to close. The other case was also that of a man
aged 40, with advanced phthisis. At first there was only a
small perforation in the membrana tympani, but in three
weeks the whole of it had melted away. The discharge
gradually ceased, and the mucous membrane of the promontory, which had been congested and swollen, assumed
a normal appearance, except that it was perfectly dry. He
was absolutely deaf from labyrinthine complications. He
lived six months afterwards and had no return of ear trouble.
The deafness remained as at first. Suppurative disease of
the tympanum in Bright's disease of the kidneys, as well as
in diabetes, is likely to run a very tedious course.

AURAL POLYPI.

In suppurative otitis, as well as in other forms of ear disease, which result in loss of substance of the membrana, the mucous surface being in a state of ulceration, the effort to repair the loss is likely to result in the production of an excess of new material, badly organized and unfit for the purpose. This assumes the form of granulation tissue, granulations and polypi. It will subsequently be seen that I hold no view pointing to the peculiar and complex structure of any of these neoplasms. They are all of the same nature, essentially, except the malignant forms, which are to be hardly classed under the heading of polypi.

The location of polyps is, first, in any portion of the tympanum, being developed from its mucous lining; secondly, from the borders of a perforation in the membrana, more likely on its mucous surface; thirdly, from any abrasion or ulceration in the meatus. The posterior and inner portion of the meatus often presents ulcerations and carious spots, communicating with the mastoid cells, which are frequently the site of granulations or polypi. After a furuncle has ruptured, the reparative action may be faulty and granulations or polypi may spring from the unhealed aperture. Fourth, the mastoid cells, especially the antrum, are occasionally the seat of granulations and polypi. Often, where a discharge continues unaccountably, a minute granulation may be found at the entrance to the mastoid antrum, or the mastoid cells may even be entirely broken down, and the cavity filled with polypi. Fifth, the Eustachian tube near its tympanic orifice occasionally is the seat of minute granulations. Any spot of unsound, necrotic or carious bone, in any of the locations here described, is likely to give rise to polypi. Sometimes the cicatrix, resulting from the repair of a membrana tympani, shows a tendency, on slight provocation, to take on inflammation, which often results in ulceration, and the formation of minute granulations. Especially is this the case where the cicatrix is adherent to the promontory. I have seen several illustrations of this peculiarity. Dr. Buck also mentions this in his book on Diseases of the Ear. One of his cases showed a granulation on a cicatricial portion of the membrana; this moved more freely on inflation than the normal portion of the

membrana, which, indeed, is the rule with new formations,
they being less resisting than the normal membrane. It
seems to be a well settled principle, that new formations
possess less vitality and power to resist disease than normal
tissues.

The appearance of the polyps.—In general, the polypus is of a
red or reddish color. Those situated deeply in the canal or in
the tympanum are of a brighter red than those projecting
far into the meatus. If extending to a level with the concha,
they may be of the color of the skin. Where they are constantly bathed in discharge, the color is not as intense. The
soft mucous polypus of gelatinous consistency is of a more
intense red than the fibrous polypus. In children, they are

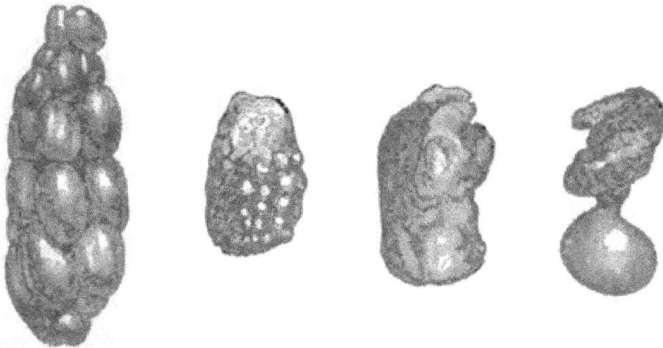

Fig. 72.—From Wilde, Showing the Different Forms a Polypus may take.

of a brighter red than in adults. In tuberculous subjects
a polyp is likely to be pale.

Its size is determined, often, by the limited space of the
meatus or tympanum.

Rarely do they extend beyond the level of the concha,
although I have seen them do so in rare instances. There
seems little tendency in polyps to cause absorption of
tissue in their neighborhood, so as to give room for greater
development, as is the case in malignant disease.

Wilde and Toynbee have figured polypi of many differing shapes and physical appearances, which has given rise
to a rather complicated nomenclature. Such words as
these appear: "raspberry polypus, or raspberry cellular;
racemose; peariform," etc., all dependent on the peculiar
shape and structure of the tumor (Fig. 72). The lobulated
polypus is very frequently met with, but it has no particular
significance. One question may be asked concerning a

polypus—is it hard or soft? The considerable amount of fibrous tissue in the former and the small amount in the latter determines the consistency.

In the Transactions of the Amer. Otol. Soc., 1874, p. 541, I described a case where moderately soft polypoid material filled the meatus, beneath which was found a material of cartilaginous hardness; beyond this was bony material occluding the canal.

The structure of polypi is about as follows: An ordinary mucous polypus was removed by me, at the Manhattan Eye and Ear Hospital, from an ear which had been suppurating for many years. Several similar polypi had been

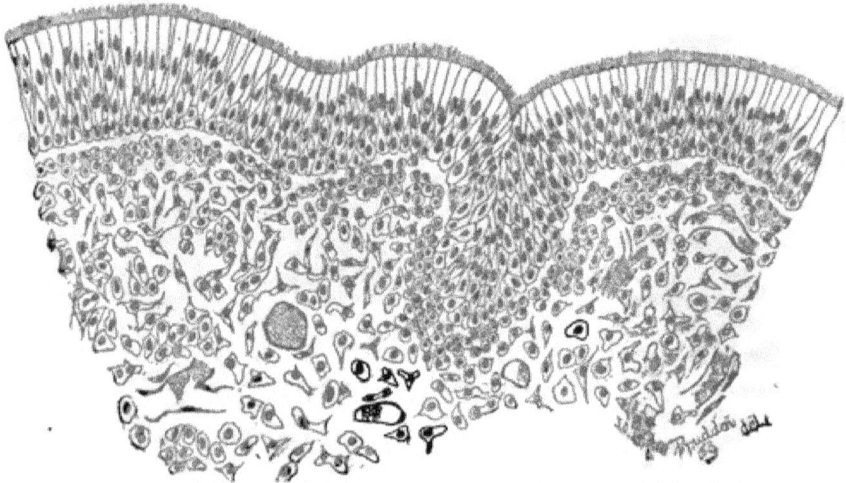

FIG. 73.—MICROSCOPIC APPEARANCE OF A MUCOUS POLYPUS OF THE TYMPANUM.

previously removed, all of which were of large size. This one filled the meatus, and at its outer portion was of a pale red color. The appended drawing is by the Pathologist of the Hospital, Dr. T. Mitchell Prudden (Fig. 73). It is done in an unusually painstaking and artistic manner; the description of the appearance is as follows:

"The tumor (about one half inch in diameter after hardening) consists of an abundant gelatinoid, scantily fibrillated basement substance, in which are imbedded numerous larger and smaller spheroidal fusiform and branched cells. Many of the cells are crowded with pigment; small and thin-walled blood-vessels are abundant, especially near the surface of the tumor. The whole tumor is covered with -

lamellated columnar epithelium, which is, in part, ciliated. Anatomical Diagnoses—Myxoma Polypoides.

As the surface of the tumor was covered with small folds and indentations, I have represented one of these. The epithelial covering is not of uniform thickness over the tumor, as is seen in the sketch."

It is not essential to our purpose to go elaborately into the structure of aural polypi.

So long as they are non-malignant, the indications are to remove them. The soft, mucous, or gelatinous variety presents in part or whole, the following appearances: they all have an epithelial covering, of few or many thicknesses; this is either ciliated, columnar, and, in a few instances, is of the trabecular variety. Internally the mass is made up of gelatinous material called mucin, numerous blood-vessels; white and red corpuscles, numerous nucleated round cells, a smaller number of spindle-shaped cells, with nuclei; only a few connective tissue fibres are present; amorphous granular matter is found. The surface of the polypus may be rendered uneven by the presence of the papillæ.

A diminution of the fluid contents of the polypus and an excess of connective tissue fibres, accompanied by few or no papillæ, places this in the category of the fibromata. The covering of all aural polypi bears a similarity the one to the other. Where the polyp is developed in the meatus, that is, in a neighborhood containing no proper epithelium, the covering of the growth will resemble that of the skin, having a coarse flat epithelium, which is sufficiently characteristic usually to enable a diagnosis of its locality to be made. It also true that a polyp of the tympanum if allowed to extend itself so as to reach the region of the concha will be covered by an epithelium of the flat, tesselated variety more nearly resembling that of the skin, and in its internal structure there will be less " mucin," fewer blood-vessels, spindle-shaped and stellate rather than round cells, and considerable increase in the number of fibrillæ which results in greater hardness than in that found in the protected tympanal portion. In a case of suppurative otitis chronica under the care of Dr. Robt. F. Weir, of New York, a polypus was found attached by a long slender pedicle to the stump of the hammer. It was removed. The case was reported by Dr. A. H. Buck in the Trans. Amer. Otol. Society for 1870, p. 75. He calls the tumor Angioma Cavernosum, according to the nomenclature of

Virchow in his treatise on tumors. It is thus described:
The entire mass . . . consists of blood-vessels, radiating
from an irregularly shaped central cavity, and separated by
a network of fibrous connective tissue, holding blood-corpuscles in its meshes.   In two or three of the sections a large
vein can be followed from the central cavity into the remaining stump of the pedicle.   In one of these, and in other
sections, the point of rupture can be distinctly traced from
the central cavity to the periphery of the polypus.   Six
weeks after removal a second one made its appearance."
He quotes Virchow as of the opinion that an *angioma* is an
independent new growth of blood-vessels.

Treatment.—The first indication is to remove the growth.
This is done by means of the snare, forceps, curette, scissors, and caustics or astringents. The snare, since the
days of Sir William Wilde, has been a popular instrument
for the removal. The best form of this instrument, which
has been called a modification of Wilde's snare, but which
really belongs mostly to the inventor, is the one devised
by Dr. C. J. Blake of Boston (Fig. 74). The cut will

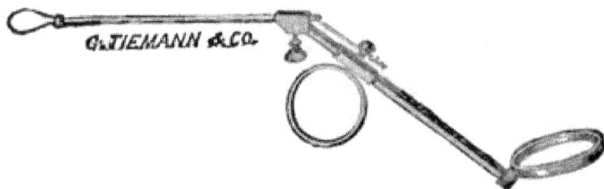

FIG. 74.—BLAKE'S POLYPUS SNARE.

sufficiently explain its mechanism. Many other modifications are in the market, but this one is quite sufficient
for all purposes. The instrument may be provided with a
wire composed of iron or silver. Fishing gimp may also be
used. If the polypus is of small size a very small wire may be
used, allowing greater neatness in manipulation and causing
much less pain to the patient. I prefer fine steel wire; it is
stronger than the silver, does not stretch as much, and is less
liable to be cut by sliding through the cylinder attached to the
instrument. It, however, is liable to rust and become brittle,
which is not the case when made of silver. The late Mr.
Hinton, of London, preferred the fishing gimp to any form
of wire. I have no experience with this, but it seems to
me to be inferior. In removing a polyp with a snare, the
principal point is to get the loop completely around the

growth. If it has a distinct pedicle this may not be difficult. It will be well to examine the polyp very carefully
and find where it is attached, whether there are few or
many points of attachment, or as to the presence or absence of a pedicle. To do this, use the forehead mirror and
speculum, or not, as indications present, and with the cotton-holder clean the ear *perfectly*. Then with a probe
passed around the polyp, a good idea of its location and
points of attachment may be gained. In fact the probe, or
cotton-holder without the cotton, is a good instrument for
diagnosticating the presence of polypi. The next step will
be to adjust the loop of the snare to the size of the polyp.
By drawing it closely upon a speculum this may be accomplished. If the polyp springs from the side of the meatus
or tympanum the loop of the snare may project directly
out of the instrument, but if it is attached to the membrana tympani or inner wall of the tympanum, the snare
may be bent so as to protrude nearly or quite at right angles to the barrel of the instrument. In enclosing the
polyp in the snare it *may* be passed directly down upon the
tumor without any trouble; but *generally* it will be necessary to pass in a probe and push the loop deeper in, so as to
reach the portion of the polypus near its attachment.
When the wire is drawn upon, it should be done gently, so
as not to allow it to slip, and cause the patient too much
pain. After the tumor is well circumscribed, then use
traction. My plan is to give a succession of very gentle
"jerks," so as to break off as much of the tumor as possible, not using force enough to extract anything except the
tumor. By this plan it may be detected if any improper
object be contained in the snare. Sometimes the latter has
caught upon the malleus handle in the effort to remove a
polypus. In a case reported by McBride in the *Ed. Med.
Jour.*, Apr. '81, p. 900, the patient having a polypus removed by the snare, felt as though the loop was caught
around the tongue. It was believed that the chorda tympani had been injured by the operation, as there was no
taste on that side of the tongue for a considerable time
afterwards; it was, however, recovered. Moss of Heidel-
berg (*Archiv. Otol.*, N. Y., 3, '80, p. 35) removed a polyp by a
snare, which caused traumatic paralysis of the chorda tympani nerve, although the latter was not seen during the
operation. There was numbness of the right half of the
tongue, with blunted taste, but the sense of touch was inten-

sified. The patient recovered completely in one month. Although the snare in general is the best method of removing polypi, there are the following objections to its use in the manner just described. When the tumor has a broad base and its violent evulsion is likely to remove some of the contents of the tympanum, as the membrana or ossicula, or wound the chorda tympani, or do violence to either of the fenestræ, it is unsafe to use it, but the polyp should be cut off by tightening the loop of the snare upon the growth without using traction upon the instrument. Naturally this course makes an incomplete operation.

The Forceps are preferred by me for the removal of polypi and granulations, for the following reasons: It is capable of removing *any* kind of polyp, while the snare is adapted only to those of considerable size. If there is no pedicle the snare does not operate satisfactorily. It is more painful usually than the forceps, unless a very fine wire is used, when it will be found too weak for a large growth. It does not remove the whole of the growth as a rule; it is liable to catch other objects than the tumor, and its mode of removal is by breaking or cutting off the polypus, while the forceps, by continuous torsion after being applied, will remove a much larger portion of the neoplasm. Let the instrument be selected which the individual surgeon can best use. The common small-sized straight nasal polypus forceps does very well in ear work, and being straight allows of more convenient handling when the tumor is removed by torsion.

Hinton's angular forceps with reverse openings (Fig. 75), so as to operate more conveniently within a speculum, is a good instrument, although complicated and expensive. Gruber has devised a forceps with reverse openings, quite similar to Hinton's, but without the compound jointed apparatus, which operates as well, and is much less liable to get out of order. The objections to Toynbee's lever ring forceps (Fig. 76) is, that the neck of the instrument near the catch is so slender as not to allow of any torsion of the polypus without twisting the instrument out of shape. A most admirable pair of forceps is one devised by Dr. H. D. Noyes, of N. Y. (Fig. 77). The lower jaw of the instrument is fixed, the upper being attached by a pivot, which acts as a fulcrum to a lever, which is moved up and down by opening and closing the handles. The leverage of the jaws is so judiciously applied as to make it very powerful in grasping an object, and it does not obstruct the view like

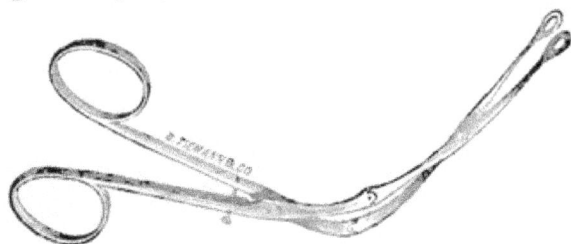

FIG. 75.—HINTON'S ANGULAR FORCEPS.

other forceps. I had a mouse-toothed forceps made for me some years since by Stohlman, Pfarre & Co. of N. Y., with a

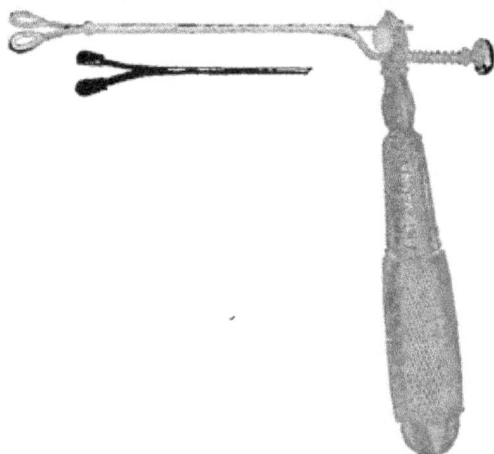

FIG. 76.—TOYNBEE'S LEVER RING FORCEPS (modified by Bumstead).

reverse opening like Gruber's, which is useful in extracting minute granulations (Fig. 78). I am in favor of attach-

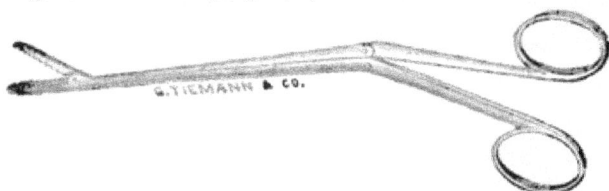

FIG. 77.—NOYES'S POLYPUS FORCEPS.

ing what instrument makers call a "magic catch" to polypus forceps. Two minute hook-like processes are attached to the inner surface of each handle, and when the instrument is closed they interlock, holding it in position. By this method the tactile sensibility is left free, and torsion may be accomplished in a proper manner. The forceps is

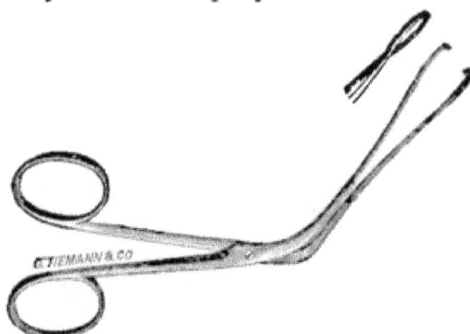

FIG. 78.—POMEROY'S POLYPUS FORCEPS.

opened again by a slight lateral movement of the handles. Dr. Buck, of N. Y., has devised an instrument for *scraping out* granulations and polypi, called a curette. The cut (Fig. 79) will give a better idea of its construction than a de-

FIG. 79.—BUCK'S CURETTE FOR REMOVING POLYPI.

scription. It is quite astonishing how readily granulations may be scraped out of the ear by this instrument. Several sizes are used, as is seen in the cut. Especially well does this operate after a polyp has been cauterized, when the re-

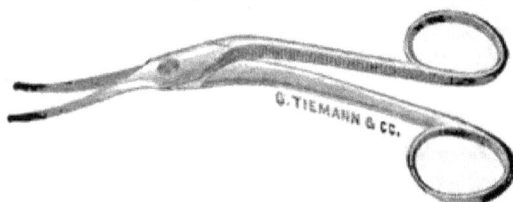

FIG. 80.—GRUBER'S SCISSORS FOR REMOVING POLYPI.

sulting coagulation or slough is easily scraped away. When the polyp is attached by a very tough and fibrous pedicle, it is often desirable to remove it by a pair of scissors (Fig. 80). *Removal by caustics and astringents* is sometimes easily done, but in other instances it is impossible. I have seen polypi grow with extreme luxuriousness, while bathed every second day in a strong solution of arg. nit. Again, the same solution may cause dispersion of polypi of long standing after a few installations. A case of my own, reported by Dr. D. Webster in the *Med. Record*, N. Y., Vol. VI., reads, "removal of a polypus of ten years' standing, by four applications of a 40-grain solution of arg. nit. used by installation," and sufficiently illustrates this phase of the question. It was a mucous polyp, but it is not certain that it had been in existence as long as the report would indicate. This, however, was the duration of the suppuration on which the polyp seemed to depend. Sometimes alum, burned or otherwise, will shrivel up a polyp very promptly. Boracic acid, used after the manner of Bezold, and described in detail under the heading of chronic suppurative otitis, will sometimes cause dispersion of these growths when somewhat minute. Dr. Seely, *Trans. Amer. Otol. Soc.*, 1871, p. 26, uses crystals of chromic acid on polyps, and finds them sometimes to disappear after 4 or 5 applications are made, at intervals of one week. This is better success, I suspect, than others have met with. Some years since I used chromic acid, but afterwards abandoned it as it was found difficult to confine its action to the polyp. Where there is dead bone, the action of cup. sulph. is very satisfactory. It may be applied in saturated solution, by means of the cotton holder, or it may be used in substance. In solution it may rubbed into the granulations more thoroughly than when used in substance; moreover, the method by the cotton on the probe is much easier to adjust to the requirements of a given case. Iodoform is sometimes useful. A bit of moistened cotton on the holder is thrust into a quantity of the agent, and then applied in the usual manner once in one or two days. An important objection to this agent is its unpleasant odor; this may be diminished by rubbing it up with balsam of Peru. Nitric acid fort. applied by the cotton on a probe or by means of a glass rod is very effective in destroying polypi. Nothing but the polypus or its pedicle should be touched, as it acts violently even on parts protected by the skin. The acid nitrate of mercury also may be used in the same manner, remembering that it is more violent in its action than even nitric acid, and is more likely to cause severe reaction. It is sometimes more effective than almost any other application.

Dr. Edward H. Clarke, of Boston, recommended the injection of 2 or 3 drops of the perchloride or persulphate of iron into the substance of the polyp, by means of a hypodermic syringe. In his book ("Observations on the Nature and Treatment of Polypus of the Ear," Boston, 1867) he says: " It causes considerable but not intense pain for ten or fifteen minutes," and in a given case caused the polyp to shrivel up, turn black, and in forty-eight hours dropped out of the ear, a black mass. It did not return. This resulted after a great many other methods of treatment had failed. I believe Dr. Hackley, of New York, has also used this method. It is a rather harsh mode of treatment; the iron is likely to irritate and to blacken the ear, making it objectionable. Dr. Clarke also recommended potassa fusa on polypi. If the parts surrounding the tumor are packed with cotton soaked in acetic acid it is safe to use it, but it is violent in its action and difficult of management. Almost any of the astringents used in suppurative otitis may act favorably in removing granulations or polypi of small size. When the growth is of considerable size, and is covered by a thick skin-like epithelium, almost all forms of astringent and caustic treatment are likely to fail: or if the tumor is fibrous in character this treatment is not likely to succeed. The use of astringents and caustics in removing these growths requires some hints. It is rare that the whole of a growth is disposed of by instrumental means, and that which remains must be managed by the method under consideration. If a polyp is of the mucous or gelatinous variety it will be difficult to grasp it with any instrument whatever. If any effective astringent is poured into the ear, the growth becomes somewhat coagulated, is more friable, having no longer a gelatinous appearance, and may readily be removed by instrumental means. This treatment also disposes of the hæmorrhage, which in some very vascular tumors is exceedingly embarrassing. Arg. nit. is the best agent for this purpose. A good plan is to saturate cotton on a holder with a strong solution (80 to 100 grains), and rub it into the growth, then remove by snare or forceps what is possible; if there is hæmorrhage sufficient to embarrass, clean the ear again by syringing, dry it, and reintroduce the arg. nit. By these means most of the growth may be removed at a single sitting. I place very great stress upon the rubbing-in process. Some (E. H. Clarke) recommend using the astringent for several

days previous to the removal of the polyp. A good plan
is to pour in a 40-grain solution of art. nit., pass in the cotton and move it about among the granulations, then stop
the ear with cotton (on which salt and water has been placed),
and allow the arg. nit to remain in the ear from 15 to 30
minutes, provided there is no pain, then attack the growth
with instruments. The arg. nit., even in pretty strong solutions, causes little or no pain to parts covered by skin,
although the contrary is stated in some of the books.
The bleeding may be quite profuse at times, but there is
never any danger. A tampon will always arrest it. Nitrate of silver has in my hands been found to act as well as
any other hæmostatic. Rholand's styptic cotton bears a
good reputation, but I have seen it fail. A malignant
growth is more likely to give trouble by excessive hæmorrhage. It is of the greatest importance to search carefully
for minute granulations in the most remote and unsuspected parts of the tympanum and its neighborhood, to
find the true cause of an existing otorrhœa. Dr. Buck, of
N. Y., has called attention to this matter, which it so well
deserves. I one time removed granulations from the mastoid cavity (the cells being broken down by caries) through
an opening in the external table. There had been a chronic suppuration of the tympanum, with granulations. The
amount of polypoid material removed at the first sitting
under ether was enough to half fill a teaspoon. There was
considerable hæmorrhage. This operation was repeated at
least four times, at varying intervals, less and less granulation material being removed at each subsequent sitting.
There was dead and denuded bone in every direction, but
the patient, two years afterwards, was entirely convalescent,
and had a useful amount of hearing, that is, could hear
ordinary conversation at three feet. The tympanic orifice
of the Eustachian tube is often the seat of granulations.
These may be managed by injections through the catheter
or by injections through the tympanum into the throat,
after Hinton's method. At the posterior or superior part
of the canal may often be found granulations, which are in
reality attached by a pedicle to a point within the mastoid
cells. These may be reached by thrusting a bent probe
armed with the cotton, saturated with the caustic, deeply
into the part, or the aperture may be enlarged by incision.
Dr. Agnew, of N. Y., has devised an instrument somewhat
resembling a "bill hook" for incising this part of the canal.

The upper part of the tympanum is a favorite seat for polypi. These are difficult of access, as it is above the upper border of the meatus. A bent probe with arg. nit. fused on the end is a good procedure, or even the cotton on a holder sharply bent will serve. A minute syringe with a curved point may reach the part by means of injections. Dr. Buck has a small glass instrument for doing this work, although it is exceedingly fragile.

The entrance to the mastoid antrum at the upper and posterior part of the tympanum is a favorite seat for granulations. One or two minute granulations in this region are quite sufficient to perpetuate an otorrhœa. When the perforation in the membrana is small, and there is un-doubted evidence that there are polypi in the tympanum, it will be proper to incise the membrane sufficiently to allow the growths to be reached and dealt with in the usual manner. It is, however, possible in some instances, by astringent injections, forced from the meatus inward, after the manner of Hinton, to reach the parts and effect a cure. Injections through the Eustachian tube by means of the catheter will also accomplish the same purpose; or an astringent may be thrown into the nostrils, the head being turned towards the affected ear, and, by performing Val-salva's operation, force it into the tympanum. A similar plan also may be adopted after this manner: about half fill the meatus with the astringent, turn the head to the opposite side, adjust the air bag by means of a nozzle tightly fitting to the meatus, then, when the patient per-forms the act of deglutition, gently compress the bag, and the column of fluid is forced into the tympanum and through the Eustachian tube to the throat. It is always desirable to adopt the milder methods of application

Politzer has introduced the use of alcohol in the treat-ment of polypi (*Wein. Med. Wochensch.*, No. 31, 1880). It is warmed, and poured into the ear, and allowed to remain ten or fifteen minutes; this is done three times a day. If it is painful he dilutes with water, but the full strength should be reached as quickly as possible. It produces a certain coagulating effect on the polyp, and turns it to a grayish color. It is more successful in the soft round-celled mucous polyp, than in the fibromata. In spongy granulations of the middle ear it does exceedingly well. It may be used in polypi which are attached to the membrana tympani; sometimes he sends the alcohol through the catheter. Dr. McBride, *Edin. Med. Jour.*, Apr., 1881, p. 900, claims to have used the alcohol treatment before Politzer published his results.   He finds that sometimes it is painful and that it should be used diluted at first.   He explains that alcohol, from its affinity for water, very readily extracts the fluid from the polypi.   He also speaks of its coagulating effect on the neoplasm.   He thinks it causes furuncles, from obstruction of the glands of the meatus, as is sometimes the case with alum.   Alcohol has been used quite extensively in New York during the last few years. Sometimes it is very successful, and again it appears to be absolutely inert.   As most remedies at times disappoint our expectations, it is well enough to have a large number to select from.   As far as my experience goes, it is at least a very harmless remedy.   As to the *recurrence of polypi.*   No subject connected with aural surgery will tax our patience and courage like the frequency with which polypi return after having been removed.   Often one would suspect malignancy from this peculiarity, yet by a sufficient amount of perseverance the growth may be removed to not again return.   I have removed a polyp in a given case more than a dozen times, with healing of the membrane and permanent cure as the final result.   It is usually stated that when these tumors recur so frequently there must be dead bone. In the main this is true; but I have seen many cases where the polyp had grown frequently and there had been no evidence of dead or even denuded bone.   It is not possible in certain cases to determine why a polyp should continue to reproduce itself in such a manner.   It is generally thought that if the base or pedicle of a polyp is thoroughly removed by cauterization, electric or otherwise, it is not as likely to grow again.   Sometimes scraping the point of origin of a polypus with a sharp instrument will prevent its recurrence.

The Consequences Resulting from Granulations and Polypi.— Deafness will result if the growth is of considerable size, from mechanical obstruction, especially where the meatus is completely filled with the growth.   If the tympanum is crowded with granulations which press upon the ossicles and the oval and round windows, hearing will be interfered with from failure of the parts to vibrate freely.   The tympanum may also be so filled with the growths as to press upon the oval and round windows sufficiently to compress the labyrinth, and they in turn the terminal filaments of the acousticus and interfere with audition.   The hyperæmia

of the tympanum and meatus, induced by the irritation of
the growths, and also by the pressure, sometimes amounts
to inflammation, may cause pain, and is sufficient to lower
the hearing, even without the presence of anything obstructive. I have seen symptoms of nervous disease result in
consequence of the pressure from polypi; undoubtedly a
certain amount of retained pus also aided in developing the
symptoms. Dr. E. H. Clarke thought that it was rare for
life to be endangered by polypi, but that it was not impossible. He cites a case where for two or three days previous
to the extraction of a polypus the patient suffered from
" headache, intolerance of light, nausea, vertigo, stupor,
and fever." These symptoms disappeared after the removal of the growth. He further states that "the irritation and inflammation induced in the ear and its neighborhood by the presence of a polypus may extend at any time
to the brain." If the canal of Fallopius has in some portions only a membranous covering, as is sometimes the
case, or if the bony wall is carious at any point, the pressure of the polyp may cause facial paralysis. Hemiplegia
very rarely has been observed as a consequence of the inflammation of the tympanum, the result of polypi, and extension to the meninges, as quoted by Burnett from Schwartze
in the *Archiv. f. Ohrenh.*, Bd. IV. p. 185. The membrane
frequently does not heal after a polypus has been removed,
although the discharge may entirely cease. The edges of
the perforation may collapse so as to touch the promontory, to which it may become adherent. The intervening
space of mucous membrane on the promontory will be covered by a thick flat epithelium, having no moisture upon it
and comporting itself like the skin. The meatus may be
narrowed as a consequence of polypi, and become preternaturally dry from atrophy or destruction of its sebaceous,
hair, or ceruminous glands; sometimes, however, it will be of
too great width. Frequently it is left in a state of chronic
diffuse inflammation, the redness and tenderness remaining for a long time. Caries and necrosis of the temporal
bone from the pressure of the polypi and the erosive action
of the discharge, may result. In any event it seems clear
that removal of the neoplasm is indicated to prevent numerous morbid changes in the ear.

THE ARTIFICIAL DRUM MEMBRANE.

This may be used in certain cases of suppuration with a perforated membrane, but very rarely when the membrane is intact. Its object is to improve the hearing and close the drum cavity, so that it may be protected from atmospheric influences. It was first used by the laity more than two hundred years ago, and afterwards by the profession. It has been made of a great variety of materials. Yearsley selected a cotton pellet for the purpose, and in his book he has much to say about his new method of treating deafness. Like all "new methods" it was rather extravagantly extolled, and after a few years fell somewhat into disuse. At the present time, however, it is being used more than formerly, and seems on the whole to be perhaps the best form of artificial ear drum. It has this great advantage that it is not likely to act as an irritating foreign body. Moreover it will be seen, by reference to previous pages, that the cotton pellet is one of the modes of dressing a perforated membrane, in the method of dry treatment. Perhaps the best known artificial membrane besides this is Toynbee's (Fig. 81); it consists of a circular disk of sheet rubber

G.TIEMANN & CO.

FIG. 81.—TOYNBEE'S ARTIFICIAL DRUM MEMBRANE.

(soft) fastened to a silver wire sufficiently long to reach to the membrana tympani. The rubber disk may require to be pared down by means of scissors, so as to accurately fit the canal. As Toynbee's disk is fastened by two round plates to the stem or handle, it sometimes becomes detached. To obviate this a mechanic in Nuremberg caused the end of the wire to be made into a spiral form so as to catch the rubber, as a cork-screw would hold a cork; the wire next to the tympanum was pushed backward through the rubber to prevent irritating the membrana. Lucae has covered the wire with a bit of rubber tubing to prevent the unpleasant rattling of the unprotected wire in the ear. Gruber uses a similar piece of rubber to that of Toynbee, but attaches to it a piece of silk thread so as to draw it out of the ear when necessary. He claims that less irritation results than with the Toynbee instrument. Beyond the use of soft rubber and cotton wool as materials for artificial drum membranes it seems unnecessary to go. *The indications for its use* are as follows: any drum cavity that needs protecting from the air, may benefit by an artificial membrane; whenever the ossicula are detached from each other, the result of destruction of their ligaments or from failure of the membrane to keep them in proper contact, or even a severance of continuity consequent on relaxation of the ligaments or absence of the incus, with a gap existing between the membrane and the stapes, the latter also may be in a state of sub-luxation from relaxation of its annular ligament, requiring the presence of the artificial membrane to keep it properly in position. The artifical membrane by moderate pressure restores contact of these parts and presents an unbroken medium for the transmission of sonorous undulations to the oval window.

Dr. Barnett in his Treatise on the Ear (Phila. 1877) presents a different explanation of the action of the membrane from that generally accepted (in a certain class of cases), which seems reasonable. Where the membrane has a large perforation with the ossicula in proper relation and the membrane is greatly retracted, in consequence of the non-resistance of a normal membrane to the action of the tensor, there is *undue* pressure on the labyrinth, and the indications are, not to press on the membrane to coaptate parts, but to *draw it outward;* this he does by tucking a bit of cotton wool beneath and behind the manubrium. The result is improvement to the hearing. He states that this may be the secret of the non-improvement to the hearing in so many cases where the artificial membrane is inserted in the ordinary manner. I had one case where a cotton pellet pressed upon an intact but relaxed and thinned membrana diminished a troublesome tinnitus, and improved the hearing. This condition, however, is exceedingly rare.

The artificial membrane is contra-indicated where there is considerable discharge, or when granulations or polyps are present. If the discharge is increased by the wearing of the membrane, it must be discontinued. Sometimes tinnitus aurium, vertigo, nausea and irritation of the chorda tympani results from its use, when it should be discontinued. Any pain resulting from wearing it, is a contraindication. If the stapes is absent, the membrane should not be worn, unless to protect the tympanum from the air. It should not

be worn constantly or at night; it is better to wear it for
short periods only, especially if the hearing is considerably
improved. The cotton pellet may, however, be worn for
days together, only it should be kept clean. If one ear has
normal hearing it is not advisable to use an artificial membrane for the fellow ear, unless it acts usefully in closing
the cavity.

There are several methods of applying the cotton pellet, it may
be pushed down upon the membrane by means of a delicate
pair of forceps, or a cotton holder may be twisted for two
or three turns into the pellet, then carried to the proper position; by turning the holder in the opposite direction so as
to disengage it, it may easily be withdrawn without distur-
bing the cotton. The Toynbee membrane may be caught in
the blades of any convenient forceps and carried to the proper
position. The Gruber plate of rubber, with its silk thread
attachment, may be inserted by appropriate forceps.

Unless the artificial membrane touches the remains of the
old membrane or the ossicula at exactly the right point,
there is no improvement to the hearing. This is determined
only by experiment. Sometimes the membrane must be
introduced several times, before the right spot is reached.
The hearing naturally is tested every time the membrane
is readjusted. The patient himself may be taught how to
introduce the membrane, and he often does it with great
skill; his sense of feeling enables him to select the spot
which before has been found to be the proper one. Frequently there will be no improvement at all to the hearing.
I believe it is not used as often as formerly. About six
years since, Dr. Spencer, of St. Louis, published an article
on the use of the artificial membrane, which contained the
opinions of a considerable number of aural surgeons in
the United States, myself among others. The verdict was
somewhat unfavorable to its use. Is is a pretty scientific toy,
and *occasionally* is very satisfactory in its effects. Dr. Blake
of Boston has used a piece of writing paper, cut of proper
size, moistened with water, and then by gentle pressure
made to adhere to a smallish aperture in the membrana
tympani. This acts both as a means of healing the perforation and as an artificial membrane, and also as a protection to the tympanum. On the development of the
dermoid layer it will be pushed off; this process may occupy
several weeks; then if the healing is not completed another
is applied. The middle ear trouble must have well subsided

to make this plan successful. Dr. Knapp in the Archiv. Otol. for 1881, No. I., p. 64, treats of the cotton disk as an artificial drum membrane. He reports the case of a patient, a woman, who wore it most of the time for twenty-nine years; the deafness previously was so great as to prevent her from mixing in society, obtaining an education, etc., and by the use of the cotton disk she was able to avail herself of all the advantages derivable from the possession of good hearing. The average opinion of the profession is in favor of using vaseline on the cotton previous to its introduction. It should be removed as often as necessary for cleanliness, but may be allowed to remain some days if not otherwise contraindicated. In the discussion at the International Medical Congress in London in 1881, a disposition was shown to increase the pressure of the Toynbee membrane from time to time, in order to maintain a maximum improvement to the hearing, although there were some dissenting opinions as to the propriety of doing this.

MASTOID AFFECTIONS.

PERIOSTITIS OF THE OUTER SURFACE OF THE MASTOID.

This is the simplest and most frequent form of mastoid disease ordinarily met with in practice.

Its usual mode of invasion is to travel outward from the tympanum, along the periostial lining of the osseous meatus, until the covering of the mastoid is reached. Then there will be redness and tenderness, swelling which pits on pressure, with possibly a greater degree of redness and swelling in the neighborhood of the insertion of the sterno-cleido-mastoid muscle. The canal must needs have exhibited some previous swelling, or redness, or tenderness on pressure. There may be some stiffness about the jaws, and the sub- and posterior auricular glands may be considerably swollen. The patient may have had pain in the ear and mastoid region which often radiates, more or less, over the whole side of the head.

If the attack is severe, some fever and elevation of temperature may be observed. The tongue may be coated and the patient feel quite ill. The rule is that pus forms sooner or later. If left to itself it becomes a formidable disease; the scalp may be undermined with pus, so that half of the side of the head may be involved in the abscess, as in the case of John Scrypes, published in the *N. Y. Journal of Med.* for Feb., 1873. The abscess had been allowed to remain some weeks without opening, and had dissected up the scalp above the level of the lower portion of the mastoid process. The diameter of the abscess was at least three inches. The neck was not involved.

It also sometimes follows the sterno-cleido-mastoid muscle, forming an abscess that extends far down the neck. In an ordinarily robust adult the external table of the mastoid is not likely to be destroyed, the disease not going beyond that of periostitis. It is true, however, that the sub-cutaneous connective tissue frequently becomes involved, when a cellulitis is added to the other symptoms. The pain is likely to be severe, like all inflammations of the periosteum, especially if there is confined pus. Occasionally the external table of the mastoid process gives way, and we have true mastoid cell disease. This is much more likely to occur in weakly subjects or those having any constitutional disease favoring destructive processes. In children it very frequently occurs, but this phase of the subject will be treated of under the head of mastoid cell disease.

I do not conclude that there is much danger to the patient's life in this form of the affection, it being a superficial matter mostly.

In the treatment it is admissible to begin with mild measures. If a poultice or leeches to the part promptly relieves the trouble, nothing more may be needed. I have seen Tr. of Iodine painted on the part until it became nearly black, discuss the inflammation. A saturated solution of Arg. Nit. will be still more effective, but it makes a hideous black crust, difficult to remove. I am compelled to admit that there is less indication for severe measures in this affection than I at one time thought. If these means fail, then the Wilde's incision must be done. A stout scalpel is introduced near the lower border of the mastoid, provided the swelling and redness extend so far; penetrate to the bone at once, extend the incision upward, parallel to the auricle and about half an inch behind, for the distance of one or two inches. Carefully note whether the knife glides smoothly over the bone. If there is any roughness, there is denuded, dead or carious bone. By pressing firmly on the part with the edge of the scalpel, a carious spot may often be broken through into the cells beneath, and a sufficient opening made for the evacuation of pus. It will be always desirable, even when no suspicion of bone involvement exists, to search the incision with a probe to see whether any rough or softened bone or fistulous opening exists.

If the abscess extends downward in the direction of the sterno-cleido-mastoideus muscle, or any of the muscles inserted into the mastoid process, a grooved director and bistoury may be used to freely lay open any burrowing sinus leading to an abscess lower down, or the abscess may simply be punctured at this lower point. Larger blood-vessels must be carefully avoided; but if an abscess is pointing, it is likely that no important blood-vessels are in the way of an incision (the abscess having a tendency to separate blood-vessels as it comes to the surface).

It will always be gratifying, in incising a mastoid swelling, to find pus, but often the surgeon will be disappointed. Where there is considerable cellulitis, the sense of fluctuation may be present without a particle of pus being subsequently found. The relief from the bleeding and the division of swollen and tense tissue, will often be very great. Again, the incision, after a day or two, may have permitted an abscess to open into it, which otherwise would have remained to inflict mischief upon the patient. Sometimes, I have by pressing firmly upon a swollen mastoid, caused the abscess to rupture into the meatus on its posterior wall, usually near the junction of the osseous and cartilaginous portions. After making the incision, I am in the habit, if no pus is found, and the general appearance of the part induces one to suspect that there is pus somewhere, to pass in a probe and push it beneath the tissues near the bone in the direction of the focus of inflammation, to determine whether a pus cavity may not be opened into, and I have often succeeded in so doing.

In a recent case, where a Wilde's incision failed to reach pus, and a day after, the abscess had ruptured into the meatus by a very small opening, I passed a probe through the incision in this direction, and easily succeeded in reaching the abscess cavity and diverting the discharge in the direction of the mastoid incision. A probe ruptures no large blood-vessels, is not likely to do harm to nerves, and after the skin and connective tissue are divided, readily penetrates the parts. Occasionally a mastoid abscess may point towards the posterior and outer portion of the meatus.

When this is the case, the incision may be made at that point. Occasionally I have seen an abscess above the auricle, somewhat trenching upon the meatus. This may be opened above the auricle, or in the upper and outer portion of the meatus, that is, where it evidently points. In this region it may involve the temporal bone and open into mastoid cells, but this is very infrequent indeed. Sometimes the inflammation, extending from the tympanum, may involve the parts in front of the meatus, when cellulitis with considerable swelling may result. This is a somewhat annoying location for an abscess. In making an incision there is great danger of dividing some of the larger branches of the temporal artery or even that vessel itself. The incision should be commenced above the swelling and quite near to the auricle, and not extend too far downward, or it may be made in the anterior portion of the meatus. With a stout probe or grooved director endeavor to open into an abscess cavity, if the incision has failed to do so. The director will exhibit a small amount of pus in its groove if it has penetrated an abscess. After the incision has been made, it will be well to use a poultice for a few days; not long enough, however, to macerate the parts, for granulations are prone to spring from these incisions, and the poultice facilitates their growth. It is my own practice to daily open the incision with a probe, moving it from side to side in the lips of the wound to prevent adhesion, and maintain a perfect opening into the abscess cavity. The method practised by many, however, is to insert a tent, and thus keep the wound open. My objection to that mode is, that while the tent is in position a confined pus cavity is formed for the time, which cannot but diminish the promptness of recovery. Any good disinfectant wash may be used, at least once a day, to cleanse the wound. This may, if necessary, be introduced by means of a syringe. If, after the first relief from an incision, the patient again has pain, with possibly fever and elevation of temperature, it is well to search the wound for some concealed pus cavity, which indeed may result from closure of the wound already made in some unobserved portion, and finding it, the patient will experience relief from its evacuation, as in the first instance. The wound should be kept open until nothing remains of the disease. Pain in the mastoid or side of the head should have completely disappeared. All proper care should be taken to prevent relapses. The matter of carious or softened bone will be considered under the head of Mastoid **Cell Disease.** Granulations at the edge of the incision will sometimes be very embarrassing: these may be clipped off with scissors or a scalpel, or removed by forceps, and the point of attachment thoroughly cauterized with arg. nit. in saturated solution or stick, or nitric acid. In making these incisions, arteries are sometimes divided, but torsion will usually arrest the hæmorrhage; if not, apply a ligature. Sometimes an aneurism of the posterior auricular artery may result from the wounding of the vessel in making the incision. Dr. Buck, in the Tr. Amer. Otol. Soc., 1873, p. 61, reports a case of aneurism of the posterior auricular artery. An incision of the mastoid was done somewhat nearer the auricle than usual, which was followed by a small jet of arterial blood. It was arrested by compression, as it seemed impossible to apply a ligature. Five days later distinct pulsation was noticed over the wound; on the next day a circumscribed pulsating tumor, the size of a hickory nut, made its appearance; on the day after, pulsation ceased and the tumor diminished in size ; an incision in the line of the former wound was made, and a blood clot was removed from a distinct cavity. Hæmorrhage recurred, and the part was stuffed with lint. Two days after, on removing the lint, a jet from an artery was observed, when the cavity was again stuffed, and a compressive bandage was applied. The lint was allowed to remain until it ulcerated its way out. There was no further trouble. Other cases of aneurism from a similar cause have been reported, but this one presents all the characteristics of such an accident.

PRIMARY INFLAMMATION OF THE MASTOID AND ITS CELLS.

This very infrequently occurs. The rule already laid down is, the disease is always secondary to middle ear inflammation or that of the meatus. A few cases on record prove that there are exceptions to this rule. In the N. Y. *Med. Record* for June 4, 1881, p. 634, Dr. E. Grüening, of N. Y., reports a case as follows: A man had, in Oct., 1880, hardness of hearing and tinnitus, with stiffness of the neck. On November 3d the doctor saw him, when he had pain in the mastoid, but no swelling, some redness. He could not hear anything but the tuning fork. *There was no disease of the ear.* The membrana tympani was perforated by a needle, and leeches were applied to the mastoid, but no relief was experienced. On November 6th Wilde's incision was done with temporary improvement. On November 12th had a severe rigor, and the next day he vomited. On the 16th of November the mastoid was opened with Buck's instrument and ten drops of pus escaped. The hearing was then much improved, and now, January 28th, 1881, the hearing is normal except that he has a little tinnitus. In a large number of cases of osteo-sclerosis of the mastoid the trouble seems to have developed so long after the middle ear affection occurred as to entitle it to the name of a primary affection. Mr. Dalby, in the *Medico-Chirurg Transactions*, vol. 62, p. 237, states that the mastoid cells become sometimes primarily inflamed, while throughout the course of the disease the tympanum remains healthy. Dr. Arthur Hartman, in the *Archives of Otology*, N. Y., 1879, p. 322, translated by Dr. J. A. Spalding, speaks of an idiopathic chronic periostitis of the mastoid process leading to osteo-sclerosis, which may be developed long after the middle ear is free from disease.

Dr. H. Knapp reports a case of "primary acute periostitis of both mastoid processes" in the report of the first otological congress, N. Y., 1876, p. 80. He quotes from Voltolini and C. J. Blake, who both agree that this affection is possible. They also furnish cases to illustrate the subject. Knapp's case is that of Mrs. H., aged 20. In March, 1876, she had pain in the left ear, but no disturbance in the hearing. Soon the skin over the mastoid region became red, swollen, and tender; not being properly treated it grew worse. The pain subsequently extended to the mastoid and neck. Dr. Knapp first saw her on March 14th, 1876, one week after the disease had commenced. She was then feverish, with loss of appetite, and in bed. The pain was in the left ear and neck. Besides the mastoid swelling previously stated, the skin on the posterior surface of the auricle was red, swollen, and raised. At the mastoidal attachment of the sterno-cleido-mastoideus muscle, for the distance of an inch down the tendon, was a swelling which was tender to the touch. Leeches, morphine, etc., did no good. On the fifth day fluctuation appeared, and the mastoid was incised to the bone, dividing the periosteum and permitting a small amount of pus to escape. The discharge ceased after the usual treatment in four days, and the wound closed—was cured in ten days after the incision.

The mastoid was sensitive to cold for three months afterwards. On April 24th, eight weeks after the beginning of
the disease in the left ear, without any known cause she
had pain behind the right ear and over the head. On the
next day there was redness, swelling and tenderness behind the ear. On the the third day worse, and leeches and
morphine were used; but in spite of these she was so much
worse in two days that an incision was made over the mastoid although there was no fluctuation. No pus was found.
She was cured in five days; did not relapse. This ear, like
the other, was free from disease, thus differing from some
other cases where at some past time the ear had been diseased. The doctor calls attention to the swelling of the
mastoidal insertion of the sterno-mastoid, and infers that
this indicates a mastoid periostitis. Voltolini's case did not
present this swelling, and he thinks it only an inflammation
of the subcutaneous cellular tissue of the mastoid. His own
case, he was at first inclined to think might have cerebral
complications, especially phlebitis and thrombosis of the
cerebral sinuses, as according to Griesinger obstruction
of the sinus near the sigmoid fossa produces a painful
œdema over the mastoid process. The independent character of the inflammation in these cases is worthy of note
and cannot be explained. This affection does not seem to
undergo resolution. The treatment is the same as already
laid down.

INFLAMMATION OF THE MASTOID CELLS

usually depends on an inflammation of the tympanum, this
inflammation extending by continuity of the muco-periosteal
lining common to both the tympanum and mastoid cells to
the latter. Any variety of inflammation of the tympanum
may extend to the cells, whether it be a mild form of catarrhal otitis or the most violent form of purulent inflammation. This process may be acute or chronic, or what perhaps
is more frequent, an exacerbation in a chronic inflammation.
Mastoid periostitis may cause destruction of the outer table,
coveing the mastoid cells, and involve this region.
    The swelling of the soft parts in this form of the disease
may resemble very closely that in simple mastoid periostitis. There is, however, a very rare form of the disease,
which has been most ably described by Bezold in the

Deutsche Med. Wochenschrift, July 9th, 1881. The mastoid becomes perforated at its inner surface in the digastric fossa. It may have been swollen over its whole surface, but on the advent of this complication the tendons of the muscles inserted into the process become so swollen as to apparently partially obliterate the original mastoid swelling. This may go on until the region occupied by the sterno-cleido-mastoid, the splenius capitis, and the trachleo-mastoid becomes deeply infiltrated and the whole side of the neck much swollen, exquisitely painful, hard, and at first with no signs of fluctuation. Somewhat early in the disease an abscess may point at the posterior part of the meatus. Pressure on the swollen muscles will cause the pus to be extruded at this opening, and also into the mastoid if that has been previously perforated. The swelling from the confined pus may extend as high as the superior semi-circular line of the occiput, as it is impossible to break through the tendon of the trapezius, which is also closely adherent to the tendons of the mastoid muscles. This is accompanied by severe pain in this region, from pressure upon the occipitalis major nerve at its point of entrance into the complexus magnus. In a somewhat advanced stage of the disease fluctuation may be detected, but the pus is at some distance from the surface and requires deep incisions for its evacuation.

In inflammation of the meatus, especially at its upper and posterior part, where the periosteum is involved, the partition wall between the cells and the meatus may give way and true mastoid cell disease result.

The ordinary *consequences* of mastoid disease are about as follows : In a certain class of cases there will be violent and destructive inflammation, resulting in the formation of pus and perhaps a carious breaking down of the osseous partitions between the cells, so that the mastoid process may contain but a single large cavity, and it may continue to destroy bone in its neighborhood until in some instances most of the temporal bone may be thrown off in a carious condition. Again the inflammation may be of a milder type, and instead of destroying tissue will result in an inflammatory proliferation, the cells undergoing a process of osteosclerosis, which may convert the mastoid into a solid bony mass, often of petreous hardness, a state of eburnation, or ivory-like. In the earlier stage of this development the proliferation material may be exceedingly soft, but ultimately the true bony quality will be developed. *The pain in mastoid cell disease* is dependent on the following causes: Confined pus is sufficient to excite severe pain here as in other parts of the body. Intense hyperæmia, causing pressure on sensitive nerves, is often a sufficient cause for pain. In those cases where there is no external manifestation of mastoid trouble, or, we might almost say, of tympanal disease, where the symptoms of osteo-sclerosis have been found to exist, there may be considerable aching pain, due undoubtedly to pressure of the proliferating material upon branches of the tri-geminus, or on sensitive filaments of other nerves supplying the mastoid cell region.

Mode of Invasion.—The mastoid is not as likely to be involved during the acute stage of a tympanal inflammation, or, at least, not until the disease has made some progress. The first hint of a commencing mastoid trouble may be in the fact that the ordinary means for the relief of the acute symptoms have failed, and the patient still has pain in spite of leeches, fomentations, douching with hot water, and the administration of opiates. Another frequent mode of invasion is during a chronic suppurative otitis, which undergoes an exacerbainto of acute inflammation, when the mastoid becomes implicated.

In a severe acute inflammation of the tympanum the periosteum of the meátus near the membrana tympani frequently becomes inflamed. This may not result in resolution; the periosteum may be separated from the bone, and the latter, deprived of nutrition, undergoes caries. If this should occur on the upper, posterior, or inferior wall of the meatus, it is likely to open into the mastoid cells. A tolerably frequent experience is to find a fistulous opening at the inner extremity of the meatus, on its posterior wall, which may give rise to a discharge, even after the membrana tympani has healed, and the tympanic trouble come to an end.

Furuncles of the meatus may sometimes involve the periosteum of the osseous meatus, and extend direct to the mastoid cells.

In periostitis of the outer surface of the mastoid, the cells are involved in a certain number of cases. The periosteum becomes destroyed, and the bone beneath dies from mal-nutrition, and a carious opening into the cells is the result. In adults this is not often likely to take place; but in children this seems to be perhaps as frequent a mode of involving the cells as any other. It is possible that the opening into the antrum, the result of a Wilde's incision, in children, may, more frequently than we are aware of, depend on the knife's crashing into the cells, from pressing too hard during the incision. In most of the cases of swelling and inflammation of the mastoid in children that I have incised, the knife has entered the antrum. Bezold, in the *Deutsche Med. Wochenschrift,* July 9, 1881, describes a rather exceptional manner in which mastoid cell disease involves the muscles of the neck. He has found that in some cases the mastoid bone at the digastric fossa, that is, on the inner surface of the process, contains numerous cells, which approach so near the surface of the bone that it is readily cut into with a knife, or even a probe may break through. In some cases there is an absence of bony covering, and the cells may be covered by the soft parts only. Numerous foramina for the passage of blood-vessels still further facilitate the passage of inflammation outward.

At this unprotected point the pus escapes, and burrows deeply among the muscles of the neck, extends upwards as far as the superior curved line of the occiput, and backward to the ligamentum nuchare, and downward as far as the lower cervical or upper dorsal vertebra. Such cases usually terminate fatally, either by exhaustion, or by involvement of the vertebræ or base of the cranium, with symptoms of paralysis; by œdema glottidis, or by burrowing of pus in the thoracic cavity. In one of his fatal cases there was, during the last weeks of life, paresis of both upper extremities, stiffness of the neck, and great difficulty in swallowing. At the time of death there were symptoms of œdema of the lungs. Deep incisions and drainage caused a few of these cases to recover. The muscles inserted into the mastoid process cover its postero-inferior portion by their tendons, and also of that of the trapezius, which is continuous with them, so as to present a resisting wall, through which it is impossible for pus to pass; hence the burrowing along the course of these muscles.

Symptoms of Mastoid Cell Inflammation.—The subjective symptoms are more valuable than the objective, and are as follows: After an acute attack of inflammation of the tympanum, the symptoms sometimes do not disappear after ordinary means of treatment have been used; there still is pain about the ear, but more especially in the mastoid region, which may radiate to almost any portion of the side of the head. There may or may not be fever, a coated tongue, with elevated temperature, vertigo, with a tendency to fall toward the opposite side, nausea, vomiting, sensitiveness to loud noises or a strong light, with sluggishness of the intellect, or preternatural activity; sometimes delirium or even convulsions. In short, there may in some cases be signs of irritation of the meninges, which do not differ from the symptoms that indicate real cerebral complications. There may or may not be considerable deafness, with tinnitis aurium. The pain is more likely to give trouble at night, and the patient may be unable to sleep. In some cases the haggard, anxious look will indicate the amount of distress he has undergone. Sometimes the fever accompanying these symptoms may be intermittent in character, with rigors, resembling that from malaria. The respiration may be preternaturally rapid, or the opposite, and the pulse may be irregular and weak, showing possibly that some portion of the pneumogastric nerve is involved in the disease. In rare instances the pupil may be sluggish or dilated, and in others contracted. Facial paralysis is occasionally seen, and very infrequently hemiplegia of the opposite side.

Difficulty of deglutition occasionally occurs. If the cavernous sinus is diseased, there may be paresis of those nerves passing in its vicinity (the 3d nerve, the ophthalmic division of the 5th and the 6th nerve). It is well-nigh impossible to diagnosticate during life as to whether the brain or its meninges are involved, as brain symptoms and those dependent on the irritation set up by the ear affection may be of a similar character. If we are dealing with a case of mastoid cell osteo-sclerosis, which may have developed from a mild catarrh of the tympanum, it is impossible, certainly, to determine the presence of this condition except by operation. A moderate amount of pain in the mastoid region and side of the head may be all the symptoms obtainable.

The patient may during the day be free from pain, but towards evening, when becoming fatigued, the pain may recur. If he has taken cold, and added to his aural catarrh, the symptoms about the mastoid are likely to be aggravated.

The Objective Signs are somewhat less equivocal. In an uncomplicated case of osteo-sclerosis, however, there may be no symptoms apparent on an examination. If there is suppuration and dead or denuded bone, there will be the following symptoms: In the upper and posterior part of the tympanum, granulations or polypi may make their appearance, which often recur, after having been removed, even in spite of the most effective cauterization. A persistent discharge from the tympanum often owes its presence to diseased mastoid cells. In the meatus a persistent inflammation of its posterior wall, in the osseous portion, may be due to mastoid cell trouble.

If there is in this region one or more fistulous openings, which often may be fringed with granulations, there is the strongest suspicion that disease of the cells is at the bottom of the difficulty. Externally, the disease may have broken through the outer table, and possibly a fistulous opening in the soft parts will indicate its position. In other cases the bone in this region may be necrosed, and no suspicion (judging from the appearance of the soft parts) of its existence may be excited. The latter, however, are frequently paler than the neighboring region, and may have an undermined doughy feeling to the touch. A case is reported by Dr. Arthur Mathewson, in the Tr. Amer. Otol. Soc. for 1878, p. 270, in which an epithelial tumor of the tympanum extended into the cranial cavity, excited a very intense otitis media and mastoid cell inflammation, which had completely destroyed the whole of the mastoid and petrous portions of the temporal bone, and the soft parts covering the mastoid showed no signs of disease. An incision into it, however, was made somewhat reluctantly, which of course revealed the true state of things.

In other cases the ordinary signs of inflammation of the periosteum of the outer table of the mastoid will be exhibited. Pus is usually found in this form of inflammation. A very striking symptom of burrowing pus will be observed in the position of the auricle, which may stand with its posterior face at right angles to the side of the head. This is easily explained, as the antrum mastoideum is just behind and almost beneath the middle insertion of the posterior part of the auricle, and as that is usually the focus of mastoid cell inflammation, formation of pus is more likely to occur there; while if the trouble is only mastoid periostitis of its outer plate, the abscess will usually point lower down. There is a strong tendency of abscesses resulting from mastoid cell disease to burrow beneath some of the muscles inserted into the mastoid process. Conspicuously is it so in the abscess which follows the sterno-cleido mastoid muscle downward. They may reach the clavicle. Sometimes the side of the neck will be swollen in the direction of the trachleo-mastoid. A frequent seat of mastoid abscess from disease of the cells, according to Bezold, is on the inner side of the process, in the digastric groove. When several of these muscles are the seat of abscess, the whole side of the neck may be intensely swollen, very hard, and painful, rendering it almost impossible for the patient to move his head. Such cases, however, are rare.

The swelling in the vicinity of the stylo-mastoid foramen may be so great as to produce temporary paralysis of the facialis. I recently saw a case of this kind at the hospital. Sometimes abscesses point on the posterior wall of the meatus, and are very painful. Another sign of mastoid cell disease is tenderness of the region when firmly pressed upon, or, better still, when percussed smartly with the finger tips. It may not excite pain in the immediate part, but if there are shooting pains in the side of the head consequent on the percussion, suspicion points to mastoid cell trouble.

Treatment of Mastoid Cell Disease.—The leading indications in the management of mastoid cell disease are:

1st. Diminish the hyperæmia and swelling of the parts by appropriate treatment.

2d. Where there is reason to believe that products of inflammation are confined in any cavity, to make a free opening so as to allow the readiest exit possible for such products.

3d. Where carious or necrotic bone exists, to cause its removal if doing harm, or use such means as shall resist the progress towards further destruction. During an attack of acute inflammation of the tympanum an uncomfortable or even painful sensation may be experienced in the mastoid region, with no external trouble which may point to inflammation of the mucous lining of the cells. This condition may be relieved by three or four leeches on the mastoid region, or by a poultice, or in some instances an ice bag. This condition is usually not serious, the inflammation never going beyond the catarrhal form. The pain is probably dependent on swelling of the lining of the cells, so that the spaces are more or less obliterated and the lining membrane may be "pinched" by the too sharp coaptation of its opposing surfaces. The general hyperæmia of the part will also result in pain, as is the case in almost

any other tissue. When the soft parts on or in the neighborhood of the mastoid are swollen, as a consequence of
the cell disease breaking through the external plate of the
mastoid, the treatment should be as follows: Any point
of fluctuation should at once be freely opened by an incision reaching quite to the bone, so as completely to divide
the periosteum. If the swelling is on the mastoid, commence
the incision at the lower portion of this bone, about half an
inch from the auricle, to avoid the posterior auricular
artery, and carry it upwards for from one to two and a half
inches, that is, sufficiently far to freely divide all the swollen
tissue. After the incision has been made pass a short
probe into it in order to find the opening into the mastoid
cells. As the antrum is the most frequent site of carious
processes, the attention may be directed more particularly
to that region.

Abscesses following any of the muscles on the neck may
be sought for, especially those burrowing in the direction
of the sterno-cleido-mastoid the trachleo-mastoid or the sple-
nius-capitis muscles. Often the collection of pus will be
found some distance from the mastoid. An abscess which
has burrowed along the course of the sterno-cleido-mastoid may point near the clavicle. There is no objection to
the use of the poultice previous to an incision, but it should
not be continued long enough to soften the parts unduly.
In making the Wilde's incision considerable force may be
applied to the scalpel, with the hope that it may break
through the softened bone and sufficiently open into the
cells. This can nearly always be done in young children.
After abscesses have been opened, a poultice may be used
for a few days and a tent introduced to keep the incision
from closing, or a drainage tube may be inserted of soft
rubber or silver. A plan I like, however, is that of passing
a probe along the lips of the incision daily to prevent closure, and at the same time press out any pus that may have
collected. A variety of disinfectants may be used to cleanse
the abscess from any foul material. In the burrowing abscesses of the muscles, it may be well to inject a solution
of acid. carbolic, a drachm to the pint of water, once a day
in the direction which the pus has taken, to encourage the
closure of the pus tract. It may, however, be necessary
after a time to lay open the somewhat fistulous canal provided it shows a disinclination to close spontaneously.
Sometimes these abscesses tax the constitutional energy of

the patient so that nourishment and stimulants must be resorted to to sustain the failing strength. The mastoid abscess requires the most careful attention. It will be well to syringe the ear with the carbolic acid solution or salt and water, with the hope that it will pass out through the opening in the mastoid. This failing, the syringe may be inserted into the latter, and the attempt made to pass the solution through the middle ear from this direction. In this condition the membrana tympani is ruptured, so that it may be easily done. The irrigation should be repeated at least once a day. If there is a free passage through the mastoid opening to the tympanum, and the pain and other symptoms subside, nothing more may need to be done. Recently I had a case of mastoid abscess in a little child with purulent otitis. The cells were opened by Wilde's incision. Water could not be forced into the tympanum through the incision, but she made a good recovery without further surgical interference. One of the striking improvements noticed was the rapid subsidence of the purulent discharge from the ear after the incision. When the abscess has already ruptured nothing more may be required provided the opening is sufficient. If not it must be enlarged. It is also desirable that the opening in the mastoid communicate with the tympanum.

Granulations are likely to form about the incision, which may be removed by snare, forceps or cauterization in the usual manner. They may also spring from the mastoid cell cavity, requiring removal. Under the head of polypi an important case of this kind is referred to. In cutting down on the mastoid a small fistulous opening occasionally may be found even where there has been no swelling of the soft parts, or an abscess has imperfectly healed and left a passage extending from the soft parts to the mastoid cavity. In such cases if there is any evidence present of confined pus or other products of inflammation, this fistulous opening must be enlarged, the carious borders chipped or scraped away, and a free communication with one or more of the larger mastoid cells made.

A case reported by Dr. C. R. Agnew, in the Tr. Amer. Otol. Soc. for 1865, will illustrate this point. Miss X., in the middle period of life, had an attack of mastoid periostitis resulting from disease of the tympanum. An incision was made down to the bone, and the latter was opened by boring with a gimlet. Pus came from the incision, and

from the perforation of the mastoid; the patient was relieved. One year after this the Doctor found a fistulous
opening in the mastoid extending through the soft parts to
the antrum. She was deaf, had pain in the temporal bone,
was nervous, anxious, and could not sleep well. It was
concluded that there was thickening of the brain covering
near the ear in consequence of basilar trouble. There was
no rough or carious bone found. She became somewhat
better, then had a relapse, when there occurred facial
paralysis, vomiting, and difficulty in articulation; had a
vacant look; concluded that the pneumogastric was slightly involved. There was pain also behind the well ear. A
half-inch trephine was used in operating, the pin of the
instrument being inserted into the fistula. The disk of bone
removed was three eighths of an inch in thickness and only
contained one small cell. This operation not opening the
sinus to the bottom, the opening was still further enlarged
to a depth of five eighths of an inch. No pus or diseased
bone was found, but the patient made a fairly good recovery. The Doctor thought that here we had to deal with
an ostitis with hyperplasia filling a few or all of the mastoid cells.

General Indications for the Opening of the Mastoid Cells.—1st.
If on cutting down upon a swollen mastoid where the bone
is roughened and the periosteum removed and perhaps
carious openings extend into the cells, the symptoms at once
completely give way, no further proceedings may be necessary unless there is a communication with the tympanum,
when it is always desirable to pass a stream of carbolized
water into this opening and through the middle ear to the
meatus. If no relief results from the Wilde's incision, then
the cells must be opened, whether the external table be
diseased or not. A perfectly healthy condition of this bone
does not contraindicate an operation. Nay, in some cases
where the external plate is exceptionally developed, the
tendency of the disease may be to pass inward or upward,
attacking a bony septum which has much less power to resist a morbid process than has the other surface. Especially
is this the case at a point internal to the antrum, near the
sigmoid flexure of the lateral sinus. Again, this tendency
to extend inward becomes greater when all of the smaller
cells have been obliterated by osteo-sclerosis, leaving the
outer plate still more fortified by accession of new bone.
2d. When the mastoid region seems perfectly free from dis-

ease, but the patient has recurring polyps of the tympanum which may be attached at or near the opening of the antrum, and no means at our disposal prevent their recurrence, it will be proper to open the cells. 3d. In a case of osteosclerosis whether any disease of the tympanum remains or not, and the patient has tenderness in the mastoid on sharp percussion, or even if he does not, but if a persistent pain is experienced in the mastoid process which may radiate to different parts of the side of the head, relief may be expected from opening into the cells. 4th. If a sequestrum is found in the mastoid cell cavity, it is generally proper to remove this, although it is often safe to allow it to separate by natural processes. The case of John Eddington will illustrate that point. It is as follows: His age was twenty months; applied for treatment July 26th. Has a purulent inflammation of the tympanum of three months'

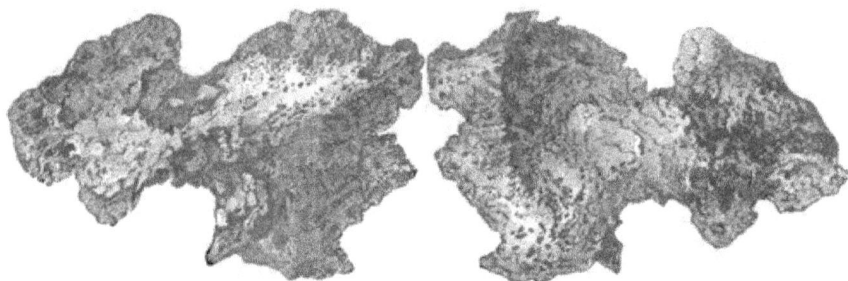

Fig. 82.—Sequestrum from the Case of J. E., Showing both Surfaces of the Bone.

duration. A pale looking boggy swelling without distinct signs of fluctuation appeared over the mastoid; this was incised without evacuating pus. In two days pus made its appearance; previous to the incision there were head symptoms which disappeared after the operation on July 28th; there was considerable cellulitis extending above and in front of the auricle. An incision was made enlarging the old wound upward and forward; bone rough and of unequal depth. Sept. 1st the boggy swelling extending still further, the incision was again made in the same direction as before; opposite the antrum the probe suddenly passed into the mastoid cells for a distance of half an inch. The patient did moderately well during the next three months, although the swelling extended to the front of the auricle. Early in the treatment, water was passed freely from the incision in the mastoid through the tympanum.

About six weeks from the commencement of the treatment facial paralysis occurred; on Oct. 1st a piece of dead bone was observed in the region of the antrum which was at first immovable, but after a week or two it could be moved about in a limited manner—it seemed to extend to a considerable distance. In one month the sequestrum was caught by forceps and moderate traction made, but as consideralbe hemorrhage resulted, no further effort was made to remove it. On Dec. 26th, or about one month after this, the mother returned to the hospital bringing the sequestrum with her (Fig. 82), which had separated spontaneously; it consisted of nearly the whole of the temporal bone with the exception of the inner part of the petrous portion, and a little of the inferior wall of the osseus meatus. The facial paralysis was less apparent than at first. One month after this a swelling extended downward from the old opening which was relieved by incision. Subsequently the child made a good recovery. This patient belonged to Dr. Agnew's clinique at the College of Physicians and Surgeons, and it is through his kindness I am permitted to report the case. He also aided me materially by his advice. During the whole course of this case there was no question as to the advisability of abstaining from any measures more heroic than those practised.

On the Different Modes of Operating the Mastoid Cells.—The instrument longest used and best known for this purpose is the Trephine. It should be of small size, ranging from something less than one fourth of an inch to one half an inch in diamater. After making an incision in the soft parts of about two inches in length, a half inch behind the auricle, the periosteum is turned back, if not already separated, and the trephine applied over a point directly behind the meatus, at a distance of $\frac{1}{4}$ to $\frac{1}{2}$ inch, but not extending above its upper wall, so as to avoid the sigmoid flexure of the lateral sinus. The boring should be done very slowly and cautiously, stopping frequently to determine by a probe or a delicate wedge shaped instrument as to whether the disk of bone has become loosened; if so, do the subsequent cutting on the side which is still immovable. If after removing it there is no communication with the tympanum, or if any fistulous opening is found, the operation may be completed by going deeper with a gimlet or hand drill or one propelled by the dental engine, or a small chisel, and open as far as seems necessary to obliter-

ate any fistula or to open a communication with the tympanum. Sometimes when the mastoid cells are obliterated
by osteo-sclerosis, a considerable distance may be reached before the disk of bone can be removed. Dr. D. R. Ambrose
of New York is reported by Dr. Roosa in his "Treatment on the Diseases of the Ear," 1873, p. 431, in a case
of osteo-sclerosis of the mastoid where a disk of bone one
inch in length was removed by a ¼ inch trephine. The Doctor
was careful to avoid the lateral sinus, and no accident resulted. In general it is not safe to go so deeply, although
if the line of the petrous portion of the temporal bone is
followed, a great distance may be travelled without reaching the brain, but the semi-circular canals would certainly
be broken up. If the region of the antrum is reached it is
ordinarily not necessary to go further. By following these
directions, the only danger to be apprehended is the wounding of or opening into the lateral sinus especially at its
anterior portion nearly opposite the antrum.

Bezold, in the *Deutsche Med. Wochenschr.*, July 9, 1881,
concludes from measuring 100 temporal bones that the mastoid portion varies greatly in size in different subjects, and
that the relations of the lateral sinus (transverse) to the antrum is by no means constant, so that in opening into the
cells according to the ordinary method there is danger of
wounding the sinus. He proposes therefore to lift the
auricle somewhat from its attachment and enter the antrum
beneath it at a point still nearer the meatus than is usual.
He quotes Hartman as having done this operation 100
times on the cadaver, and that he arrived at about the same
results that he himself had. Hartman calls attention to
the fact that it is possible to wound the dura mater at the
middle fossa of the base of the cranium when the perforation is made at a point a little too far above the upper border
of the auditory canal. Other points have been selected for
entering the mastoid. A fistulous opening is usually
selected as a point of entrance for the instrument. Dr.
Buck in his book on the diagnosis and treatment of diseases of the ear published in Wood's Library in 1880, reports
a case where he opened the mastoid at its lower portion,
hoping to take advantage of gravity in more perfectly
draining the cells, but he found little or no communication
between this and the upper part of the process; probably
on account of closure of the cells from inflammatory pro-

liferation. Subsequently the mastoid required to be opened over the antrum in the usual manner.

Bezold on the contrary, recommends perforating the lower extremity of the mastoid so as to completely open both the inner and outer surfaces at the digastric fossa. He regards this as sufficient for the drainage of both the mastoid and the tympanum. He reports a case of successful performance of this operation on a peasant, aged 73, who recovered in fourteen days.

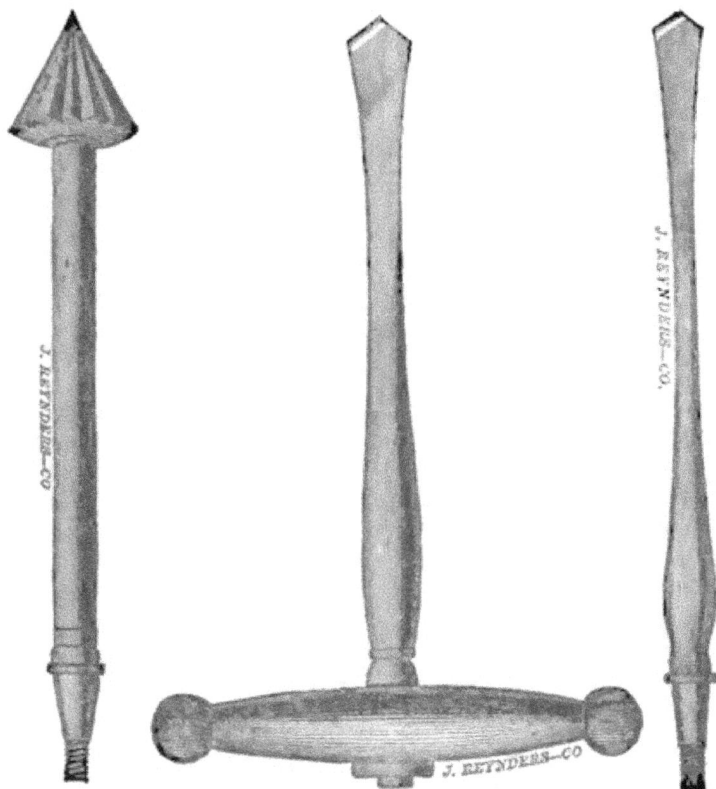

FIG. 83.—BUCK'S DRILLS FOR OPENING THE MASTOID.

Some form of hand drill is just now very much in vogue for opening the mastoid. The one figured here is known as Buck's (Fig. 83). I believe it to be the best instrument for the purpose I am acquainted with; the inventor of these drills prefers the larger one to the smaller. To prevent it suddenly going too deeply he is in the habit of extending one of the fingers of the hand holding the drill so as to rest it

against the side of the head.   This drill has a wide application.   A great variety of shapes may be used, and the dentist's engine may be utilized for the purpose of driving them.
A cross-cut burr drill is a very convenient instrument for
excavating bone, enlarging fistula, etc.   An ordinary gimlet may sometimes be used.   Dr. Buck also uses a pyra-
midal borer which is in no danger of suddenly penetrating
too deeply.   (See instrument at the left hand in the figure.)
   The gouge is much used, and is a very good instrument.
It is usually better to push it in by the pressure of the
hand than by the use of the hammer, as is the custom with
many of the continental surgeons.
   Schwartze uses what is called a sharp spoon, an instrument somewhat resembling a gouge.   He also uses a chisel,
a gouge, and a borer.   When the bone is eburnated a hammer may become necessary to drive the chisel with sufficient
force.
   Whether using a gouge or drill, it is well to frequently
pass in a probe or a finger to observe what progress is be-

FIG. 84.—STOUT KNIFE FOR CUTTING THROUGH THE EXTERNAL TABLE OF THE MASTOID.

ing made.   If the lateral sinus has its bony wall chipped
away, the sensation of a soft body revealed by the finger or
the touch of the probe will give a hint of this fact, or if the
meninges have been reached, this test will make us aware
of the fact.   After the external table has been removed,—
which will vary from one twelfth to one fifth of an inch in
thickness, unless it should be associated with osteo-sclerosis, when its thickness may be much greater,—the cells may
be broken up, if it is thought necessary, by a stout probe or
steel director.   The lips of the wound in the soft parts may
be kept apart by an assistant using retractors for the purpose.   Dr. Burnett, of Philadelphia, figures in his book on
the ear a stout knife for opening into the mastoid.   I take
great pleasure in alluding to it, for I am assured that very
many cases may be operated on by means of this simple
and effective instrument (Fig. 84).
   Carious spots on the mastoid process may be scraped
until healthy bone is reached.   Buck's instrument, devised
for this purpose is a good one.   Occasionally the lateral
sinus is opened in the effort to reach the mastoid cells.

The resulting hemorrhage is much like that from an extensive phlebotomy, the blood welling out rather excessively.
A tampon of lint is usually sufficient to arrest it, and it is
not as serious an accident as one might suppose. It
only occasionally destroys life. Dr. Knapp recently operated on a painful mastoid where there was no tympana-
complication, in which the cells had undergone a certain amount of osteo-sclerosis, with success, but the lateral sinus was opened. The hemorrhage was readily controlled by a tampon and no ill results followed. He used
a narrow chisel and a mallet occasionally in the operation.

Where the mastoid trouble appears at the meatus, or a
swelling on its posterior wall causes us to suspect complicity
with the mastoid, it will be proper to open into the cells
through this region. A drill and a stout probe are the best
instruments for this purpose. This should be followed by
washing out the aperture with a syringe having a bent nozzle, similar to Schwartze's. Bezold and other authorities,
however, do not think this plan as effective as opening from
the outer surface. In the complication described by Bezold,
where the abscess points at the meatus, it certainly would
be proper to enlarge the opening by appropriate means.
If the patient is tuberculous or very much reduced it would
hardly be proper to open the mastoid. Sometimes erysipelas
supervenes upon an operation on the mastoid; the whole
side of the head may be excessively swollen, extending to
the eye of the same side, so as to close it, and the patient
have intense fever and be very ill indeed. Where a mastoid
periostitis recovers and relapses frequently, it is a good rule
to open the mastoid, as usually some explanation will be
found for the relapses. In a subsequent section will be
found a study of the symptoms accompanying fatal cases
of mastoid and tympanal diseases, when certain points not
fully developed here will be discussed.

The after treatment of operations on the mastoid should be con-
ducted for the most part on general principles. Sometimes
the patient may suffer from the shock of the operation, requiring anodynes and possibly stimulants. The erysipelas
occasionally seen may be treated on general surgical principles, being careful to include iron, quinine and even
brandy if necessary for support. If there is much reaction
the patient should be kept in bed for from one to two weeks,
and occasionally, if there is intense injection of the parts,

iced applications may be cautiously used as recommended by Schwartze.

The wound should be kept open until suppuration has ceased and the patient is well advanced in convalescence. A tent may be introduced to keep the wound open, similarly to that which is done in simple incisions of the mastoid process, only being careful to pass the tent into the antrum as far as possible. It should remain in from half a day to one day, and should be removed whenever signs of confined pus appear. I still adhere to my plan of daily passing in a probe to break up any adhesions that may have taken place in the soft parts, preferring this to the tent for reasons already given. A drainage tube of soft rubber or silver may sometimes act admirably in keeping the wound clean.

Irrigation of the parts should be begun at once, and at first practised at least twice a day; afterwards once a day may suffice, and as the discharge grows less it may be done once in two or three days. Acid, carbolic, three or four grs. to the ounce of water, with or without an equal quantity of chloride of sodium, is a good injection to use. Any of the agents used in an ordinary chronic suppurative otitis may be appropriate here; at first mild injections may be used, always warm. These should be thrown into the part with great gentleness, for fear of causing head symptoms. Clark's douche is a good apparatus for this purpose. Inject the Eustachian tube with air frequently. Occasionally for some reason not immediately apparent no fluid will pass from the antrum to the tympanum. After a while, however, there is usually no difficulty. If carefully done, a probe may be inserted into the open antrum and an effort made to cause this to communicate with the tympanum should it seem desirable. If at any time during the treatment the patient should have pain or fever, carefully examine the wound to see whether pus is anywhere confined. If the soft parts close so as to obstruct free irrigation they should be incised. Granulations and polypi often give trouble, especially in preventing egress of fluids from the tympanum. These may be dealt with as already mentioned in the section on polypi. Schwartze and others use drainage tubes both in the antrum and in the mastoid, or pass a single tube into the mastoid, causing it to emerge at the meatus; Schwartze recommends the use of a rubber tube at first, and when the parts are too much closed then one made of lead, which should be bent at its outer extremity, and fastened by a piece of tape passing around the head.

I do not think it ordinarily necessary to use tubes in the mastoid, but they certainly are necessary in the meatus. Instead of using a sponge tent for dilatation as recommended by Schwartze, where the canal is too much narrowed, I use the rubber tube as directed in the section on chronic suppuration of the tympanum. This may also act favorably after the manner of a drainage tube.

The duration of treatment is extremely variable; some will recover in a few weeks, while others go on for one or several years. Schwartze states that the average duration of treatment in his cases was from nine to ten months. It has been thought not amiss to refer to 86 cases of operations on the mastoid done by H. Schwartze, of Halle, perhaps the best authority now living on this subject.

(*From the "Archiv. für Ohrenheilkunde," B.* 14, *S.* 202. *H. Schwartze.*)

"SUMMARY OF FIRST FIFTY CASES OF SURGICAL PERFORATION OF MASTOID PROCESS (MASTOID BONE.)

Cured.....................................35 cases = 70%
Not Cured........................... 5 "   = 10%
Died...10 "   = 20%
                                            —
    Total...............................50 "·

By cure, I intend the certain and permanent cessation of the purulent processes; but by no means in all cases a functional and anatomical restitution to a normal condition. Cure has resulted after a duration of after treatment from one month (case 26) to two years (cases 23, 45, 46). The average duration of after treatment is to be estimated at from nine to ten months. Of the thirty-five cases which have been put down as cured in the above sense, I must except four cases (14, 17, 30, 35), because they did not remain long enough under obsertion to give assurance of a definitive and perfect cure. Of the remaining thirty-one cures, twenty-two were chronic cases, and nine were acute. The average duration of acute cases was from six to seven months; only a single chronic case lasted ten months, so that between the two no essential difference appears. In the five uncured cases (2, 9, 10, 11, 22), the grounds of failure in cases 2 and 10 lay in the unsuitable method of operation; in case 9, of total sclerosis of the mastoid process, and in cases 11 and 22, in inadeqate after treatment. In the fatal cases the causes of death were: meningitis (case 1), six weeks after the operation; meningitis of the convexity (case 19); tubercular meningitis (case 40), six weeks after the operation; pyæmia (cases 7, 36). In case 7 it was present before the operation; abscess of the cerebellum (case 44), which existed, undoubtedly, at the time of the operation; anæmia (case 50), nine weeks after the operation; epithelioma of the temporal bone (case 18); pulmonary tuberculosis (cases 4 and 6), six and twelve months after the operation.

"In six of these ten fatal cases death was certainly independent of the operative interference (cases 4, 6, 7, 18, 40, 44); in three the connection is not certainly proved (cases 1, 36, 50); and only in one case (19) can the fatal termination be held with certainty as the direct consequence of the operation (traumatic meningitis of the convexity). Of the doubtful cases, case 1 relates to a child of ·15 months who died with symptoms of meningitis after otitis-media-acuta with formation of abscess on the mastoid, (process), some six weeks after the perforation of the carious cortex. No post-mortem. The probable cause of the fatal issue lies in the inadequateness of the operative procedure.

"The small opening which was made with the "grooved" director gave no sufficient exit to the pus; through prolonged retention of pus there developed purulent meningitis.

"By means of wider opening of the mastoid, the use of the sharp spoon and disinfecting cleansing, death might perhaps have been avoided. In case 36, where with total sclerosis of the mastoid the canal of the facialis and the external semicircular canal were injured by a faulty use of the drill, the decomposition of the thrombus in the lateral sinus was not caused by the operation nor by the faulty performance of it, but very probably only by the circumstance that notwithstanding the perforation of the mastoid; on account of the excessive stenosis of the auditory canal and the closure of the Eustachian tube, the retention of pus was not relieved.

"In case 50, death occurred nine weeks after the operation, after the patient had successfully passed through a severe pyæmia, and was already in full convalescence, apparently in consequence of collapse. No post-mortem.

"The single instance in which we must regard death as the

direct consequence of the operation relates to a case of abnormal development of the temporal bone, where instead of the antrum the middle cranial fossa was opened, and unfortunately a splinter of bone perforated the dura mater, causing traumatic meningitis of the convexity."

" From the history of the above cases, the fact appears that it is possible by the perforation of the mastoid to bring the most difficult and most tedious cases of caries of the temporal bone to a perfect and permanent cure, even when the petrosal portion is involved in the carious destruction; that by this means are not only the direct dangers to life definitively removed, but that also the indirect consequences to the general condition which the bone disease so often induces disappear after the operation, provided they are not already too far advanced (pulmonary tuberculosis). The patients, who before the operation were reduced mentally and physically to a condition of chronic invalidism, recover in a few months, even when through terrible sufferings lasting many years they had been brought to the verge of the grave. [I refer only in this therapeutical relation to the most remarkable case, (c. 42).] The favorable influence of the operation on the cure of a secondary pulmonary tuberculosis, in cases 5 and 6, is undeniable. In case 5, after the cure of the ear difficulty, epilepsy disappeared permanently which had lasted for ten years. Facial paralysis disappeared two days after the operation in Case 33, and in the third week after the operation in Case 30; on the contrary it remained unchanged in Case 4, in which six months after the operation death occurred from pulmonary tuberculosis. In regard to the hearing in the 31 cases of cure, it may be mentioned that eight times (cases 3, 5, 13, 21, 25, 27, 29, 39) it became perfectly normal ; in the remaining cases, as far as we have reliable data, the hearing distance for the watch ranged from 2 to 34 cm. (2 m. normal). In only one case was contact of the watch with the auricle requisite. This difference depends, as we may readily understand, on the extent and location of the derangements in the sound-conducting apparatus caused by disease before the operation. In 11 cases (3, 5, 13, 15, 21, 24, 25, 27, 28, 39, 43) cicatrization of the perforations of the drum-membrane was observed. In the four cases (33, 34, 42, 45) in which after perfect cure of the caries absolute deafness remained, it was to be assumed, from the results of the tuning-fork applied to the cervix before and after the operation, that the carious process had already attacked the bony labyrinth. In case 33 there was positive proof for this assumption in the necrosed and extruded cochlea; in case 45, where the disease affected both sides, there was proof in the deaf-mutism.

"As regards age in the fifty cases, there were

From 1 to 10 years ....12 cases.
" 11 to 20 " ....16 "
" 21 to 30 " ....13 "
" 31 to 40 " .... 1 case.
" 41 to 50 " .... 3 cases.
Over 50 " .. .. 5 "

The youngest patients were in the second year (cases 1, 32, 40); the oldest 78 years.

"Second series of fifty cases of surgical perforation of mastoid" (Casuistic zur chirurgischen Eröffnung des Warzenfortsatzes). H. Schwartze, zweite serie Archiv. für Ohrenheilkunde. B. XVI., XVII., XVIII., being a brief statement of the salient points of each case.

"Case 51. Otitis media purulenta chronica with abscess on the mastoid and fistula. Dilatation of the osseous fistula with the gouge and hammer. Scraping the mastoid antrum. Cure in ten months.

"Case 52. Otitis media purulenta chronica with abscess on the mastoid. Perforation of the mastoid with grooved director (Hohlsonde). Permanent cure in nine months.

"Case 53. Otitis media purulenta acuta with abscess on the mastoid. Dilatation of an osseous fistula with the grooved director. Death in eight days of miliary tuberculosis.

"Case 54. Otitis media purulenta chronica after scarlet fever. Sinus-phlebitis. Metastatic pyæmia. Opening of the antrum with the gouge. Death.

"Case 55. Chronic suppurative process of middle ear with fistulous opening into the meatus and inflammation of the mastoid. Opening the antrum with the chisel. Cure in two months.

"Case 56. Chronic suppurative process of middle ear. Fistulous opening in the cortex of mastoid. Cutaneous covering unchanged. Dilatation with gouge. Drainage. Cure in twenty-one months.

"Case 57. Development of acute abscess in the mastoid cells without perforation of the drum membrane. Opening of the mastoid with the chisel. Cure in seven weeks.

"Case 58. Central caries of mastoid with fistula into meatus. Opening the mastoid with chisel. Cure in fifteen months.

"Case 59. Periostitis syphilitica chronica. Opening of mastoid with the chisel. No escape of pus. Cure in nine weeks.

"Case 60. Otitis media purulenta acuta with secondary periostitis and fistulous opening through the cortex of the mastoid. Multiple polypi in auditory canal. Opening of mastoid with the chisel. Cure in three months.

"Case 61. Chronic purulent disease of middle ear, lasting four years. Repeated abscesses on the mastoid. Fistula in the cortex. Dilatation. Scraping out the antrum with the sharp spoon. Cure after two years.

"Case 62. Otitis media purulenta acuta with abscesson mastoid. Minute fistulous opening in the cortex. Dilated with gouge. Cure in three weeks.

"Case 63. Chronic purulent disease of middle ear with polypi and facial paralysis. Opening the mastoid with gouge. Subperiosteal abscess of the mastoid abscess on the carious petrous portion beneath the dura mater. Death from meningitis.

"Case 64. Otitis media purulenta acuta with implication of the mastoid. Repeated incisions. Perforation with chisel. Cure in nine months.

"Case 65. Otitis media purulenta acuta with involvement of the mastoid. Antrum opened with chisel. Cure in three months.

"Case 66. Scrofulous caries of mastoid. Antrum opened with the chisel. Cure in one year.

"Case 67. Otitis media purulenta chronica. Purulent periostitis with superficial caries of the mastoid. Scraping. Cure after four weeks.

"Case 68. Otitis media purulenta chronica with abscess and fistula into the auditory canal. Antrum opened with chisel. Cure in six weeks.

"Case 69. Otitis media purulenta chronica. Fistula in mastoid. Dilatation with chisel. Antrum filled with cholesteatomata. Masses cleared out by means of the sharp spoon. Cure in nine months with cicatrized fistula in the bone.

"Case 70. Caries necrotica with fistulous perforation of the mastoid. Scraped out with sharp spoon in 1869. Apparent cure. Relapse after eight years with facial paralysis and cerebral symptoms. Mastoid opened with chisel. Death from meningito purulenta diffusa after twenty-two days, in consequence of necrosis of the labyrinth.

"Case 71. Otitis media purulenta chronica. Fistula in mastoid. Dilatation with chisel. Scraping out the mastoid and auditory canal with sharp spoon. Cure in eight months.

"Case 72. Otitis media purulenta chronica. Secondary periostitis of mastoid. Carious opening through cortex. Chiselling and scraping out the antrum. Drainage. Permanent cure in two months.

"Case 73. Otitis media purulenta acuta with empyema of mastoid. Antrum opened with chisel. Cure in five weeks.

"Case 74. Otitis media purulenta chronica with necrosis of mastoid. Extraction of sequestrum. Dura mater laid bare. Cure. Death after thirteen months from pulmonary tuberculosis—long after the wound from the operation had healed.

"Case 75. Otitis media purulent aacuta with neucrotic caries of mastoid. Extraction of sequestrum. Dilatation of fistula. Death four weeks after of catarrhal pneumonia.

"Case 76. Otitis media purulenta chronica with necrotic caries of mastoid. Sharp spoon. Death in twenty-four days of tubercular meningitis.

"Case 77. Otitis media purulenta chronica duplex with caries of left mastoid of ten years' duration. Opening the antrum with chisel. Scraping. Drainage. Cure in two years.

"Case 78. Otitis media purulenta chronica. Fistula. Sequestrum of cortex removed. Dilatation of fistula with chisel. Removal of granulations with sharp spoon. Drainage. Cure in two years.

"Case 79. Otitis media purulenta chronica for seventeen years. Sudden symptoms of brain irritation and chills. Disturbance of speech. Subperiosteal abscess on the mastoid. Fistulous opening in cortex. Dilatation and drainage. Death eight days after the operation, of abscess of brain, with meningitis purulenta (?). No post-mortem.

"Case 80. Otitis media purulenta chronica with abscess as large as one's fist, behind the ear, and fistula leading to antrum. Dilatation of fistula. Drainage. Death next day. No post-mortem.

"Case 81. Otitis media purulenta chronica of 12 years' standing with central caries of mastoid. The sound outer wall chiselled through and a fistulous canal kept open. Cure in one year.

"Case 82. Otitis media purulenta chronic aafter pneumonia. Abscess above and behind the ear and burrowing into the lateral and posterior regions of neck. Secondary superficial caries of occiput. Opening antrum with chisel. Cure in four months.

"Case 83. Otitis media purulenta chronica with necrotic caries of mastoid operation. Cure in a year and a quarter.

"Case 84. Chronic periostitis with sclerosis of mastoid and widely extending firm-walled fistulous passages leading to the lateral and posterior regions of the neck. Cure in two and a half months.

"Case 85. Otitis media purulenta chronica lasting four months. Necrotic caries of mastoid. Use of chisel and sharp spoon. Cure in four weeks.

"Case 86. Otitis media purulenta chronica duplex and secondary periosteal abscess over the right mastoid. Fistulous opening in cortex. Dilatation with chisel. Sharp spoon. Drainage. Death in four weeks of purulent meningitis."

UNCLASSIFIED DISEASES.

SOME CONSIDERATIONS REGARDING FATAL CASES OF SUPPURATIVE OTITIS WITH OR WITHOUT MASTOID COMPLICATIONS.

Fatal complications arise—

From the disease extending to the brain, causing inflammation of the meninges, brain substance, blood-vessels, and sinuses.

Purulent inflammation of the mastoid cells, where the mastoid ruptures at the inferior portion, at or near the digastric fossa, or in the outer part of the osseous meatus, causing the pus to burrow beneath the muscles of the neck, may result fatally from the exhaustion incident to the purulent process.

Inflammation may extend from the ear to the brain whenever any of the bony partitions which separate the tympanum, the mastoid cells, or the labyrinth from the brain become necrotic or carious, or when these bony partitions become perforated from inflammatory processes; carious processes in the meatus sometimes extend to the brain. Inflammation may extend to the brain from the ear by continuity, there being a free interchange between the ear and the brain by means of numerous vessels, nerves, foramina, etc. If the oval or round window be destroyed by ulceration, there is then essentially a continuous opening into the brain through the continuity of the peri-lymph and subarachnoidal spaces. Septic material may be carried from the ear to the brain by means of blood-vessels, causing abscess of the brain-substance, phlebitis, thrombosis, etc.

Necrosis and caries are the direct results of inflammation; the former being rather the consequence of inter-

rupted nutrition from disturbances in the blood supply
during the inflammatory process; the latter is the result of
a true ostitis communicated from the soft parts, which results in ulceration, a breaking down and exfoliation of the
bone as sequestra.   Both of these are more likely to occur
in the strumous and the syphilitic, but still more likely in
infants and young children.   Pus confined in any cavity
may attack the bone, especially if there is considerable
pressure.   On the other hand, any part or the whole of the
temporal bone may be removed by carious processes without destroying life.   My own case of John Eddington,
recorded under the heading of mastoid diseases, fully illustrates this.   Numerous other cases of a similar nature to
be found in literature abundantly prove that life is not
necessarily lost by destruction of the temporal bone by
caries.   The meninges, it is true, are laid bare, but it is
probable that the inflammatory process has so thickened
them as to cause them to be a sufficient protection to the
brain from the inflammatory processes.   I am inclined to
the opinion that children bear a greater amount of bone
exfoliation from ear disease than adults.   Abscess of the
brain is thought by some of the best authorities to be
caused by ear affections in from one half to two thirds of
the cases, it never being a primary affection.   There may
be no diseased tissue between the abscess situated deeply
in the cerebral substance and the diseased ear, but which
may, nevertheless, have depended on the ear for its cause.
The explanation of this is not easy.   Abscess of the rectum
may cause a cerebral abscess, although the latter is rarely
metastatic; a brain abscess, as has been observed by the
late E. H. Clarke, M.D., of Boston, and Geo. P. Field, of
Glasgow, and others, sometimes gives no notice of its presence by disturbance of function.   If, however, it approaches
the surface, a meningitis is excited which rapidly destroys
life.   The abscess is usually situated in the posterior part
of the brain and is generally single.   Its remoteness from
the motor tract will explain why paralysis is so infrequent.

   According to Mr. Field in the London *Lancet*, June 5,
1880, an abscess of the brain sometimes remains for years
unsuspected, and the patient may die of some other disease; the pus-cells undergoing fatty and granular degeneration, and all that remains of the abscess may be a fibroid
sac, containing chalky material, the fluid portions having
been absorbed.

The tendency of the disease to extend to the brain is increased whenever, from any cause, products of inflammation fail to find a free outlet.   This may occur in tympanal affections where the membrana stoutly resists perforation, and matter is confined in the tympanum.   Polypi and granulations may so crowd the tympanum and meatus as to prevent egress of inflammatory products.   Impacted cerumen sometimes acts very obstructively.   Bony growths in the meatus, or even occlusion from other causes, may so act.   Foreign bodies in the meatus or tympanum may produce similar results.   The roof of the tympanum is exceedingly thin, and it is sometimes found partially wanting, which naturally offers a feeble resistance to destructive processes.   The inner and upper wall of the mastoid antrum is thin, lies in apposition with the lateral sinus, and offers much less resistance to the invasion of disease than the outer or cortical portion of the mastoid cells.   The latter also is frequently much thickened by osteo-sclerosis, making it quite impervious to disintegrating inflammatory action, which, if destroyed by caries, might allow escape of the purulent material.

The ear affections which destroy life are, as a rule, chronic, or a new inflammation is lighted up on an old process.   A few exceptions to this rule, however, occur.   The following case may illustrate an exception.   It is reported by G. S. Ryerson, M.D., etc., in the Canada *Lancet*, Nov., 1881.   The patient was a child who was comatose when first seen; ptosis of right eyelid, divergent squint, and dilatation of both pupils occurred; recovering from scarlatina. Had been screaming with pain in the left ear for some days; membrana tympani bulging; was incised, and bloody serum escaped, giving relief.   April 5th, five days later, a purulent process was inaugurated; pains in the arms and legs.   On April 8th, pain and tenderness in the mastoid, pain in the head; mastoid trephined and a brownish grumous fluid escaped; periosteum detached and discolored; this gave great relief.   Next day very little ptosis, all the symptoms relieved except profuse and offensive discharge; no fever.

April 10th, could not see well; pupils dilated but moderately movable; optic disks swollen, with enlarged and tortuous arteries and veins.   A small abscess of the arm and one of the auricle was opened.   April 12th, screamed with

pain all the time; discharge less; fever and delirium. On
April 21st severe rigors developed and the child died.

Post-Mortem.—Dura mater thickened and adherent to
calvarium. On division, much serum escaped; brain congested, more on left hemisphere; ventricles full of serum;
degenerated pus beneath the pia mater over each superior
lobe; lateral sinus full of blood clots and pus; caries of roof
of tympanum; dura mater extensively detached, beneath
which was considerable pus.

The case was remarkable as being an abscess of the
brain following *acute* inflammation of the ear, and for its
remoteness from the organ. Toynbee reported only one
case of abscess of the brain consequent on acute inflammation of the ear. In this instance, the pus was carried by
the veins, as there were symptoms of pyæmia. (*Am. Jour.
Otol.*, '82, No. I., p. 75, signed G. B.)

*Some of the more Frequent Symptoms observed in Fatal Cases
of Ear Disease.*—True brain symptoms in cases which result
fatally are not certainly diagnosticated from the signs of
brain irritation found in many cases which recover. It
will always, perhaps, be a vexed question whether a patient with any considerable brain disease has ever recovered. Certain minor brain troubles have, without doubt,
terminated favorably. Brain symptoms where paralysis is
included, usually point to fatal brain lesions. With few exceptions, the trouble that terminates in death is engrafted
on an old case, and appears as an acute exacerbation, which
may repeatedly be relieved, but recurs from time to time,
and terminates only in death.

One symptom, which is the most important of all, is *pain ;*
this may be in any part of the ear or head, and often is of
the most excruciating character, defying all known means
of relief, except, perhaps, the temporary palliation afforded
by ether or chloroform. Its location, in the ear or head, is
by no means a certain indication of the exact situation of
the lesion.

When the neck is involved in burrowing abscesses, the
pain will be more likely to be in that neighborhood, and in
that part of the spine adjacent to the swelling.

Pain may occur in the most remote parts: the legs and
arms or the chest; these facts are not so easy of explanation. Exacerbations of the pain are more likely to occur
at night; the haggard and emaciated appearance of the patient gives evidence of the gravity of the disease.

Fever nearly always accompanies these processes, although there may be many remissions, especially in the earlier and later developments of the disease. This will often simulate an intermittent form, and large doses of quinine will frequently be beneficial.

The temperature in the earlier period of the progress of a case may not be elevated, but later on it is very likely to be; it varies, some days being normal, and others elevated. At the last it may be below the normal.

The *pulse* is a valuable indication. At the outset it may not be changed in character, but subsequently it may become very rapid. It is not likely to be of the full, bounding variety, but occasionally it is. In the later stages it may be weak and threadlike. If there is much pressure upon the brain and medulla oblongata, it may be intermittent, slow, and perhaps dicrotic. The peculiar pulse of extreme exhaustion may be often observed. *The temperature* of the body is not usually greatly elevated as in some sthenic forms of fever, although it occasionally is.

A profuse perspiration is often an important symptom, and later on, the cold, clammy perspiration of complete exhaustion may be present. *The tongue* may show no marked signs of disease in many instances, while in others it may exhibit the usual signs of fever (dry, rough, coated, etc.). We are constantly confronted by contradictory symptoms.

The bowels in only a few instances show the marked constipation characteristic of brain disease, and only occasionally the involuntary evacuations of the last stages of brain trouble are observed.

The intelligence is often most remarkable, and depends on the fact that only certain parts of the brain, usually those located posteriorly, are involved in the disease. Sluggishness of the intellect is occasionally seen, with inability to articulate certain words. Most of the cases end with comatose symptoms.

Delirium is not uncommon, especially among children. Convulsions are not infrequent, and are sometimes accompanied by opisthotonos. Sleeplessness is an important symptom. *Rigors* are a valuable symptom, showing the pernicious action of confined, and possibly absorbed pus upon the system. *Dizziness or vertigo*, when persistent, is a bad symptom, the patient being inclined to fall in a direction opposite to the affected ear.

Nausea and vomiting are always bad symptoms, especially

if they recur from time to time. Where a patient has previously heard a tuning fork by conduction in the affected ear, and subsequently fails to hear it, there is evidence that the disease has reached the labyrinth or the brain. *Paralysis* is an important symptom, especially that of any of the nerves passing near the lateral sinus, this being accompanied by ptosis, divergent squint, dilated pupil, etc; intracranial trouble may then be strongly suspected, and indeed, almost proven.

If the medulla oblongata become involved, paralysis of the hypoglossus ensues, when, upon the patient making an attempt to show the tongue, it will protrude towards the paralyzed side. The arm opposite to the affected ear, or even the whole side, may become paralyzed, or in extreme disorganization of the brain general paralysis may ensue. Any disturbance in the functions of the pneumogastric nerve is an unfavorable symptom.

Tinnitus aurium coming on suddenly in an acute exacerbation is sometimes a noticeable symptom. A sudden arrest of the discharge, in a chronic case, with febrile symptoms, often precedes a fatal issue, although in itself considered, it may not be of grave moment. I can give no special reason for the following observation, but in looking over an extended record of fatal cases, it appears that those who have a considerable swelling below and in front of the meatus, frequently do badly; herpetic eruptions about the face and neck are often observed in the grave cases; so also is exophthalmos with redness and swelling of the eyelids.

Some of the more important *post-mortem appearances* are as follows: Mucous lining of the tympanum swollen, infiltrated, and congested; tympanum filled with polypoid material or malignant growths or cheesy tuberculous material with or without epithelial masses; cholesteatomata or mucus, pus, blood, serum, etc., may fill the tympanum; there may be total absence of mucous lining of the cavum with necrotic discoloration of the bony wall. Ossicula all present, if membrane is intact, but the ligaments may be partly destroyed by ulceration, or may be found sclerosed. In other instances, membrana and ossicles may be swept away, leaving, however, for the most part, the base of the stapes, which resists very strongly destructive processes. The chorda tympani nerve is frequently destroyed by ulceration.

The mastoid cells are sometimes found completely closed by an hyperostotic process, and converted into a bony ma-

terial of ivory-like hardness. The process of suppuration
may have swept away all the cell divisions, leaving one
vast cavity in the mastoid process, which may open by
a large aperture into the tympanum. It may be extensively infiltrated with pus and other inflammatory products
or completely filled with polypoid material. Fistulous
openings may often be noticed in the mastoid process,
usually communicating with the antrum, but occasionally
found in the lower and inner portion of the mastoid process, and sometimes communicating with the meatus; where
the perforation of the mastoid is in the digastric fossa, the
burrowing abscesses may cause the bones of some of the
cervical vertebræ to be attacked by caries. The roof of
the tympanum is frequently perforated by carious processes, or it may be thickened, and the region in juxta-
position to the inner wall of the mastoid cells may be
destroyed, resulting in an opening into the lateral sinus.
Occasionally the thin anterior wall of the tympanum may
present a carious opening, communicating with the carotid
artery, when sudden death from hæmorrhage is likely to
result. Occasionally the whole of the temporal bone is
destroyed by caries, and an immense cavity formed, which
may be filled with fœtid purulent material. This condition
is more likely to result from malignant disease. The
vestibule, semicircular canals and cochleæ are sometimes
filled with a red, solid, flesh-like granular mass, which project into the aqueductus vestibuli and the internal auditory meatus. This material is composed of a net-work of
connective tissue fibres interspersed with round cells. In
the meshes of this stroma, larger cells have been found.
Fat granules and epithelial cells have been found in the
vestibule. The Eustachian tube sometimes has its calibre
much widened by destructive ulcerative processes, or it
may be blocked up with inflammatory products.

The dura mater may be hyperæmic, more especially that
portion in the vicinity of the diseased ear. The veins may
be much injected, and some of the vessels embolic. It may
be thickened and abnormally adherent to the calvarium.
It may also be detached from the bone (in the petrous
portion it is normally adherent), discolored, and with pus
formations beneath it. Sometimes it is found in a state of
pachymeningitis. Growths on the dura near the petrous
portion have been observed. There may be a purulent
meningitis, with or without sloughing of the meninges.

Foci of inflammation may sometimes be seen, and in a single instance, one of these was found at the lobulus of the pneumogastric nerve. Sometimes the dura is adherent to the arachnoid. In one case, the dura over the orifice of the aqueductus vestibuli, on the posterior part of the petrous bone, was bulging with pus which had reached this point from the vestibule, and the auditory nerve was in some places discolored, and in others disorganized by the purulent process.

The pia mater sometimes shows injected veins with other signs of hyperæmia. It may be infiltrated with serum; pus is occasionally found beneath it. It may be adherent to the gray substance of the brain.

The Arachnoid.—Pseudo-membranous deposits are sometimes found upon the arachnoid, while in others it may be in a state of purulent infiltration, or swollen with œdema.

The cerebral lobes, as well as those of the cerebellum, may be more or less intensely congested. Where considerable intra-cranial pressure has existed, the convolutions of the brain may be compressed and flattened. Sometimes the surface of the arachnoid will be covered by a puriform lymph.

Abscesses are found in the *brain substance* usually at or near the middle or posterior part, and may be some distance from the surface. The *ventricles* may be broken up by an invading abscess. The contents of these abscesses may be of the greatest possible variety, from normal, not ill smelling pus, to the most fœtid and degenerated purulent material. Hæmorrhagic spots are sometimes observed in the gray substance. In other instances, the brain appears "pale, flabby, and diffluent," from degenerative changes. Collections of pus are sometimes seen along the walls of the vessels of the convexity. In extensive disease of the brain, blood-vessels are often obliterated. The vertebral arteries have been found wanting. The ventricles, more especially the lateral, often contain an excess of serum, which may be clear, or rendered turbid by grayish red flocculent material. Sometimes they are filled with pus, which may have broken down their walls. The septum lucidum may be destroyed by purulent processes. The membrane of the pons varolii has been found covered with a greenish deposit, and the substance broken down nearly to the walls of the fourth ventricle.

The sinuses are very frequently attacked, notably the

lateral, which is in such close proximity to the antrum of the mastoid. Occasionally the petrosal and transverse sinuses may be involved. In rare instances the superior longitudinal sinus may be attacked. The sinuses may be covered by a tenacious thick deposit on their walls, or bathed in masses of pus. In their interior may be found coagula, or fluid-blood, or pus. Thrombi are perhaps more frequently seen adhering to their walls. The greenish red thrombi are more recent, and the yellowish, gray, or black are of greater age.

These may break down into a puriform material. The walls of the sinuses may be found destroyed by suppurative processes of the dura mater, and obliterated, their site being occupied by a reddish, fleshy granular material, representing the ordinary products of inflammatory proliferation.

The jugular vein sometimes contains clots, which after a time become dense, firm, granular, and of a brown color. It is also found with similar changes to those seen in the sinuses.

The carotid has been found infiltrated with purulent exudations. Both jugular vein and carotid artery have exhibited thickening of their walls. Metastatic abscesses have been found in the spleen, kidneys, liver, lungs, etc., dependent on the brain disease. Pleurisy, with effusion of serum and pus, has been noted as secondary to abscesses of the brain. Peritonitis with submucous ecchymoses in the intestines has been observed in autopsies of fatal ear cases. In very rare instances the upper wall of the meatus and the roof of the tympanum have become carious, and involved the brain, without attacking the tympanum at all. Dr. J. O. Greene reports such a case in "The Tr. Am. Otol. Society" for 1871, p. 70.

In the report of the first congress of the International Otological Society, held in New York, September, 1876, I reported a case of abscess over the mastoid region, extending to the squamous portion and involving the meninges and brain substance without attacking the tympanum.

MALIGNANT DISEASE IN AND ABOUT THE EAR.

I am thus indefinite in this designation, because in a large number of cases it is impossible to determine the exact point from which a given tumor springs. It is known that the meatus auditorious externus may be the starting point of a malignant growth; at the internal auditory meatus, on the sheath of the auditory nerve, or the dura mater lining the meatus, is a favorite seat for these tumors. Virchow has stated that the acoustic nerve is more likely to take on malignant disease than any other cranial nerve. As a rule, the fibrous sheath of this nerve is attacked, but one case, herewith recorded, shows that the fibres of the nerve itself may become involved in the disease. Other cranial nerves also become affected, notably those in the vicinity of the Gasserian ganglion. Some of the cases herewith recorded show the tumor to have arisen from the mucous lining of the tympanum. Moos states that the submucosa of the tympanum was the starting point of a malignant tumor he reported. The variety of tumor usually found, is for the most part that denominated sarcoma, with the subdivisions of Fibro-Sarcoma, Melano-Sarcoma, Osteo-Sarcoma, Myxo-Sarcoma, Round-celled Sarcoma, Spindle-celled Sarcoma, Chondro-Sarcoma, Glyoma or Glyo-Sarcoma, and occasionally a Chondro-adenoma, with a few cases of Epithelioma of the meatus or middle ear, and occasionally carcinoma of the acousticus may be found.

The appearances on inspection of malignant growths in this region are as follows: In the meatus or tympanum they are likely to be darker colored than other growths, although their appearance is not diagnostic. They bleed more profusely than benign growths and are more painful; the discharge is likely to be very offensive, and the tumor invariably reappears after removal—sometimes as soon as one or two days. They are likely to extend to the mastoid region and to involve the post and pre-auricular glands which may result in abscesses. These often have a dark color and a more or less boggy feel.

A fungous appearance of any tumor about the ear casts suspicion on its benignity. The variety of tumor which is most malignant has the appearance of a mucous or a gelatinous polypus. *Before the tumor has made its appearance externally*, its presence may be suspected by paralysis of certain muscles, the nervous supply of which has been interrupted by the tumor. For instance, it is not uncommon to see a paralysis of nerves from the 3d to the 8th or 9th, with the exception generally of the 4th. As a consequence there will be strabismus with ptosis, or the eye may remain in the centre of the palpebral fissure when the 3d and 6th nerves are paralyzed.

Deglutition is often interfered with in consequence of the paralysis ; sometimes making it well-nigh impossible for the patient to swallow food ; in other cases the food may fall into the larnyx or pass out through the nostrils. The invasion of the pharynx by an aural tumor may also produce the same result. Paralysis of most or all of the branches of the facial is not uncommon.

Ataxic movements of the arms and legs are frequently seen, more especially, however, of those on the diseased side. Faulty co-ordination is sometimes observed in walking. Epileptic attacks are not infrequent, as well as other forms of convulsions. The optic nerve is sometimes inflamed or simply hyperæmic, or in a state known as choked disk. When the tumor has approached the region of the eye, there is likely to be exophthalmos with occasionally an injected eyeball. The vision is not often lowered; the pupil may be dilated or contracted. Anæsthesia of the side of the face is sometimes seen. In a few instances a neuro-paralytic inflammation of the cornea with occasional destruction of that tissue is observed; spasm of muscles of the eyeball previous to becoming paralyzed has been observed. The pulse is frequently elevated, but not always. The respiration is sometimes quite frequent but by no means always so; the temperature is not often elevated, and sometimes in the later stages it may be even lowered. Towards the termination of the disease, sordes may be seen on the lips, gums and teeth, and the patient is likely to pick at the mouth and nose. There is occasionally an ichorous discharge from the nostrils. In the soft varieties of sarcoma, the tumor may slough to a considerable extent, but will soon reproduce itself. In autopsies, the dura mater is infrequently found to be penetrated by the tumor, though it may be thinned by the pressure exerted upon it by the growth. The malignant quality of the tumor is often very slight at its commencement, but it seems to be aggravated in proportion as efforts are made to remove it. Occasionally the tumors seem entirely innocuous in the commencement.

The subjective symptoms are of great variety. In some of the worst cases, very few symptoms may be observed. A tumor may press on the brain, with apparently no functional disturbance. The intellect is almost always clear until the latter stage of the affection, and in some cases to the very last. There is a variable amount of pain in the head, ear and throat, but this symptom has no great significance. The patient is sometimes irritable, loses appetite, and does not always sleep well. Vertigo, nausea, vomiting, rigors, delirium and coma are frequently seen towards the termination. The affection is more frequent in women than men, and in children than in adults.

The duration of malignant disease about the ear is exceedingly variable. In some cases the patient has lived six or seven years. In others a few weeks has been sufficient for a fatal termination. The duration seems to depend on the degree of malignancy and possibly on the amount of operative interference to which the tumor is subjected. *The causes* are often difficult to determine. Heredity is by no means a constant factor in developing the affection. Syphilis sometimes seems to have caused it. Traumatism occasionally appears to be the starting point of the growths, but a suspicion will creep in that a predisposition to the disease already existed, which was called into activity by the injury. The same remark applies to the influence of cold in developing the affection.

Sex evidently has something to do with the causation of the disease, women being more subject to it than men. *The results of the disease* are sufficiently exhibited in the accompanying cases. I do not remember to have heard of a case which terminated favorably, unless it were a mitigated form of epithelioma.

The treatment is sufficiently detailed in the cases hereafter reported. If a high degree of malignancy is discovered by the microscopic examination, a judicious surgeon will refrain from operating. Similar affections confined to the interior of the eye may be arrested by enucleating; but about the ear there seems no means of drawing so distinct a line between the diseased and healthy tissues. Symptoms must be treated as they arise, on general surgical principles, a prominent indication being to render the patient as comfortable and free from pain as possible.

A brief statement of the case reported by me in the *Am. Jour. Otol.*, April, 1881, is as follows: a girl six years of age

applied to me on April 17th, 1880, on account of a convergent squint in the left eye which had existed for ten weeks.
External rectus of the left eye completely paralyzed and
internus in a state of spasm, drawing the eye sharply inward. Hearing of right ear, for the watch, 24 inches; of
the left, watch in contact with the auricle. In the right are
symptoms of middle ear catarrh. The left shows a minute
granulation protruding through a perforated membrane.
Tuning fork heard best in the left ear. Pulse and temperature normal. After a consultation, the diagnosis was between basilar meningitis and malaria. Quinia was first
administered; afterwards large doses of pot. iodid. Faradism was used on the affected muscle. The granulation
was removed by forceps.

After the first week there was pain in the left eye and
ear, darting towards the throat. Iodine to the temple, a
blister to the mastoid and atropine in the eye relieved the
pain for a time. Patient is too quiet, sleeps too much, but a
paroxysm of pain readily arouses her; bowels constipated.
In eight or nine days complete paralysis of the motor oculi
occurred and the eye remained in the centre of the palpebral
fissure, with a drooping lid. Pulse occasionally rose to 130,
but the temperature was normal. The pain soon returned
and was not relieved by an opiate; patient very irritable.
Optic nerve decidedly hyperæmic. On May 26th there was
a little twitching of hands and feet, more noticeable in the
left. On the 27th the pulse ranged from 130 to 84; will
cry out if the eye is touched, although it is not reddened.

Mouth drawn to the right side, and she talks a little
" thick." Tongue protruded towards the left. Respiration about keeps pace with the temperature. Mind clear
and always has been. Sometimes she behaves quite naturally. Not inclined to take nourishment unless urged. May
31st, darkish coating on lips and gums; picks her mouth
and nose.

June 1st. Return of pain in the eye, and there is an
offensive discharge from the ear; found a soft, gelatinous
mass filling the meatus; removed it by forceps, and some
hæmorrhage resulted.

June 2d. Nausea; lid did not droop as much on account
of paralysis of orbicularis.

Being somewhat better, she was sent to her home in the
country. Carbolic acid solution was used in the ear;
iodide of potass. continued with attention to the nutrition.

It was now believed that all the symptoms depended on a malignant growth in the tympanum or its neighborhood. After a few weeks, the ear again filled with a material similar to that which had been removed, This was followed soon after by a tumor just behind the lobulus and another beneath the tragus.

These tumors grew with alarming rapidity, and on her return to town on Aug. 15th they had extended down the neck for nearly five inches (Fig. 85). The tumor was here

FIG. 85.—EXTRA-CRANIAL PORTION OF THE TUMOR REPORTED ON PAGE 303.

and there bloody with spots which had undergone decomposition, and it being warm weather, flies were constantly alighting on it, rendering the patient restive; it was very offensive. The patient was taken home. After a time, deglutition became difficult on account of the paralysis. There was some pain about the back of the neck and in the region of the tumor; usually slept moderately well.

Occasionally considerable portions of the tumor sloughed away, noticeably diminishing its size, but subsequent growth soon restored it to nearly its former size. After death the tumor shrunk about one third. The appetite was good towards the last, but the food required to be placed in the right side of the mouth to enable it to be swallowed. Near the termination of the disease, there was some bronchitis. The patient died from inanition and exhaustion. There was never any well marked cachexia. All pain had disappeared for several weeks before death.

No hereditary tendency to malignant disease could be discovered. When seventeen months of age, the child fell down-stairs, but without apparently doing serious injury. There was no history of syphilis in the family. The autopsy showed that there were two tolerably distinct tumors, which were outgrowths from the tympanal tumor. The early development of the tumor was evidently from the neighborhood of the tympanum, as the first symptom of the disease was paralysis of the abducens, then of the third, and soon afterwards of the facial, these passing in the neighborhood of the growth. As the paralysis of certain muscles of deglutition did not occur until some time after that of the facial muscles, it may be inferred that the anterior portion of the facial, in the hiatus fallopii, where the petrosal branches are given off from the intumescentia gangliformis, was invaded by the tumor at a later date. At this time, the tumor had destroyed the inner bony wall of the tympanum and neighborhood sufficiently to involve the cavernous sinus.

The intra-cranial portion of the tumor is distinctly lobulated, and has in only one or two places penetrated the dura mater. The absence of brain symptoms seems extraordinary, in view of the fact that the brain was much pressed upon by the intra-cranial portion of the tumor. The cases herewith reported, although called different varieties of sarcomatous tumors (with few exceptions), seem in the main to exhibit the following characteristics, to wit: they develop more or less slowly at first, but after operative measures have been resorted to the development becomes much more rapid, and they seem to take on a more malignant quality.

Autopsy by Dr. J. A. Andrews, formerly assistant surgeon to the Manhattan Eye and Ear Hospital, forty-eight hours after death; body greatly emaciated. The dura mater was pale but otherwise normal. Superior longitudinal sinus contains a long yellowish red cord-like clot. Pia mater raised by a clear fluid. The removal of the brain displays a growth beneath the dura mater, filling the left middle cranial fossa and a portion of the anterior and posterior fossæ. Dura mater overlying the growth is thickened and adherent to it. Surface of the growth nodular; one of the pedunculated projections of the tumor passes under the left anterior ciinoid process, displacing the nerves and vessels in this region. On incision the tumor presents the appearance and consistency of fat. It envelopes the lower two thirds of the petrous bone, which is found to be carious in every part. The tumor extends forward as far as the optic foramen, and from thence to the anterior condyloid foramen; behind it is bounded by a horizontal line touching the anterior condyloid foramen. The entire petrous portion is carious; this process terminates just behind the temporal suture; the squamous portion is separated from its articulation with the wing of the sphenoid; through this gap the tumor projects outward upon the face. Posteriorly, the caries extends beyond the temporo-occipital articulation to the foramen magnum. Behind the mastoid process there is a carious gap at the occipito-temporal articulation through which projects the portion of the tumor behind the ear. The superior petrosal and lateral sinuses are obliterated and the underlying bone is carious. Carotid canal and middle ear obliterated. Brain healthy, except the lateral ventricles, which contain ʒ ij of clear fluid. *The microscopical examination* was made by T. Mitchell Prudden, Pathologist to the Manhattan Eye and Ear Hospital; and is as follows: "The intra-cranial tumor, after preservation in Müller's fluid and alcohol, measures 8 ctm. in length, 6 ctm. in breadth and 4 ctm. in thickness, and was attached to the inside of the temporal bone and covered above by dura mater. It was mostly soft and gelatinous, except in the central portions, where it was firmer and distinctly nodular. Within were small irregular cavities, some of them filled with blood. The dura was in general somewhat thickened, except over the apex of the nodules, where it was thinned, and in two instances was perforated. On the inside of the temporal bone at its upper portion was a broad line of jagged, irregular osteophytes. The cells of the tumor were fusiform, branching and spheroidal and in some parts crowded together, in others separated by granular or fibrillated intercellular substance.

It was very vascular in many parts; the walls of many of the smaller vessels were considerably and irregularly pouched. Numerous nerves passed through the substance of the tumor apparently unchanged. In other places, single medullated nerve fibres were seen surrounded by the new growth.

At the upper and inner side of the tumor and occupying about one fourth of its bulk, was a circumscribed and somewhat denser portion, which, in addition to the above described structure, presented spheroidal and ganglionic shaped cells and non-medullated nerve fibres. The ganglion cells, which were finely granular, had large vesicular nuclei and sharply defined nucleoli; they were frequently pigmented near the nucleus, and in many cases connected with non-medullated nerve fibres. They were mostly enclosed in cellular capsules and were surrounded by the tumor tissue. They were as a rule, smaller than the space within the capsule, and in some cases the nucleus was at the end of the cell from which the process passed off. The largest had a diameter of 0.035 m. m., the smallest 0.015 m. m. In many places they were closely crowded together, and in others widely separated by tumor tissue. Everywhere among them non-medullated nerve fibres were found in abundance (Fig. 86). Associated with but

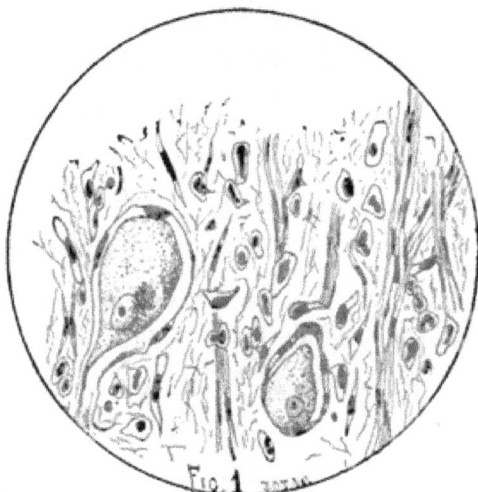

FIG. 86.—MICROSCOPIC APPEARANCES OF A MYXO-SARCOMA.

separated from these ganglion cells were found cells or cell groups so peculiar as to deserve notice. They were of various shapes, but in general more or less spheroidal, more coarsely granular than the ganglion cells, and usually smaller. The largest measured 0.0375 mm., the smallest 0.0075 mm.

They had a large irregular shaped nucleus, and usually a well-defined, large nucleolus. From some of them, distinct, narrow processes were seen, passing off from one

end.   Many were devoid of capsules, but a large proportion
were enclosed in a shrunken structureless non-nucleated
capsule, which generally was larger than the cell itself.
Among these cells were seen elongated spindle-cells, with
rod-like nuclei.   (Fig. 87)   These seemed to represent gan-

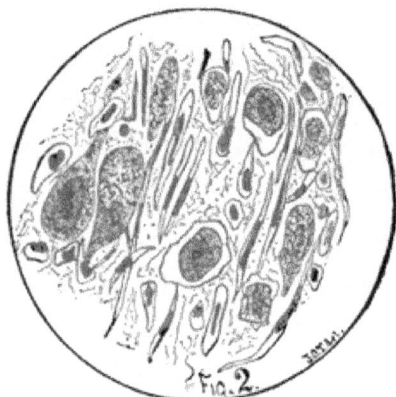

Fig. 87.—Microscopic Appearances of a Myxo-Sarcoma.

glion cells, immature or imperfectly formed.   On the inner
border of the tumor was a pedunculated nodule, about 4
m. m. in diameter, projecting upwards, and a small nerve
passed into its substance near its base (Fig. 88), the fibres

Fig. 88.—Microscopic Appearances of a Myxo-Sarcoma.

of which became separated by the tumor tissue.   Before
the nerve entered the nodule it was perfectly normal.   The
fragment of the petrous bone, besides the carious condition, exhibited granulation tissue, partaking in part of
the nature of the tumor itself.   The dura mater, at points
where it was most closely adherent to the tumor, showed
on its inner layer infiltration with cells similar to those
in the tumor.   The growth evidently involved the Gasserian ganglion; whether originating here or in the connective
tissue of adjacent nerves, could not be determined.
     The rare occurrence of the new formation of ganglion

cells in tumors, together with the meagreness of our knowledge concerning the development of cerebro-spinal ganglion cells under normal conditions, would suggest the propriety of a simple objective record of the structures in question without definitely attributing to any portion of the tumor the character of a true neuroma.

The *anatomical diagnosis* would accordingly be *myxo-sarcoma*, involving or possibly originating in the Gasserian ganglion and the nerves in its vicinity, with partial destruction of the petrous and other portions of the temporal bone and the soft parts within them.

The subjoined cases, representing tumors similar in kind in this locality, may further elucidate the subject.

Dr. C. A. Robertson, in the Tr. Am. Otol. Soc., 1870, p. 35, reports a case of fasciculated sarcoma of the tympanum:

A woman, æt. 40, had tinnitus aurium and deafness. Five years afterwards there was an offensive discharge, accompanied by pain and paralysis of some of the branches of the facial nerve. The removal of a polypus resulted in excessive hæmorrhage. Could not hear a watch in contact; pre-auricular glands swollen; no further record.

Wilde reports, in his text-book on the ear, a case of malignant disease of the meatus, probably also involving the tympanum.

A woman, æt. 50, whose brother died of cancer, had a growth in the ear, which had been in existence for many years; she had an unhealthy look; there was a fœtid discharge. She was giddy, had nausea, and could not rest. The tumor was painful; it filled the meatus and was never wholly removed; showed a tendency to bleed, and often reproduced itself after partial removal; presented a fungous appearance. A fluctuating tumor of the mastoid was opened, from which a dark-colored fœtid matter was evacuated. There was facial paralysis and abscesses along the mastoid muscles. The patient had rigors and convulsions, with excruciating pain, followed by coma. Post- and infra-aural regions enlarged. The integument of the mastoid gave way, and a fungoid mass sprouted from this region; fœtor intolerable. Death in three weeks from the first appearance of the tumor. The malignant quality of the disease was evidently aggravated by efforts made to remove it. No autopsy.

Another case from the same source, p. 297:

A boy, seven years of age, had a polypus of the meatus, which was removed several times, but reappeared in a day or two; not long after this he had an epileptic fit, which was followed by an abscess over the mastoid region. This abscess cavity communicated with the meatus. A fungus almost immediately sprung from it; the parts in front and around the ear became swollen and were boggy to the touch; repeated attacks of epilepsy were followed by death. The autopsy revealed an osteo-sarcoma involving the petrous and mastoid portions of the temporal bone. The petrous portion was enlarged and softened, and presented a fungous appearance; the internal ear was obliterated; the portion of the brain lying on the tumor was unaffected.

Dr. Hartman, of Berlin, in a translation by Dr. Knapp in the *Archiv. Otol.*, March, 1880, reports a case of round-celled sarcoma of the tympanum. O. J., æt. three and a half years, was first seen in Oct., 1878. Four weeks previous to this he had a discharge from the ear, accompanied by pain and inflammation. Two weeks later a tumor in the meatus was observed the size of a pea; this, with others found in the canal, were removed as far as possible, and the parts cauterized; the tumor returned in a few days, appearing to start from all sides of the tympanum and inner end of the meatus. Cauterization was again done, but the tumor reappeared. In October, most of the mass was again removed and the galvano-cautery applied. After this it reappeared and the surroundings of the ear became swollen, and an abscess below the ear, communicating with the meatus, was incised. Other tumors formed in the vicinity of the tympanum and meatus. These masses were found to be round-celled sarcomata; they pushed the auricle outward. In February they were the size of a goose egg; general health declined. There was bronchitis, anorexia, diarrhœa, emaciation and headache. On March 21st, bilateral convulsions occurred, lasting several hours. A prominence in the right side of the fauces interfered with mastication and deglutition. Death occurred on March 28th; convulsions reappeared two days previously accompanied by coma.

Autopsy.—The auricle was on the crest of the tumor. The latter consisted of a number of lobes, varying in size from a walnut to a hen's egg. There were ulcerated places communicating with pus cavities. The upper and posterior walls of the meatus, roof of tympanum and a part of the squamous portion, destroyed; at the latter point the external tumor communicated with another in the cranial cavity. The mastoid cells were filled with the tumor. The petrous pyramid was easily separated from the mastoid process. Meatus auditorius internus free from disease. This seemed malignant from the start, which was exceptional, as this class of tumors was usually preceded by otorrhœa and polypi, indicating an inflammatory origin. (Most of the cases herewith reported seem to have been malignant at the commencement.) He also quotes a case from Wishart in the *Ed. Med. and Surg. Jour.* A child three years of age had a severe pain in the ear for some weeks with a discharge; a tumor appeared surrounding the ear, soon ulcerated, and discharged fœtid bloody matter; hæmorrhages frequently occurred and the child died in fifteen weeks from the commencement of the disease. The tumor was as large as the child's head. The petrous portion of the temporal bone, the zygomatic process, and the condyloid process of the lower jaw were destroyed. The tumor extended into a large orifice in the squamous portion of the temporal bone, forming a depression in the middle lobe of the brain, the latter being in other respects quite sound. Dr. Hartman thinks the tumor he reported sprang from the sub-mucosa of the tympanum. Mr. James Hinton, in his book "The Questions of Aural Surgery," 1874, says: "Now and then, but rarely, a true malignant growth simulates a polypus and demands caution in treatment. - The general cachexia, the pain, the livid color, the great tendency to bleed, and the swelling which is always present around the ear, serve to distinguish the malignant growth. A soft, persistent swelling about the attachment of the auricle, attended with great pain, I have never known except in malignant disease." Wilde, it will be remembered, speaks of the livid color of the malignant growths in the ear as being characteristic. My own opinion, however, is that the malignant nature of the growth is determined with certainty only by the microscopic examination.

It is true that the appearance of the tumor may lead to a suspicion of its nature. Very infrequently is there any true cachexia in this form of growth. Mr. J. W. Hulke, F.R.C.S., in the *Medical Times and Gazette*, Jan. 20th, 1877, reports a case similar to the preceding: J. W., æt. 39. The tumor seemed to arise from the tip of the petrous portion of the temporal bone. It had involved and destroyed the

Gasserian ganglion, the third, fourth, and sixth nerves. It also extended to the posterior part of the orbit through the sphenoidal fissure; the cavernous sinus was destroyed, together with the eighth and ninth nerves at their foramen of exit from the skull. No brain symptoms. The patient being syphilitic, a gummy tumor was diagnosticated. The eye-ball was inflamed and protruding; there was great pain. The tongue was protruded to one side; there was squint, myosis, etc.

Dr. George P. Field in the London *Lancet*, Dec. 8th, 1877, reports a case of sarcomatous tumor in connection with the auditory nerve, in a woman aged twenty-nine years. She has brain symptoms; was admitted to the hospital Jan. 29th, 1877, and died April 29th, 1877. A tumor the size of a Maltese orange was attached to the posterior surface of the right petrous bone, above the internal meatus. It resembled the cerebellum in appearance. It arose from the dura mater lining the meatus, and seemed to unsheath the auditory nerve. His diagnosis was round-celled sarcoma of the auditory nerve, of moderately rapid growth. He mentions two other cases, somewhat similar to this one.

A case of fibro-sarcoma of the auditory nerve is reported in the *Archiv. Ophthal. and Otol.*, Vol. III., p. 135, by Boettcher, of Dorpat. Prof. Moos in the *Archiv. Ophth. and Otol.*, Vol. IV., 1874, p. 482, reports a case of sarcoma of the auditory nerve, in which there was fatty metamorphosis and partial destruction of the organ of Corti. It was a spindle-celled sarcoma. The cells were abundant, vessels were numerous, and the intercellular substance was very little developed. In one place the tumor was harder and transparent and in another softer and gelatinous. The patient was a woman forty-nine years of age. The duration of the disease was from August, 1870, to August, 1871. On July 24th, (1871) she had permanent headache, dizziness, weakness of sight, ataxic movements of the left arm (left ear affected), anæsthesia of left half of the face, so that the food fell out of that side of the mouth; swallows her food the "wrong way;" fluids regurgitated into the left nostril, rarely into the right; converging strabismus of left eye; left frontalis has a permanent spasm. Left arm makes ataxic movements, but only under impulses of the will; lower lip tremulous on the left side. August 1st, temperature steadily below 37.5°; diarrhœa continued, patient weaker. On August 11th, swallowed food the "wrong way," fell back with a rattle in her throat, turned black and died in a few minutes. There were also eye symptoms. He quotes Landifort, Lévêque-Lascuorce, Brückner, Bötticher, Forster, and Voltolini, as having described this kind of tumor.

The causation of these tumors in general was here stated to be mechanical injuries and syphilis. This patient slept all night by an open window, and cold might have caused the trouble, or at least, as Moos remarks, the tumor might have been present before the exposure, but have been excited to a more active development in consequence of it.

In one hundred cases of ear disease treated by Dr. Kirk-Duncausen, and reported in the *Edinb. Med. Jour.*, March, 1878, one of malignant growth was recorded.

It occurred in connection with purulent otitis media; the malignant growth distended the auditory canal and implicated the auricle. It was painful, recurred often after removal, and finally proved fatal. It was not stated what variety of tumor it was, but its history resembled that of sarcoma.

Dr. Stevens in the *Arch. Ophthalmology and Otology*, July, 1879, reports a case of spindle-celled sarcoma of the acoustic nerve. The nerve fibres were lost in the substance of the tumor. The patient was a girl, aged seventeen years. External recti were paralyzed, pupils sluggish. Vision $\frac{20}{70}$ in each eye. Hearing was lost in the left ear and impaired in right. Both optic disks were choked. Headache in frontal and occipital regions; signs of meningitis. Duration of the disease, several years. Towards the fatal termination her gait lacked co-ordination.

In the *Arch. Otol.*, Vol. VIII., No. 4, is reported a case of parotidean and intra-tympanic tumor, by Dr. Knapp. J. H. W., aet. 37, consulted Dr. K. on account of sudden deafness in the right ear. Below and in front of the ear was a tumor the size of a hen's egg, which the patient first noticed six or seven years previously. During the last six months it had grown more rapidly. Left ear has a chronic otorrhœa. The membrane of the right ear was bluish red, convex and on pressure with a probe yielded as though a soft substance were behind it. Membrane punctured on May 8th, which penetrated a tumor and resulted in considerable hæmorrhage. This tumor was touched with a probe and presented the appearance of a soft fleshy mass.

The parotidean tumor was thought to connect with the tympanal tumor through the Gasserian fissure.

May 22d. Great pain in the ear for a day and a night. The aural tumor occupies half the meatus.

May 29th. Pain relieved; an abscess has formed and pus escapes on pressing the tragus.

June 20. The tumor fills the meatus. Dr. K. being absent from town, Dr. Buck took charge of the case.

Dr. Sands removed the tumor of the parotid on June 28th. This was found to have no connection with the aural tumor. The latter was also partially removed at the same sitting with Dr. Buck's assistance.

July 28th. The incision for the removal of the parotid tumor has healed; pain in the ear continues; aural tumor increased perceptibly in size.

This was mostly removed by incision and was followed by considerable hæmorrhage. Other efforts at removal were made, but resulted in so much hæmorrhage as to necessitate plugging the meatus with cotton wool to arrest it. There was little or no fibrous element in this tumor, and it seemed quite similar to the one I reported in the commencement of this article.

Microscopic examination by Dr. W. H. Porter, of New York. A detailed description is given, the conclusion of which is that both parotidian and aural tumors were essentially a chondro-sarcoma, or chondro-adenoma. Dr. Buck described it as a chondro-adenoma. On the same day another effort was made to remove the tumor, but hæmorrhage interrupted it. July 26, the rongeur was used to remove the growth, and red-hot needles were thrust into it to arrest the hæmorrhage. Soon after this an abscess formed below the mastoid. Patient returned to his home in the country. On Oct. 25th he came to town and a consultation of all the surgeons who had previously attended him was held. The tumor below the auricle increased in size; meatus filled with a growth having fistulous openings in it, discharging pus; facial paralysis since five days; decided not to operate. Patient returned to his home. The tumor continued to grow until its dimensions were as follows: seven inches, eight inches and five inches in its different diameters. The ear rested on the outer surface of the tumor. No mental disturbance was observed during the course of the disease. Knapp refers to Schwartze for the literature of the subject, who had found twenty-one cases of malignant disease of the organ of hearing.

Dr. Harlan in the Philadelphia *Med. Times*, Dec., 1873, reports a case of round-celled sarcoma in the ear of a girl three years of age. Two months before examination she had a bloody discharge from the ear, with pain on swallowing. There was swelling about the ear, and the face was drawn to the right side; meatus externus filled by a firm, globular polypus. Below and behind the auricle was a fluctuating swelling. Polypus removed and an incision behind the ear resulted in a discharge of sanious pus; subsequently the growth was twice removed. Twenty-eight days after the first examination the tumor was of the size of a hen's egg, bright red, lobulated, and having a granular surface; complete ptosis of left eyelid resulted; conjunctiva became congested, the cornea was hazy and the lower portion infiltrated; sanious discharge from the left nostril. Seventeen days later an ulcer penetrated the cornea. Breathing through the mouth and nostril had become laborious. In one week from this she died, apparently from exhaustion.

Double convergent strabismus appeared some time before death, but nearly disappeared at a later date. On autopsy the bone at the base of the tumor behind the ear was found eroded, inner tympanic wall destroyed. The inflammation of the eye was regarded as neuro-paralytic.

THE EFFECTS OF QUININE ON THE EAR.

Inasmuch as quinine is administered largely for the treatment of malaria, and as the latter condition is known to act injuriously on the ears, it is somewhat difficult to arrive at an exact conclusion as to the effects of the drug on the organ of hearing.

One fact comes out prominently enough—that on administering quinine in large doses, and sometimes even in small ones (3 to 5 grs.), the patient will have a buzzing or ringing in the ears with deafness and a sensation in which his own voice sounds hollow and unpleasant. The voices of others will seem to be muffled and to have an unnatural pitch. The Eustachian tubes may be more or less obstructed, crackling sounds will be heard in the ears from the interchange of air between the throat and tympanum, the tubes being easily forced in some instances, while in others inflation becomes difficult and the patient may have a feeling of fulness

in the ear cavities.   The throat is also frequently congested.
These symptoms are temporary provided the dose be not
repeated, but disappear in a few hours or a day; but if the
drug is constantly repeated the symptoms may continue
and cause permanent harm to the ear, according to Kirch-
ner in the *Berliner Klinische Wochenschrift*, No. 49, 1881
(*Am. Jour. Otol.*, Vol. IV., No. 1).   He also concludes that
the hyperæmia of the tympanum may go on even to
hœmorrhage, with hyperæmia of the labyrinth and injury
to the ultimate nerve fibres.   The cause of the hyperæmia,
he states to be vaso-motor disturbance, which in extreme
cases may result in paralysis of the vessels and exudation
into the various parts of the ear.

Weber-Liel, *Monatsch. f. Ohrenheilk.* No. 1, 1882, in
conjunction with Gruber, instituted the following experiments on twelve medical men.   One gramme of quinia mu-
riatica was given.   In the course of two and a half hours the
temperature of the meatus as well as that of the whole body
fell to an average of 56° Cent.   No hyperæmia of the membrana or meatus was observed, but on the contrary in five
cases the slight injection previously present had disappeared.
Roaring and buzzing or ringing sounds were always produced in from one to one and a half hours, gradually disappearing in twelve hours.   After from two to three hours, a
decided diminution of hearing showed itself, which continued until the disappearance of the subjective noises.   The
greatest loss of hearing occurred at the period of lowest
temperature.

Dr. H. N. Spencer and others, St. Louis Medico-Chirurgi-
cal Society, published in the St. Louis *Courier of Medicine*,
Nov., 1880 (*Am. Jour. Otol.*, Vol. III., No. 2), have discussed
this subject.   Dr. S. some years since gave eighty grains of
quinine to a gentleman attending medical lectures.   In a
few minutes there was a general congestion of the vessels
of the membrana tympani, which gradually subsided.
There were deafness and tinnitus, which disappeared as the
influence of the drug passed off.   No anæmia succeeded
the hyperæmia.   He believes it possible that in some cases
quinine may permanently injure the hearing, but he has
seen no case where lasting injury has resulted from quinine
alone, other explanations of the deafness always being
found.   In the Tr. Am. Otol. Soc. for 1875, Dr. Roosa records the following experiments concerning the effects
of quinine upon the ear.   He quotes Dr. Hammond's article

on the subject in the *Physiological and Medico-Legal Journal*, Oct., 1874, in which also is recorded the literature of the subject to that date. In this was related the experience of Dr. Hammond, who administered a dose of quinine to himself, Dr. Roosa making the observations. Dr. R. also repeated the experiment on others. Dr. H. took ten grains of quinine at 8.30 P.M. At 10.30 P.M. the right drum-head was very much injected along the malleus handle and upper margin; left less injected. Both optic papillæ were pink, left more so than right; face flushed, eyes suffused, ocular conjunctiva decidedly injected. Slight headache, tinnitus in both ears, auricles burn and are decidedly blushed; lobe of the right so congested as to resemble an ecchymosis. Most of these symptoms had much diminished by 11 P.M.

Ten grains of quinine were given to Dr. E. T. E., æt. 24. The experiment was similar to the last except that there were no symptoms but slight congestion of the drum membranes.

Another observation was made on Dr. C., æt. 25. At 10.16 A.M. fifteen grains of sulphate of quinine were administered. At 11 A.M. a vessel is seen along the malleus handle of the right membrane; slight vertigo. At 12.30 there is a sense of heat and tingling over the whole surface of the body; some fulness in the ears and head. Handles of both mallei injected, hands tremulous, sounds of a high note in the ears, auricles feel warm. At 12.30 the injection of the mallei, the vertigo and tremor are disappearing. At 12.50 mallei still injected, motions of the jaws cause peculiar and unpleasant sense of vibration in the ears. Dr. R. refers to a paper published in the *Am. Jour. of Med. Sciences*, Oct., 1874, to show that in that paper he had expressed the opinion that hyperæmia and not anæmia of the ears was caused by the administration of quinine. He refers to two other cases in his experience, one behaving somewhat similarly to those already reported, and the other showing no congestion of the ear whatever; the latter case, however, was one where quinine had been habitually taken and the patient was anæmic. Dr. R. expresses the opinion that the tinnitus and deafness "following the use of quinine depend upon congestion of the ultimate fibres of the auditory nerve in the cochlea, and that the redness of the drum-heads is merely an index of the former condition." In the *Trans. Am. Otol. Soc.*, 1872, p. 57, Dr. Roosa also reports a case where an inflammation of the tympanum, followed by im-

paction of cerumen, occurred when the otitis, "although to
a certain extent dependent upon the naso-pharyngeal catarrh," was "chiefly caused by the use of quinia. By looking at the history and observing how promptly and invariably the pain in the ears occurred in several instances after
the use of the agent (quinine), we are forced to the conclusion that quinine was the exciting cause of the aural inflammation." He remarks that this case illustrates the
effect of quinine upon the ear, "which I am inclined to
suspect is sometimes an inflammation of the conducting
portions, as well as of the acoustic nerve or labyrinth." Opposed to the ideas here presented except those of Weber-
Liel, are those of Knapp ($Z. f. O.$, Vol. X., p. 279)—Politzer's
reference—who, "in cases of blindness and deafness caused
by large doses of quinine, observed excessive paleness of the
disk of the optic nerve with almost complete invisibility of
the retinal vessels; and he believes that a similar state in
the cochlea causes deafness." The practical point regarding these matters seems to be to avoid quinine in cases
where it is plainly obnoxious, and select some other salt of
Peruvian bark. I have found the sulphate of cinchonidia
reasonably effective though perhaps not as much so as
quinine, and it causes very little disturbance to the ear functions in doses sufficient to accomplish results. Those who
have confidence in dextro-quinine may use that remedy, as
it is not likely to effect the ears unpleasantly. All the other
salts of Peruvian bark are less likely to injuriously affect the
ears than quinine.

SYPHILITIC INFLAMMATION OF THE MIDDLE EAR, THE LABY-RINTH AND ACOUSTIC NERVE.

I treat these subjects under one heading for the reason
that in practice most of the cases met with come under the
designation of middle ear affections with or without labyrinth complications, the labyrinth often suffering in its
functions whether it be the seat of organic disease or not.
Moreover, the difficulty of always stating how far the labyrinth is involved in a given case, is sometimes well-nigh insurmountable. In one class of cases few symptoms may
be elicited except those of ordinary middle ear disease.
There is some discrepancy in the statements of different
authors as to whether the Eustachian tubes are pervious in

specific otitis media. Most agree that the tubes are pervious, and that there is little or no improvement to the hearing after inflation. My own impression is that there are at least numerous exceptions to this rule, the middle ear affection exhibiting symptoms quite similar to those found in ordinary middle ear disease. Vertigo, nausea, vomiting, unsteadiness of gait, etc., are laid down as characteristic symptoms of syphilitic otitis, especially where the functions of the labyrinth are disturbed. As so many other ear affections have these symptoms, they cannot be regarded as altogether peculiar to this disease. The vertigo is likely to be a persistent symptom. Rapid or sudden loss of hearing is certainly a characteristic of specific otitis. The hearing may be hopelessly spoiled in a day, or it may be a year or two before profound deafness results, but in either event the hearing is lowered much more rapidly than in a nonspecific affection. Tinnitus aurium is present in a considerable number of cases, although in some instances there is a surprising absence of subjective sounds in the ear.

This disease, on the whole, is a somewhat painless one, although specific neuralgias about the head are frequent, the pain running down the spine or either arm; as this is often nocturnal, it has been regarded as characteristic of syphilis. As in other grave ear affections, the patient may be despondent, and suffer from a treacherous memory, with loss of appetite, constipated bowels, and other signs of systemic disturbance. Sometimes, though not often, optic neuritis accompanies this affection. Paralysis of the facialis is a somewhat frequent symptom. Sometimes both nerves are paralyzed, when the patient is unable to open his mouth.

The facial is much more likely to be diseased in its passage through the hiatus Fallopii than in ordinary middle ear trouble. Paralysis of the facialis from a lesion at the inner auditory meatus is not an uncommon occurrence. Hemiplegia and even paraplegia may accompany this form of otitis. Epileptic seizures are not of very infrequent occurrence. Double vision with strabismus is sometimes seen, more frequently in consequence of paralysis of the abducent muscle. Many other symptoms dependent on syphilis might be enumerated. Ordinarily there is no discharge from the ear, but occasionally a suppurative otitis will be modified by the constitutional dyscrasia. Thus Buck, *Am. Jour. Otol.*, Vol. I., No. 1, mentions a case where the suppurative process resulted in carious bone at the inner end of the meatus externus, and another where two perforations occurred in the membrana, the latter seeming to melt down by a destructive process resembling that sometimes seen in tuberculous suppuration of the middle ear. It was the opinion of Buck, in the case of the carious bone, that as there was a free outlet for purulent products, that the caries would not have resulted had it not been for the syphilitic influences. The throat affections accompanying the middle ear disease, besides having nothing characteristic of syphilis, may have all the peculiarities of syphilis, as warty or gummous growths, deep ulcerations, perforations of the uvula and velum, etc. Many authors, among whom are C. H. Burnett (*Amer. Jour. of Otol.*, July, 1881) and Samuel Sexton (*Amer. Jour. of Otol.*, October, 1880), conclude that similar changes are likely to occur in the middle ear to those in the throat, even when inspection of the membrane gives no hint of a diseased tympanum. Both ears are usually attacked, but sometimes one remains healthy while the other may suffer from well-marked syphilitic disease. Little can be said about the appearance of the membrana. It very frequently does not differ in appearance from that in non-specific disease. Generally a dry looking membrane moderately reddened, with little or no cerumen in the canal, possibly with redness at the inner end of the meatus, is all that may be ordinarily seen, or there may be no hyperæmia of the membrane at all. Signs of infiltration of the membrane, giving it a greater degree of opacity than would result from a non-specific inflammation, will cause one to suspect constitutional disease. The most important aid, however, to diagnosis is the history.

Pathology.—It is extremely difficult to determine in all cases whether the lesions are in the middle ear, in the labyrinth, or in the acoustic nerve. I believe in a majority of cases that the middle ear will be found diseased. It is true that the Eustachian tubes will be pervious in a large number of cases, and that no improvement to hearing results from inflation. It is interesting to note the impression Sir William Wilde had of this class of cases. He says in his book on the Ear (Amer. Ed., p. 255): "This inflammation does not end in muco-purulent discharge from the tympanum, the surface of the membrana tympani, or the sides of the auditory canal; nor have I seen lymph effused upon

the membrane, as in the more violent and painful forms of
otitis; but from its brownish red color in the very early
stage, from a yellowish speckled opacity, which is generally observable in it on the subsidence of the redness, and
from the intense degree of thickening and dulness which
were present in some cases, which were evidently the result
of syphilitic disease, I am inclined to think that the lymph
is largely effused between the laminæ or upon the inner
surface of the membrana tympani.   Two of the worst cases
of non-congenital deafness I ever saw appeared to have
been the result of syphilitic inflammation, and in both there
was great thickening, opacity, and insensibility of the membrane.   I am also inclined to think that syphilis has played
a more extensive part in the production of deafness than
we are aware of."

On page 256 he reports a case the exact counterpart of
the description just quoted.   The patient had undergone
the usual local treatment without a particle of improvement.   Afterwards mercury was given by the mouth.   "This
mode of administering the mineral disagreeing, we were
obliged to discontinue it, and substitute inunction in its
stead.   The deafness and the appearance in the ear remained unaltered until the morning on which salivation
was produced, and then hearing was restored almost mirac-
ulously, and the next day the redness and vascularity in the
ears had almost disappeared."   There was no return of the
deafness.   I will not attempt to assert, in the absence of
the tuning-fork observations, that this was a case of middle-
ear disease alone, but I infer that it was.   I also conclude
that an ordinary otitis would not have improved so suddenly under the influence of mercury.   Following Wilde's
opinion that there is infiltration within the tympanum,
Sexton (*Am. Jour. Otol.*, Oct., 1880, p. 308); says "As regards the pathology of this syphilitic invasion of the middle ear, it may be surmised that granuloma, or circumscribed small round-cell infiltration, takes place within the
tympanum—that the invasion is rapid, and that the conduc-
tive apparatus is prevented from performing its movements
by the particular manner of fixation that occurs.   It is
doubtful if our present means of pathological study could
definitely determine its present seat in all cases."   C. H.
Burnett quotes with approval this extract, except the last
sentence, in a clinical contribution to the *Am. Jour. Otol.*,
July, 1881.   He also, finding a syphilitic wart on the uvula

of the patient in question, concluded that there was a similar condition in the tympanum.   Sexton states, *Tr. Amer.
Otol. Soc.*, 1878, that "the sudden deafness of syphilis has,
beyond doubt, its principal seat in the conducting mechanism, for in all such cases which I have examined *the vibrations of the tuning fork, when placed on the head, are transmitted
through the tissues to the auditory nerves, and are heard; and the
patient can hear distinctly his own voice*."   Buck states, *Amer.
Jour. Otol.*, Jan., 1879, that "the mucous membrane lining
the niche for the round window, or the membrana tympani
secondaria itself, may become congested and very much
swollen, without narrowing of the Eustachian tube, without
the exudation of fluid into the middle ear, and without the
slightest change in the condition of the drum membrane.
. . . The swelling of these parts, and especially of the
secondary tympanic membrane, cannot take place without
producing pressure upon the fluid of the labyrinth."   He
still further discusses the subject of interference of the
function of the acousticus from middle-ear syphilis, and even
of organic changes secondarily induced by the middle-ear
affections, and ends by saying that "In the present state of
our knowledge of aural pathology, therefore, we are hardly
justified in using the expression 'labyrinthine disease,' except in those cases where demonstrable lesions are found in
this part of the ear at the post-mortem examination."   Dr.
F. R. Sturgis, in the *Bost. Med. and Surg. Jour.*, June 3, 1880,
p. 532, speaks of the "infiltration of the tympanum con-
joined with absence of vascular congestions" as characteristic of syphilitic otitis media, and "not, I think, common
to the ordinary forms of middle-ear trouble."   He speaks
of one case of old suppurative otitis media which had become convalescent and presented the usual changes incident to that disease.   The right ear subsequently became
attacked by syphilis; when the membrana "became succulent, looking as if it were infiltrated with fluid and air;
exudation had taken place . . . which diminished very
considerably in size as the syphilis improved."   Buck further states that in one class of cases (loc. cit., p. 39) the
drum membrane showed a tendency to breakdown under the
syphilitic inflammation, often causing more than a single
perforation, similar to that which frequently may be seen
in the velum in syphilitic subjects.   In his case No. XII. he
concludes that "the development of caries and facial paralysis with no evidence of interference with the free escape

of the pus formed, was certainly not in harmony with the ordinary course of a non-syphilitic otitis media purulenta acuta."

Numerous cases on record exhibit paralysis which appears concurrently with the deafness. In this class of cases it would not be unreasonable to infer a hyperæmia of the acoustic and facial nerves at the internal meatus, with infiltration; or a periostitis in that vicinity, or a gumma—a concatenation of symptoms perhaps not dissimilar to what occur in specific neuro-retinitis. Dr. F. M. Pierce, in the *Tr. International Med. Congress*, London, 1881, Vol. III. p. 399, says: "The evidence of syphilis attacking the middle ear is mainly of a catarrhal character, with a marked prevalence of anomalous auditory nerve symptoms, and in adults these symptoms are suggestive of acquired or congenital syphilis as a predisposing cause." I am firmly of the belief that the naso-pharyngeal and middle ear catarrh of syphilis very often does not differ in appearance from that dependent on other causes, but that its peculiar nature is revealed by the rapidly curative effect of mercurial treatment. This is, however, denied by some. I am inclined to believe that any peculiar appearance of the throat which is characteristic of syphilis may be repeated in the tympanum, the inflammation travelling up the Eustachian tube to involve the middle ear. Roosa says (Text-book, 4th Edition, p. 521), after recognizing the fact "that syphilitic affections of the middle ear are perhaps more common than those of the labyrinth," goes on to say that "we have good reason for believing that we may have hyperæmia and periostitis of the labyrinth, as well as gummata." Sometimes the paralysis extends so that the patient becomes hemiplegic. Occasionally both facial nerves become paralyzed, as in one case reported by Sexton (loc. cit.), where the patient was unable to open his mouth.

Diagnosis.—If only catarrh of the middle ear exists without any of the characteristic signs of a specific lesion, the syphilitic history, the obstinacy of the affection under ordinary treatment, the proneness to lowering of the hearing more rapidly and to a greater degree than in non-specific catarrh, and lastly the wonderful effect frequently resulting from constitutional treatment, will point with considerable certainty to the true nature of the affection. In other cases the throat will exhibit so many of the signs of specific trouble as to at once place the surgeon on the right

track of investigation. The appearance of the membrana
is not a valuable guide on the whole. If there is only moderate redness of the membrana, the inner end of the meatus
reddened, dry looking, and the membrane has a degree of
opacity or is infiltrated out of proportion to the amount
of inflammation present, and there is absence of any considerable pain in the ear, suspicion may be entertained of
specific trouble.

Vertigo, nausea, vomiting, unsteadiness of gait, etc.,
are very frequently met with, but are not diagnostic of
syphilis, or even of labyrinthine disease. The rapid or sudden loss of hearing, apparently without sufficient local
cause, is a valuable diagnostic of syphilitic otitis. False
hearing or double hearing proves that Corti's organ is
functionally affected; so also does the failure to hear higher
tones of the piano while the lower ones are heard. Sensitiveness to harsh or loud sounds, the unpleasant and discordant perception of harmonics, proves that the region of
the cochlea is functionally affected, but not that the disease is located in that region. The tuning-fork is of diagnostic value in eliminating middle ear complications; but
even here the exceptions of intermittent bone conduction
will often make an exact diagnosis well-nigh impossible.

I conclude that sudden and absolute deafness without
middle-ear symptoms is the most certain diagnostic symptom
of labyrinthine or nervous trouble. Sudden facial paraly
sis and deafness, as Roosa (l.c.) has mentioned is diagnostic of
disease located in the internal ear or the acoustic nerve.
Again, a pervious Eustachian tube, with absence of disease
of the membrana tympani, does not prove absence of middle-
ear complication. The greatest trouble in diagnosis will be
in those cases where the middle ear is known to be diseased,
to determine how far the acoustic nerve or labyrinth is involved. In most of the published cases of ear syphilis the
tuning-fork and other tests have not been made with
sufficient thoroughness to determine all that may be known
of the exact locality of the lesion. This matter will test
expertness of diagnosis more than most subjects.

The prognosis is of the gravest nature. Much turns on
the commencement of treatment before profound deafness
has overtaken the patient. After that period little can be
done, although a few cases are reported where recovery
has resulted, as Wilde puts it, in an almost miraculous manner, where the deafness had become profound. In most

of the cases of non-implication of the middle ear with sudden and extreme deafness, the prognosis is very unfavorable. Where little is at fault except the middle ear, a very hopeful prognosis may be given, as the mercurial influence is often wonderful in relieving these symptoms. The presence of a number of paralytic symptoms renders the prognosis nearly hopeless.

The conditions under which the affection is developed.—Naturally, the syphilitic poison is capable of inducing disease in any part of the hearing apparatus, but we also have the element of exciting causes to consider. For instance, a syphilitic subject takes on a simple catarrhal otitis. This often follows the law of a specific affection, and the patient may suddenly become permanently deaf. Again, a patient may have had an otitis from which he has become convalescent, when, on syphilis attacking the organism, a relapse may occur in consequence of the syphilis. The condition of the mouth may act as a predisposing cause to specific otitis—bad teeth, diseased gums from wearing badly fitting artificial teeth, the vulcanite plate, etc.

Treatment.—All middle-ear and throat symptoms may be treated according to the rules already laid down under the heading of Chronic Otitis Media, with the hope that the symptoms may not require specific treatment. As numerous adhesions are likely to form within the tympanum, it is well enough to try exhaustion, with alternate condensation of the air in the meatus externus, by means of Siéglè's tympanoscope or its equivalent, as practised by Pinckney, Sexton, Brandeis, and others, with a view to rupture them or break up any anchylosis of the ossicula. I have practised this method, however, with indifferent success, on the whole. Local means proving unsatisfactory, resort may be had to mercury and iodide of potass. The form of mercurialization most effectual is by inunction. The older methods may be employed, or the more modern one of oleate of mercury—a half drachm of a twenty-per-cent solution, rubbed daily into the skin, or any convenient part of the body. It is well not to delay the treatment too long. It should be commenced with considerable energy, as valuable time may be lost; avoid salivation if possible. The late Dr. F. J. Bumstead often used the blue pill, sometimes continued for months together, without ptyalism, and with great success. In children, inunction may be used, but the hyd. c. creta, given internally, will be perhaps a better method. It is often desirable to give a tonic—iron or some of the bitter tonics—while the inunction is being applied. Some administer the iodide of potassium during the inunction. I prefer the use of the mercury for from one to several weeks, then alternate with the potass. iodid. When the specific symptoms verge upon the tertiary, and sometimes in the secondary stage, the iodide of potass will accomplish wonders, provided it is exhibited in sufficiently large doses. The neurologists have made innovations in this respect. I have seen two hundred grains of the iodide given daily, at the Manhattan Eye and Ear Hospital, to a child of six, for symptoms of basilar meningitis. It was perfectly well borne. Roosa (l. c.) reports cases of specific otitis, where the amount of the iodide administered daily reached the enormous quantity of three hundred and sixty-nine grains. He also observed that several cases showed improvement only when the larger doses had been administered. It usually seemed to be well borne.

It is good practice, as recommended by A. Seessel, M.D., of the Manhattan Eye and Ear Hospital, to take the iodide well diluted with water during the meals and in divided portions, as in this manner it may be better digested. Hyperæmia and opacities of the membrana tympani sometimes disappear with great rapidity under proper constitutional treatment. Mercurial baths often succeed well in those who are too feeble to bear treatment by other means. As long as the hearing improves, the treatment should be continued. In any event, if there is good reason to believe that syphilis is at the bottom of the trouble, a thorough course of mercury and iodide of potass. is justifiable, although there may be no improvement in the hearing for weeks together. All possible means for supporting the patient's strength should be resorted to. In accordance with the general principle, when there is a tendency to destruction of tissue, iodide of potass., tonics, improved diet, and perhaps stimulants, may be recommended.

The paralytic symptoms may be treated by electricity. On the whole, Galvanism is more likely to benefit than Faradism. I do not believe that electricity will benefit the hearing, and therefore do not recommend it for that purpose. In accordance with the general principle of maintaining the patient's health, it may sometimes be proper to send him to a more congenial climate, giving him the benefit of change of air and scene.

AFFECTIONS OF THE EAR DEPENDENT ON HEREDITARY SYPHILIS.

For the most part these are " mixed " cases, although the labyrinth is more frequently involved than the middle ear. Ordinarily the affection follows that of a parenchymatous keratitis or kerato-iritis in a subject of syphilitic parentage. The patient usually, although not always, shows the notched and ill-developed teeth described by Hutchinson as characteristic of hereditary syphilis. The ear affection does not occur nearly as frequently as the eye lesion, there being six or seven eye cases to one of the ear. Ordinarily the ear symptoms come on at about the time the eye is convalescent or soon after, although it may precede the eye symptoms, or there may have been none of the latter, or the eye symptoms may alternate with those of the ear. An important characteristic of the affection is the more or less sudden loss of hearing—much more so than could be accounted for had the patient only middle-ear disease. The hearing may be mostly abolished in one day, or some months may elapse before the patient reaches his extreme point of hardness of hearing. Girls are more subject to this disease than boys, in the proportion of three or four to one. In boys a period somewhat after the age of puberty is the one in which the attack is more likely to occur. In girls the attack more frequently occurs at puberty. In Kipp's six cases (*Tr. Am. Otol. Soc.*, 1880), the ages ranged from six to twenty-three years. Where the hearing is restored it is at a period subsequent to that of the convalescence of the accompanying eye affection. At the commencement of the attack, vertigo, nausea, vomiting, unsteadiness of gait, frontal or occipital headache, possibly sleeplessness, deafness, as previously stated, tinnitus, although the latter is not constant, absence of pain about the ear, and in some instances head symptoms occur. The ear symptoms may vary with the progress of the eye affection—that is, may be more declared when the eye affection is progressing favorably, or may show improvement during a relapse of the eye trouble. Sooner or later both ears are likely to be attacked, whether simultaneously or one following the other after a brief period, does not exactly appear where there are middle-ear complications. Naturally the symptoms of that affection will be added to those already enumerated, although it may be remembered that vertigo, unsteadiness of gait, nausea, vomiting, and pseudo head symptoms may result from middle-ear affections alone. As a rule, the Eustachian tubes are found to be pervious, and little or no improvement to hearing results from inflation. Bone conduction was found by Kipp (loc. cit.) to be unaccountably good ; but he concluded that as his patients were not of a high order of intelligence, that they often mistook the feeling of vibration of the tuning-fork for actual hearing. I have often been able to make this differentiation myself. I conclude that the suddenness of the loss of hearing, especially when there. is absence of middle-ear symptoms, amounts to proof of labyrinthine or nervous involvement. Another point which Kipp calls attention to is the fact that even when middle-ear trouble is present the restoration of the hearing is delayed long after complete recovery of the middle-ear disease ; whereas, if it were only the latter which has caused the deafness the restoration of the hearing would occur before all of the middle-ear symptoms had subsided. According to Roosa (Text Book on Diseases of the Ear), if the labyrinth is the seat of the lesion the patient will hear the upper tones of a piano at some period of his disease badly, while the lower tones are more likely to he heard well. Without further argument, I conclude that the lesion on which much or most of the deafness depends is principally in the acoustic nerve or labyrinth. I am aware of the great difficulty of determining in a mixed case how far nervous involvement has taken place, and am ready to admit that there is much reasonableness in the conclusions of such authors as Sexton and C. H. Burnette, that middle-ear trouble is often the sole lesion in the deafness of syphilis. Under the heading " On the Use of the Tuning-Fork" will be found a brief discussion of intermittent bone conduction, from which it will be seen that some of the cases reported as labyrinthine or nervous deafness have not been recorded with perhaps sufficient carefulness to eliminate this source of fallacy. I am inclined to think that with more exact observation a greater number of middle-ear causes of deafness and fewer of labyrinthine in this class of cases will be found. Naso-pharyngeal catarrh very frequently accompanies this form of ear disease, and is undoubtedly in many instances a consequence of the constitutional dyscrasia. There is evidence enough to show that this catarrh often passes into the tympanum and from thence to the labyrinth, through the oval and round windows. As far as I am able to observe, this catarrhal process differs very little from that dependent on other causes, except that in the labyrinth there seems to be a stronger tendency to hyperæmia and proliferation, and that in the pharynx the tendency to destruction of parts is greater than in other forms of catarrh. Witness the frequent perforation of the uvula and deep ulcerations often found elsewhere in the upper pharyngeal space; purulent inflammations of the tympanum have been noticed by Schwartze and others as a characteristic of hereditary syphilis, but these seem to make a good recovery.

The Prognosis is exceedingly unfavorable, many cases going on from one relapse to another until almost or quite total deafness results, if indeed the patient does not become hopelessly deaf in a few days from the commencement of the attack. A few cases taken early have completely recovered. It is thought by some (Knapp) that those in the higher walks of life, with ample facilities for carrying out treatment, do much better than others. This from recorded cases does not seem to have been true in a large sense. Whatever treatment is adopted, most patients will go on to hopeless deafness. The best to be hoped for is that the affection may be principally of the middle-ear, when a more favorable prognosis can be made. As this matter will often be quite difficult to determine with positiveness, a hopeful prognosis may be made and treatment persevered in for some months.

Treatment may be directed to any middle-ear symptoms which may be present, not forgetting that many authors (Sexton, Burnette, Buck, and others) claim that thickening of the lining of the tympanum, causing the stapes to become fixed in the oval window and the membrane of the round window thickened, besides other changes which seriously interfere with free vibrations and consequently simulate nervous affections, may depend on the specific character of the inflammation. These symptoms may often be relieved by the sorbefacient power of mercury, so that the middle-ear symptoms may require constitutional treatment as well as the nervous symptoms. The principal hope of relief to the internal ear will be from the use of mercury and iodide of potass. Dr. Kipp (l. c.), however, does not believe in the use of mercury. My own view of the matter is, to begin by placing the patient under the influence of mercury, in the shortest space of time and in the most thorough manner, not, however, depressing the patient by ptyalism. After this the use of the iodide of potass. in large doses, increasing the size of the dose, if no amelioration of symptoms appear. Neurologists have shown that a basilar meningitis that has resisted ten or twenty gr. doses of the iodide of potassium three times a day often improve on reaching fifty or sixty grs. To obviate the tendency to disturbance of the stomach, the dose may be administered with the meals, largely diluted with water, and taken in divided portions as wine would be drunk under similar circumstances. At all times tonics will be indicated, and may even be given during the inunction treatment. Tröltsch recommends the iodide of ammonium in his text-book on the ear (Dr. Roosa's translation). Iodide of iron has long been a favorite remedy. Strychnia and electricity have been used extensively, but without effect. Most of the recent text-books on the ear treat of this subject with more or less fulness.

THE DEAFNESS OF BOILER-MAKERS, SHIP-CALKERS AND OTHERS WHO ARE EXPOSED TO SIMILAR INFLUENCES.

This affection has long been regarded as principally belonging to the internal ear, although latterly a feeling seems to be creeping into the minds of aural surgeons that middle ear disease may exist as an important complication. Roosa, "Diseases of the Ear," fourth edition, p. 509, says, "The impairment of hearing is generally attributed to some lesion of the labyrinth, probably of the cochlea; for the chief symptoms are loss of hearing and tinnitus aurium; there is no vertigo or staggering in the gait. Superadded to this serious trouble, tympanic or middle ear catarrh is very frequently present, but these must be regarded as purely co incidental. Boiler-makers are constantly exposed to sudden and marked changes of temperature, and hence often catch cold, intensifying and increasing by this means the aural affection." Buck, "Diagnosis and Treatment of the Diseases of the Ear," p. 403, regards this affection as in a general way labyrinthine, but in some cases he has examined he has inferred from the thickened and toughened appearance of the membrana that the deafness was owing, in part at least, to the great changes in temperature the workmen were exposed to. He adds, however, that the condition is more likely dependent on the prolonged irritation consequent on the concussion of the air upon the drum membrane. Burnett, in his "Treatise on the Ear," p. 403, says: "Boiler-makers' and telegraph operators' deafness may be due as much to nervous exhaustion from continuous shock as from catarrhal disease, but the latter is generally found to have a part in the train of symptoms." In the Tr. Amer. Otol. Society, 1882, p. 34, Dr. Holt, of Portland, Maine, gives his experience in the examination of forty boiler-makers, and concludes that they had middle-ear disease to the exclusion of labyrinth affection, but admitted to me in a conversation that labyrinth complications might ultimately occur. He observed that if the workman was right-handed the left ear would be worse; on the contrary, if he were left-handed the right ear was the worse. The same observation was made of ship-calkers. He concluded that the violent vibration of the middle ear mechanism induced by the noise of the hammering was sufficient to excite inflammation of the tympanum. Inasmuch as these workmen were exposed to great varieties of inclement weather and sudden changes in temperature while in a perspiring condition, the purely catarrhal influence was an important agent in producing the inflammation of the middle ear. The evidence of its being principally middle ear trouble was as follows: Bone conduction was never in a single instance destroyed; the tuning fork when placed on the central portion of the head or on the teeth was always heard better in the worse ear; the discrepancy between the length of time the tuning fork was heard in the air and on the teeth was less than in a healthy ear, that is, hearing by bone conduction was comparatively better than that by aërial conduction; when both ears were stopped, the tuning fork on the teeth was heard about equally on either side and for about the same length of time as in the case of ears of perfect hearing. No observation was made of the exceptional condition of intermittent bone conduction. In nearly every instance there was catarrh of the upper pharyngeal space, but the Eustachian tubes were almost always pervious to Valsalva's method of inflation. Holt concluded that it was only a matter of time as to whether a boiler-maker would ultimately become deaf, especially if he had any tendency to catarrhal otitis when entering upon the work. Roosa (l. c.), p. 510, concludes that those who work inside of boilers as riveters lose their hearing completely,

As to the *management* of such cases, it seems that insurmountable difficulties stand in the way of affording any considerable relief. The first indication would seem to be to stop the ears so as to diminish the intensity of the sonorous impression. Holt has tried filling the ears with cotton and found it to be inoperative, for the reason that it makes them tender after a while, and in some cases caused excoriations in the meatus. He has also tried ear pads and vulcanite ear stoppers, but without success. As the contact of air to the hearing apparatus seems essential, would it not be well to wear covers for the ears made of fine wire gauze on the same principle as in its application to hearing trumpets where it is desirable to diminish harsh and discordant vibrations? It is well-nigh impossible to make any suggestion to prevent the ever operating catarrhal influence which seems so potent a factor in developing the affection. The labyrinthine disease sooner or later following the middle ear affection is mostly secondary to that and does not respond to treatment. If the patient is removed from the noise and din of the shop before he has become profoundly deaf, recovery may take place in some cases, and in others the disease may remain stationary.

DISEASE OF THE LABYRINTH DEPENDENT ON MIDDLE-EAR AFFECTIONS.

In acute inflammation of the tympanum there may be a simultaneous hyperæmia of the labyrinth or it may occur some days after the tympanal affection has made its appearance. Although it is still a matter of dispute as to whether any considerable number of tympanal blood-vessels supply the labyrinth, there is a sufficiently close anastamosis of labyrinthine and tympanal vessels to afford a ready mode of transmitting inflammatory processes from the tympanum to the labyrinth.

"A serous saturation of the labyrinth" may result from hyperæmia of this region (Politzer on "Diseases of the Ear," English translation, p. 250). In such cases various symptoms pointing to disturbance of function in the labyrinth may be noted, together with weakened or abolished bone conduction. The ear may become extremely sensitive to all sorts of sounds; there may be vertigo, unsteadiness of gait, nausea, vomiting, and even convulsions, and yet the patient may make a good recovery. Some years since I had a patient with acute inflammation of the tympanum, in which the hearing was nearly *nil* for the voice, and bone conduction was almost obliterated, showing undoubted evidence of labyrinthine trouble. After two months, however, the hearing was almost completely restored. Besides this serous saturation of the labyrinth, hæmorrhages may occur, according to Moos, *Archiv. Ophth. and Otol.*, Vol. III., p. 118 (Burnett's reference), when the most undoubted signs of labyrinthine disease will appear, followed by partial or even complete loss of hearing.

Toynbee, "Diseases of the Ear," English edition, p. 356, reports a case where total deafness resulted from a middle ear catarrh contracted from sleeping in the open air. It was as follows: A farm laborer, æt 28, eighteen months previous to his call on the doctor slept in an open cart while riding during the winter months. Intense pain came on between the right temple and ear, which was relieved by veratrine ointment. About three weeks after this attack, deafness made its appearance, at first only for a day or two, when it would disappear, but again return. This state continued for a few days, when he became totally and permanently deaf. He complained of loud noises in the head and great heaviness and sleepiness; some relief to the head symptoms occurred after the drum membranes ruptured and the discharge appeared.

In *purulent inflammation* of the middle ear in the acute form, where the tympanum becomes completely filled with inflammatory products, with considerable swelling of the membranes of the labyrinthine fenestræ the consequent pressure upon the labyrinth may occasionally abolish bone conduction and produce great disturbance in the functions of the internal ear. It is, however, somewhat temporary unless the inflammatory process should reach a high degree of intensity, when disorganization of the labyrinth and permanent mischief may result. Sometimes the disturbance of function takes on the form of double or false hearing, as in a case reported by Roosa, "Diseases of the Ear," fourth edition, p. 492. P. A. S., æt 25, a pianist, had an acute suppuration of the right middle ear. In addition to the usual symptoms he noticed that in sounding a note on a musical instrument that the proper tone was heard by the affected ear and simultaneously a tone a half note above was also heard, which, however, soon died out, leaving the true tone distinctly audible. This symptom lasted about three weeks.

In chronic purulent disease of the middle ear the pressure from the collected pus may disturb the functions of the labyrinth sufficiently to cause a roaring in the ear, nausea, giddiness, faintness, dimness of vision, etc., as in a case reported by Burnett in his "Treatise on the Ear," p. 509. These symptoms, however, disappeared in a few minutes. According to Tröltsch "Diseases of the Ear in Children," English translation by Green, should the pus or inflammation in the tympanum break through the delicate structures which close the oval and round fenestrae, a direct passage is formed into one of these labyrinthine cavities, and the dura mater is then only separated from the disease by a thin perforated lamella of bone through which the minutest fibres of the nervus acousticus are distributed to the structure of the cochlea and vestibule. On page 143 Schwartze is quoted as stating that even in the highest degree of inflammation of the tympanum it is only exceptionally that a simultaneous hyperæmia of the labyrinth occurs. According to Schwartze, "The Pathological Anatomy of the Ear," English translation, secondary inflammation of the labyrinth is found with middle ear disease more frequently in the purulent form, with or without caries, the labyrinth cavity being filled with pus and the membranous structures destroyed. The parts are involved through perforation of the labyrinthine fenestræ or by a fistula in the labyrinth wall of the tympanum, or even without the existence of any direct communication with the tympanum. The polypi which so frequently accompany a chronic purulent inflammation of the tympanum may do serious harm to the labyrinth by pressure. When the membrane of the round window is pressed upon, there may be loss of power to hear the upper notes of the piano, the fibres of Corti which vibrate in unison with such notes being located according to Helmholtz, in the cochlea, near the round window.

In scarlet-fever the middle-ear inflammation so often accompanying the fever, as is well known, may be of the most serious nature, the labyrinth often becoming involved secondarily to the middle-ear disease. Toynbee (l. c.), p. 395, enumerates 98 cases of acquired deaf-mutism where scarlet-fever was the cause in thirty-six instances.

It is not necessary to affirm here that in most cases of acquired deaf-mutism that the labyrinth is diseased. Ordinarily we have the suppuration form of inflammation of the tympanum to deal with in scarlatina, but in a case reported by Toynbee (l. c.), p. 361, it would seem that total deafness may result without suppuration of the tympanum. The case was that of a man æt. 20. He had an attack of scarlet-fever at four years of age; since that time he has been totally deaf in his right ear. The membranæ were observed to be only a little more opaque than natural. Toynbee states on p. 399 that "It is clear, therefore, that in a majority of cases of deafness from scarlet-fever, the effects of very active disease were apparent; this disease was usually catarrhal inflammation of the tympanic mucous membrane, ending in an ulceration of that membrane, which extended to the labyrinth." Field ("Diseases of the Ear," p. 249), refers to a girl æt. 10 years who was totally deaf from an attack of scarlet-fever when she was two years of age. On p. 309 he speaks of scarlet-fever as a cause of deaf-mutism, and quotes Hartman to the same effect. Tröltsch (l. c.), p. 158, refers to scarlet-fever as a cause of deaf-mutism from labyrinthine complications. Schwartze (l. c.), p. 161, quotes Moss to the effect that the otitis media of several varieties of fever, including scarlet-fever, when of the milder varieties, may not go on to suppuration, but may result in a small cell infiltration of lymphoid corpuscles in the membranous labyrinth.

Labyrinthine disease resulting from the several varieties of *chronic inflammation of the tympanum* may be developed in a variety of modes. Perhaps one of the most persistent and frequent cause of labyrinthine trouble is the long-continued pressure exerted on the labyrinth waters, as a consequence of the middle-ear disease. This pressure may induce anæmia of the labyrinth, which if continued for a sufficient time will result in nutritive changes, of the nature of atrophy.

The mechanism of the pressure is somewhat complex; the most frequent condition perhaps is the impaction of the foot-plate of the stapes in the oval window from closure of the Eustachian tube and collapsed drum membrane. Cicatricial contraction of adhesions within the tympanum may draw the membrana inward. Thickening of the membrane of the round window may cause pressure on the labyrinth. At the base of the stapes ossific proliferations are often found causing the area of the vestibule to be encroached upon, and which results in pressure upon the labyrinth. Politzer states (l. c.), p. 246, "the progress of the deafness (in adhesive affection of the middle ear) depends, therefore, principally upon retrogressive alterations (contraction, calcification) which are taking place slowly or quickly in the middle ear, and on the participation of the labyrinth sooner or later in the diseased process." This deafness may, however, only in exceptional cases be absolute.

Hinton, in "The Questions of Aural Surgery," p. 235, quotes Gruber to this effect: "Gruber lays great stress upon the frequency with which labyrinthine affections are secondary to tympanic ones, and observes that the secondary nerve affections may remain though the tympanic affection disappears." Hinton makes out a very convincing argument to prove the tympanal causation of many labyrinthine affections, as in a case on p. 259. B. A., æt. 42, had signs of chronic middle-ear disease. On Nov. 7, 1871, there was the usual tuning-fork observation made, which characterized middle-ear disease only. On the third of the following June, the tuning-fork, instead of being heard better on the worse side, was heard better on the better side. In November he suddenly fell ill, had nausea and vertigo, but was better on the following day. He subsequently had several similar attacks, each of which lasted about half an hour. Hinton thinks the vertigo, etc., indicated the period at which the labyrinth first became involved.

Inasmuch as all these symptoms may result from tympanal disease alone, the statement does not quite amount to proof. Hinton also concludes, from a series of observations he had made, that nervous deafness in after life often depends on possibly slight and unregarded ear affections during childhood or youth. He quotes the case of G. T., æt. 42, who had gradually lost his hearing at fourteen. Had earache as a child; hears some kind of sounds better than others. Bone conduction very much diminished. He cites the case as an example of labyrinthine disease, depending on tympanald isease in childhood. He reports another case bearing upon the same point. Miss F. S., æt. 26. At five years of age she had a severe earache in the right ear, which produced almost total deafness. The clicking of the nails could only be heard close to the ear, the voice scarcely understood at all; bone conduction gone, even when the meatus was closed. He could find no cause for the deafness besides the tympanal trouble in childhood.

Dalby, in "Diseases and Injuries of the Ear," p. 204, speaks of a chronic aural catarrh, which in a given case has somewhat lowered the hearing. If there is no catarrhal relapse he thinks the hearing should remain stationary, but "on the accession of some additional circumstance, which does not reawaken the catarrh, but without doing this, or in any way interfering with the conduction of sound to the labyrinth, induces a decrease in the hearing power." In such cases he concludes that the labyrinth has become involved secondarily, and in consequence of the tympanal disease.

Tröltsch on the Ear, Roosa's translation, p. 508, seems to hold the opinion that in a case of chronic middle-ear disease, with "complete immobility of the ossicula, or calcification of the fenestra rotunda, that an atrophy of the acoustic expansion with retrogressive metamorphosis of the fatty or colloid degeneration may occur, as a consequence of the deficient specific excitation of the nerve." He also agrees with propositions previously laid down, regarding the effect of calcified membranæ on the integrity of the labyrinth.

Schwartze (l. c.), p. 161, also fully agrees with this proposition, although he states that "the anatomical proofs of this are, however, as yet very few." I have for years noticed in old cases of chronic aural catarrh, in which the tympanal symptoms showed no signs of activity, that the hearing would gradually lower in an unaccountable manner, and the bone conduction become weakened. I have no doubt but that a considerable number of cases of old otitis, with gradual but certain loss of hearing, must be recorded as "mixed cases," in which the labyrinth trouble is secondary to and consequent upon the middle-ear affection. I believe it is held by most aural surgeons that the deafness of boiler-makers, and other affections similar to these, depend on a middle followed by an internal ear affection. A considerable number of ear diseases, dependent on various fevers, are first a tympanal affection, which subsequently extends to the labyrinth. In all elderly or old subjects there is a tendency to atrophic changes in the acousticus. Toynbee (l. c.) found atrophy of the acoustic nerve thirteen times in subjects between sixty and ninety years of age. Politzer states, however (l. c., p. 175), that patients of fifty years of age or older often have weak bone conduction, without necessarily having disease of the ear.

On the *diagnosis* of labyrinthine complications.—If one ear only is affected, there is usually little or no difficulty in determining as to whether the functions of the labyrinth are disturbed. Naturally, the tuning-fork placed on any central point of the head, or on the upper central incisors, will be heard better on the affected side if there is no nerve trouble. The tuning-fork, by aërial conduction, will usually be heard longer than by bone conduction, but not as long as in a healthy ear. Closing the ear will increase the distinctness and duration of the bone conduction. It is true that in some portions of the central line of the skull the tuning-fork may not be heard well (Politzer, 1. c.), but if it is heard in any portion, it is enough for our purpose.

In a decided case of labyrinthine deafness there is no difficulty—the tuning-fork will be heard badly or not at all in any part of the central line of the skull or incisor teeth, and if there is only a moderate middle-ear deafness remaining, the aërial conduction will be more distinct and prolonged than by bone conduction. Stopping the affected ear will in all cases increase the bone conduction, as well as add to the duration of the impression, provided there be any bone conduction in the beginning.

Great care needs to be exercised to discriminate between feeling the vibrations and actually hearing them. An unintelligent patient will not respond to the finer tests. In a doubtful case the patient is uncertain as to his ability to hear better on one side; the tuning-fork may be applied to the canine teeth, or even the molars, or to any point on the opposite side of the head, or on the mastoid process. If he still insists that the opposite ear hears the tuning-fork the best, there is no longer any doubt.

Again, if both ears are stopped by the fingers (not too tightly) and both nerves are in a healthy condition there will be very little difference in the duration of the hearing in each ear, provided that from the affected ear is eliminated the exceptions of intermittent bone conduction.

If the bone conduction is weak in a given case, and the aërial conductions are considerably stronger, evidence accumulates in favor of nervous trouble. It is needless to observe that the ear should be placed, by inflation and other means, in such a condition that the best hearing for the watch, or click of the finger-nails, and voice, may be obtained previous to the tuning-fork test.

A large-sized tuning-fork with clamps should be selected, as in feeble nerve reaction no impression would be received from a small one.

By pushing the clamps towards the opposite extremity of the instrument, a great variety of tones may be produced, and if any fibres of Corti have become unresponsive, this test would elicit that fact. If any exhaustive test be applied, the whole range of a piano scale would be required to be called into requisition. Where both labyrinths are involved the differentiation method of testing will not yield results (comparing one with the other).

The first proposition will be to determine how much of the affection is middle ear and how much labyrinthine. This is done by noting the difference between the length of time and distinctness of the bone conduction, as compared with the aërial. As has been before stated, if the aërial conduction bears the same or similar ratio to the bone conduction as in the normal ear, it is fair to infer that the trouble is mostly labyrinthine. The duration of the hearing through the bones, as compared to that in the normal ear, will give a hint as to the amount of impairment of function. Then the principal problem will be, first, to determine how much the deafness depends on middle-ear disease and how much upon labyrinth disease. Second, to find whether few or many of Corti's fibres have been injured by the disease, by noting the degree of perfection with which different tones are perceived. The diagnosis of labyrinthine disease by other signs is not as certain or valuable as with the tuning-fork.

Vertigo, nausea, vomiting, unsteadiness of gait, etc., being sometimes seen in middle-ear disease, are not of very great value as diagnostics; neither is sudden and profound deafness, as it somewhat infrequently results in this class of cases. Improvement of the hearing distance by inflation of the tympanum is ordinarily diagnostic of a certain amount at least of middle-ear disease, but it does not exclude labyrinthine disease, as Politzer has stated (l. c., p. 350) that some slight increase or distinctness of the hearing after inflation often occurs in pure labyrinthine disease. Sometimes by stopping both ears with a hearing trumpet, and comparing the bone conduction with the aërial conduction in both ears, (the tuning-fork being held in the latter instance before the funnel of the trumpet) an exact knowledge of the condition of the hearing may be obtained. Naturally, by closing both ears with the hearing trumpet, the bone conduction in each ear is approximated to a similar condition, unless there is considerably more labyrinthine disease in one ear than in the other.

DOUBLE HEARING PARACUSIS DUPLICATA (MOOS).

This may be binauricular or monauricular. Knapp (Arch. Otol.,vol. i.) defines the affection as diplacusis. The binauricular variety consists in the hearing of a tone normally with one ear, while in the other it is heard one or several tones higher or lower. In diplacusis monauricularis the double hearing is confined to the affected ear. Thus, as in the person of Von Gumpert, quoted by Turnbull from Moos' article in the Klinik der Ohrenkrankheit, Wien, 1866, Von G. could distinguish the true as well as the false tone in the affected ear. The true tone appeared close to him, while the pseudo tone seemed several yards away. This was sometimes a third, then a fourth, and finally an octave above the normal tone. This double hearing also existed for words. Von Wittich's famous case in his own person, from the Königsberg Med. Chronicle, vol. iii. (Turnbull's quotation), illustrates the condition of binauricular double hearing. A tuning-fork placed upon the glabella sounded a half tone *higher* in the affected ear; this was the case, however, with moderately high tones only. Knapp's case (l. c.) differed from this in the fact that the diseased ear heard the tone *lower*, instead of higher, as in Wittich's case. The explanation of this double hearing is best made by Knapp (l. c.),although it has some similarity to Moos' explanation. It is this: it depends upon the faulty tuning of some of the fibres of Corti's organ. In certain cases the fibres are too tense, vibrating in unison with a tone above that received in the normal state, while in other instances they are relaxed, so that a number of vibrations fewer than the normal suffice to induce sonorous activity. Consequently, when the fibre is tuned to a note above its normal pitch it conveys to the acoustic centre its own proper note, and the sound perceived will be lower than that of its normal fellow. To illustrate: Suppose a sound of 350 vibrations per second is perceived by both ears. The normal ear having a fibre which co-vibrates with a sound produced by 350 vibrations per second, this impression will be carried to the acoustic centre. But the fellow ear has a fibre of 300 normal vibrations per second, and from its morbid tension it is tuned to 350 vibrations, so that when the 350 vibrations reach each cochlea the healthy ear carries the impression of these to the sensorium, while the opposite ear only carries the impression of 300 vibrations, and consequently hears the sound just so much lower than its fellow. When there is relaxation of the fibres of Corti an opposite condition exists; that is, the diseased ear would perceive sounds of a higher pitch than normal. The result of this faulty tuning is to make all musical sounds discordant, and very annoying to a trained musical ear.

There are still other forms of double or false hearing of the monauricular variety. The patient finds that the affected ear at first hears a note at its normal pitch, but directly there appears a second sound like an echo, but with no perceptible alteration in its pitch.

Double hearing is *caused* by some pathological changes in the cochlea. No autopsies have revealed what that exact condition is.

Undoubtedly there is frequently intra-labyrinthine pressure, which in a majority of cases is secondary to middle-ear disease. In Wittich's case the double hearing came on four weeks after an attack of purulent otitis media, and remained unchanged in pitch until it disappeared altogether. In Knapp's case the double hearing resulted from a suppurative inflammation of the tympanum caused by the use of cold water by means of a nasal douche. Three days after the discharge set in the double hearing was noticed. For one week after, the pitch of the tone heard in the affected ear remained unchanged; but subsequently, and until the end of the third week, the pitch was scarcely half a tone different from the sound ear, whereas previously it differed by two tones.

In one of Moos' cases the double hearing seemed to have depended for its exciting cause on the administration of chloroform, although the patient at the time had a chronic middle-ear catarrh.

Glauert—Berlin. Klin. Wochenschr., No. 48, 1881 (Arch. Otol., vol. xi.)—reports a case of monauricular diplacusis in which the affection resulted from an acute exudative middle-ear disease. All tones from middle "D" up were heard double. This disappeared when the ear affection was cured. In the Arch. Otol., vol. xi., Moos reports a case, translated by Spalding, of double hearing during the exhibition of iodide of potassium. The patient was a man 40 years of age, who had a nervous asthma but no ear disease. For this gr. x. of potass. iodid. were administered daily for about ten days, when, some unpleasant symptoms of iodism resulting, the drug was discontinued.

The double hearing was only for the lower "D" in the treble clef, downward to the distance of five notes. This continued for three days, the iodide having been discontinued. There was an iodide exanthema over the shoulders and on the extremities, with violent sneezing and excessive lachrymation. Moos concludes that a process similar to that on the skin was going on in the cochlea, and that minute petechiæ on the zona pectinata caused an abnormal tension which resulted in the double hearing. Little need be said regarding *treatment*. Finding the *cause* of the trouble furnishes sufficient suggestion for treatment. It may be worthy of note that diplacusis is rarely or never diagnosticated in a patient who has not a good musical ear. So far, the affection has not been observed in any who could not lay a reasonable claim to exactness in appreciating musical tones.

DISEASES OF THE ACOUSTICUS AND LABYRINTH DEPENDENT ON AFFECTIONS OF THE BRAIN AND ITS MENINGES.

In hemorrhagic pachymeningitis the nerve becomes involved at the inner meatus, the inflammation extending to the internal ear; and purulent formations, hemorrhages, organized and serous exudations, with pigmentary, fatty, and colloid degenerations, occur within the labyrinth which are very likely to destroy the hearing. Atrophy of the nervous expansion within the labyrinth almost inevitably results, unless the parts become wholly destroyed by the purulent process. A case reported by Moos in the Arch. Otol., vol. ix., well illustrates the subject. It was that of a man, æt. 49, who died in the Marburg Insane Asylum of a hemorrhagic pachymeningitis, and who, after repeated so-called paralytic attacks, inside of a year's time had completely lost his hearing. The right petrous bone was found in the condition usual to purulent middle-ear catarrh, the left in that found in simple chronic catarrh. The alterations in the labyrinth consisted in microscopically demonstrable hemorrhages, with numerous transformations of the blood into pigment. As a consequence of the hemorrhages there was inflammation. This led in part to a hyperplasia of the connective-tissue elements, with here and there spots of fatty degeneration. Along with this was found a partial atrophy of the epithelial tissue and nerve, the latter being more highly atrophied, with abundant colloid formation. This atrophy was followed to the ganglion cells, and the trunk of the porus. acoust. internus on one side up to Rosenthal's canal, and on the other to the terminal nerve fibres of the crista of the ampullæ, etc. He showed also the connection between the alterations in the labyrinth and those in the intra-cranial process, presenting drawings exhibiting the similarity of the alterations in the vascular territory of the internal auditory artery, to those found in the region of the meningeal artery, as has also been described by Kremiansky and Rindfleisch.

In inflammations of the brain substance or of the meninges, at any point, there may be disturbances in the function of the labyrinth, as in any event there is likely to be pressure upon the sub-arachnoid space, and this, as is well known, communicates with the intra-labyrinthine fluids through the aqueducts of the vestibule and cochlea. Disease of the labyrinth, however, is more likely to result where the cerebral affections involve the base of the brain, whether in the immediate region of the acousticus, at the inner meatus or more posteriorly, where the nuclei of the auditory nerves are placed. A basilar-meningitis, with the formation of abscess, is very likely to involve the acousticus, causing infiltration of this nerve with pus cells. The acousticus, from its soft and comparatively loose texture, is peculiarly liable to inflammatory infiltration. In acute meningitis, according to Politzer (Text Book, Eng. translation), the deafness is perceived immediately after the return of consciousness, between the third and eighth week of the disease; or it is developed more or less rapidly after convalescence. Children are more likely to become quite deaf, with unsteadiness of gait. In adults there is less likelihood of profound deafness, but tinnitus is often a troublesome symptom. Deafness in this affection is mentioned somewhat infrequently by authors on general medicine or neurology—Watson, Hammond. There is usually no subsequent improvement in the hearing. Sometimes a previous affection of the tympanum is aggravated by an attack of meningitis; thus Roosa, in his Text-book on the Ear, reports the case of a man, æt. 27, who had congestion of the brain,

resulting in hemiplegia of the left side, from which he soon
recovered. The hearing of the right ear was *nil;* of the
left, $\frac{18}{48}$. Bone conduction, however, seemed better on the
right side. Another case from the same author was a man,
æt. 27, who had an attack of acute basilar meningitis.
After recovering consciousness the hearing was found to
be nearly destroyed. Hears the watch laid upon the right
ear, but not on the left. Tuning-fork heard indistinctly in
the right. There was a roaring noise in the left ear. Another case from the same author: H. S., a man, had acute
meningitis and inflammation of the internal ears. He became totally deaf in one ear, and nearly so in the other.
Tinnitus and giddiness; H. D. R. watch in contact, left *nil.*
Bone conduction only in right. Treated with bromide and
the iod. of potass. and the constant current. In about five
weeks the right ear was restored to normal hearing. The
other ear could, after a time, occasionally hear the ticking
of the watch. Tröltsch states, in his book on diseases of
the ear in children, translation by J. O. Greene, M.D., that
"hyperæmia of the labyrinth is, on the other hand, observed with hyperæmia and passive congestion within the
skull, with some of the general febrile diseases, such as
typhus and acute tuberculosis. . . . Under such circumstances, ecchymosis or even hemorrhage may occur in the
membranous tissue of the labyrinth."

A specific basilar meningitis, with or without gummata,
often enough involves the labyrinth, but this subject has
been considered under the topic of Syphilitic Affections of
the Ear. The same remark applies to brain affections, the
result of traumatism, which see under the head of Fractures
at the Base of the Brain, Concussions, etc. Deafness from
cerebro-spinal meningitis has already been spoken of.
Certain grave forms of fever, as is well known, result in
more or less hyperæmia and inflammation of the brain and
its meninges, this process occasionally extending to the
acoustic nerves or the labyrinth. The meningitis, which is
secondary to necrotic or carious bone, has been considered
under the head of mastoid affections and caries of the
petrous and squamous portions of the temporal bone. *In
tubercular basilar meningitis* the inner ear, or auditory nerve,
is sometimes involved, although affections of the optic
nerve are more likely to complicate this disease. Lucae,
quoted by Politzer, reports a case of a boy four years of
age, who had become totally deaf in the course of a tuber-

cular basilar meningitis. He had hemorrhagic inflammation in the semicircular canals and vestibules of both ears. Dr. Roosa (l. c.) reports a case as follows: Carrie X., æt. 4, had an attack of acute hydrocephalus. On recovery she was found to be deaf; the deafness still continued; there was no evidence of disease of the tympanum or pharynx.

Chronic hydrocephalus sometimes causes profound or total deafness, and the first thought arising is, that pressure on the brain and nerves, from the effusion, is the explanation of the loss of hearing. Many cases, however, where the head is enormously enlarged and translucent, show no defect of the hearing. Speaking of this affection, Watson, in his " Practice of Physic," third American edition, says: " But far more important effects of the disease are those which relate to the three great functions of the brain. The child is soon found to be deaf, or dumb, or palsied in one or more of his limbs; or idiotic, or all three."

Apoplexy, whether dependent on hæmorrhage into the brain substance or any of the cavities within the calvarium, or serous effusions (serous apoplexy), or softening of brain substance, or fibroid degeneration, is likely to result disastrously to the organ of hearing, but is by no means certain to do so. Naturally, if the apopletic lesion is near the acoustic nerve or its origin, it is not difficult to comprehend how extensive lesions of this nerve may result. Localized inflammations of brain substance may involve the acousticus. According to *Moos*, Politzer's quotation (l.c.), if the hæmorrhages occur in the pons and cerebellum the ear is more likely to suffer. According to Itard, Oppolzer, Andral v. Tröltsch and Nothnagel, subjective noises are often the forerunners of apoplexy.

Tumors of the brain have perhaps been sufficiently disposed of in another portion of this book. It may suffice to mention here, that any form of cerebral tumor is capable of destroying the hearing by involving the acousticus or its origin. If the tumor be malignant, it is quite possible for it to penetrate even the petrous portion of the temporal bone, as in a case of my own reported elsewhere, and totally destroy the hearing.

DEAFNESS DEPENDENT ON CEREBRO-SPINAL MENINGITIS.

This affection is usually regarded as dependent on a lesion situated in the labyrinth. It is developed in about ten or fifteen per cent of the cases of cerebro-spinal meningitis.

It makes its appearance early in the disease—from three to fourteen days, but in some cases as late as six months from the commencement of the meningitis. In the last instance, however, there may be evidence of the presence of a previous inflammation of the ear. The deafness may be profound in the outset, but often it is only moderate; but it is likely to increase to the degree of absolute deafness after a few weeks. There is little or no pain, but tinnitus, vertigo, vomiting, great sensitiveness to discordant sounds, etc., are quite frequent symptoms. The evidence of labyrinthine complications is reasonably good, but in some cases not conclusive. Buck, in his " Diagnosis and Treatment of Diseases of the Ear," reports a case from Moos of absolute deafness from cerebro-spinal meningitis, where the labyrinth was perfectly healthy, but the lining of the middle ear was intensely congested. It is not necessary to demonstrate, in this connection, that middle-ear trouble may, in some instances, give apparently all the signs of labyrinthine disease.

The condition of the brain does not seem to have any immediate connection with the ear affection, although many authors have concluded that an opposite statement was the correct one.

In a general way, if the ear symptoms were dependent on the cerebral trouble, evidence of involvement of neighboring nerves would exist, which is not the case: for instance, the portio-dura is unaffected, even with its intimate connection with the acousticus. There are middle-ear and throat complications. A grave form of suppurative otitis is not infrequently seen. Absence of middle-ear trouble, and the usual symptoms of disturbance of function, point undoubtedly to labyrinthine complication. Exactly what this condition is, is hardly known, as too few autopsies have so far been made to determine what changes have been produced. In some cases there is evidence of anæmia of the labyrinth, while in others hyperæmia seems to be present. Dr. Knapp, in the Tr. Amer. Otol. Soc. for 1873, made the autopsy of a patient who had become deaf from cerebro-spinal meningitis. No lesion was found, except that one acoustic nerve was softened by suppuration, while the facial was apparently normal. The other acoustic nerve had not suffered, but numerous pus cells were found in the inner auditory canal. Lucæ and Heller are quoted as having recorded autopsies where the affection under consideration exhibited a suppuration in the labyrinth. In the *Arch. Otol.*, Vol. I., N. Y., in a translation by Aub, appears the following statement from Moos: "In the pathological specimen examined by myself in connection with Prof. Luschka . . . consisting of the pons, cerebellum, medulla oblongata, and portion of the spinal column of J. Schwartz, we found the acousticus nerve, up to its exit from the skull, so completely imbedded in the mass of exudation, that Prof. Luschka felt justified in supposing that the inflammation and exudation following the course of the nerves might easily, in some cases, extend into the labyrinth, and thus produce deafness." This specimen enabled him to corroborate the fact that the exudation extended from the base of the brain, through the hiatus Majendie, into the fourth ventricle, and there principally covered the striæ acousticæ. "Adding to this the fact that the immersion of the acoustic nerve in exudative masses alone may produce deafness, we cannot be astonished at the frequency of this symptom in the epidemic cerebro-spinal meningitis."

The autopsy of Case I. of purulent cerebro-spinal meningitis is as follows: "Fresh hæmorrhagic and numerous encephalitic foci in the brain. Croupous pneumonia of the left side; splenic tumor; diffused swelling of the kidneys. In both tympanic cavities much pus. In the vestibules numerous pus cells, a large quantity in the ampullæ; the cochleæ very red, and filled with pus cells. The vessels of the membranous portion of the lamina spiralis much injected; its peripheral half filled with pus, the inner half of its surface less so. N. N. *Acoust.* and *facialis* of both sides surrounded by pus in the *meat. audit. int.* . . . very few pus-cells were found between the fibres of the facialis, while those of the acousticus and its ganglion cells were densely surrounded by them. In both nerves the fibres were well preserved, the vessels filled to bursting and their walls thickened." The next case was similar to this, with the addition of dotted ecchymoses in the peripheral portion of the lamina spiralis. He infers that purulent inflammation of the inner ear and tympanic cavity may accompany the disease in question. "We have to deal with a well-marked inflammation of the labyrinth. This may occur simultaneously with the changes in the meninges of the brain and spinal column, or, following the course of the neurilemma, it may advance into the labyrinth." It is a well-known fact that purulent choroiditis is often a consequence of cerebro-spinal meningitis. Dr. Knapp concluded (l.c.) that the anæmia of the labyrinth and inflammation of the tympanum found in these cases is recovered from, but that these are accidental complications, and that the affection which leads to permanent loss of hearing consists of "*a homogeneous extension of the original morbid process, namely, a suppurative inflammation in the labyrinth.*" He places it in the same category with purulent choroiditis and inflammation of the different joints so often found in cerebro-spinal meningitis. It is now a well-known fact, that the arachnoid cavity and that containing the peri-lymph and the endo-lymph are in free communication with each other, and it is easy to see how readily disease may extend through these channels of communication. The fact that the arachnoid is so frequently found affected in cerebrospinal meningitis, gives color to the theory of extension of the morbid process direct to the endo-lymph and perilymph cavities. Moos, in the *Arch. Otol.*, N. Y., 1881, No. 4, regards this affection as a neuritis descendens in a majority of cases, there being "a slow encroachment of the inflammation from the interior of the cranium into the labyrinth, along the peri-neural vessels of the auditory nerve." Prof. Lucæ, quoted by Burnett in his "Treatise on the Ear," found "the hemispheres, base of brain, pons and medulla affected by a purulent inflammation of the pia mater in a case of cerebro-spinal meningitis. The microscopic examination traced the purulent inflammation along the auditory nerve to the cochlea. Purulent inflammation of the sacculi, ampullæ, and canals of the membranous labyrinth was also found; along their vessels were masses of pus-cells and free blood-corpuscles; the vessels were intensely congested and much thickened; the semicircular canals also showed occasional ecchymoses. The tympanic cavities, except a slight injection, were normal The fibres of the facial nerve were subjected to microscopic examination, but were found to be normal. In the ampullæ and sacculi were here and there deposits of fat and chalk." Prof. Lucæ concluded that it was probable the disease began first in the brain and then passed to the ear. In the same article it is stated that Heller found, in a case presenting similar disorganizations in the labyrinth, purulent inflammation of the middle ears.

So-called Ménière's disease sometimes appears to be labyrinth disease dependent on cerebro-spinal meningitis of so slight a type as not always to be recognized. Thus, J. Gottstein is reported, in the *Am. Journ. Otol.*, Vol. III., No. 4, p. 329, as stating that in many of the cases of the neuropathic form of Ménière's disease the affection depends on cerebro-spinal meningitis, the meningitis being overlooked on account of its presenting so few noticeable symptoms. Many cases of the so-called otitis labyrinthica present the clinical picture of a mild cerebro-spinal meningitis, and the strongest point of all is, that they occur during an epidemic of that disease. He shows by statistics that in places and at the time when the cerebro-spinal meningitis was epidemic, so-called otitis labyrinthica was frequent, and as the epidemic diminished the cases of this aural disease diminished.

Buck, on Diseases of the Ear, speaks in a similar manner of an epidemic of cerebro-spinal meningitis in New York city and parts of Connecticut, where the disease was so slight as to be often overlooked, but ear complications were frequently noticed.

Prognosis.—There is great diversity of opinion as to the patient's liability to recover the hearing. Some affirm that nearly all who are attacked become hopelessly deaf, while others make the statement that a large number have the hearing partially or completely restored. In the *Arch. Otol.*, N. Y., Vol. I., Moos quotes Dr. Fluegel in the report of an epidemic of cerebro-spinal meningitis in the district of Naila, where among about 300 cases, five remained deaf, six hard of hearing, one deaf and blind, three deaf and unable to walk, and one blind in one eye. Dr. Bauer, quoted also by Moos (l. c.), reports 109 cases of cerebro-spinal meningitis. "Hallucinations of hearing were always present, complete deafness in seven cases of recovery, and in six of death." In the *Arch. Otol.*, N. Y., 1881, No. IV., in a translation by Dr. Spaulding, Moos found 59.3 per cent deaf-mute, 31.4 per cent deaf without deaf-mutism, 7.8 per cent permanently hard of hearing, while only one patient (1.5 per cent) escaped without any subsequent affection of the hearing. I have not seen much of this disease, but the above figures are more nearly in accordance with my experience than other less favorable statistics, except that the number of cases having ear complications is excessively large.

Where suppurative otitis media is present the patient may be profoundly deaf, but subsequently recovers. In such cases the suspicion will exist that there was no labyrinthine complication, which indeed, under such circumstances, cannot be certainly diagnosticated; where extensive purulent processes have gone on in the labyrinth, it certainly could not be expected that recovery of the hearing would take place.

Treatment, for the most part, is of little avail. Where a severe suppurative inflammation exists, it should be managed according to the rules already laid down, with the hope that there may be no labyrinthine disease present. In the absence of all middle-ear complications, treatment will very infrequently accomplish anything. If simple congestion or inflammation of the labyrinth is suspected, general remedies acting on the cerebral circulation may be administered, such as pot. brom., ergot, pot. iodid., etc. Later on, if products of inflammation exist which may be influenced by sorbefacient remedies, then mercury, iodide of potass., etc., may be administered. At a still later stage strychnia and galvanism may be used, with a view of stimulating diminished innervation of the labyrinth. It is well enough to persevere in treatment for a few months after the first attack, as a certain number of cases have unexpectedly improved. If, however, the deafness is complete from the first, and there is no change after one or several months, treatment becomes quite hopeless.

MÉNIÈRE'S DISEASE.

This is an affection of the labyrinth, generally secondary to diseases of the tympanum, the brain or the spinal cord, or it may be the result of traumatism. A few cases undoubtedly occur where no other lesion can be detected than that of the labyrinth. Ménière held the opinion that the affection was principally of the semicircular canals, apparently ignoring the fact of cochlear involvment, although the hearing was very frequently totally abolished, and in those cases where it was partially intact, there was inability to hear certain tones, usually the higher or lower tones or both, a fact pointing to impairment of some of the fibres of Corti, within the cochlea. This latter observation has been much dwelt upon by Knapp, in the *Arch. Opthalmology and Otology.*, Vol. II. Probably the grounds for Ménière's opinion were based upon the famous case of the young woman who caught cold while menstruating; suddenly she became deaf, with symptoms of extreme vertigo and frequent vomiting. She died in five days from the commencement of the attack. The autopsy showed a bloody, plastic-looking exudation into the semicircular canals, a trace of which also was found in the vestibule. No other tissue was diseased; the hearing was not carefully tested.

The symptoms of Ménière's disease are so numerous, and sometimes contradictory, as to be difficult of adequate and lucid statement. The more frequent manifestations are those simulating brain symptoms. The patient is suddenly attacked by a vertigo, with or without nausea, and vomiting, often falling in a faint, although he does not always lose consciousness; the face may be pale, and the skin bathed in perspiration, the symptoms often resembling sea-sickness. The attack may last only a few minutes, or a day or two. The vertigo, however, may continue with sufficient intensity to keep the patient in bed for several weeks. Pains in the back and head are complained of, with frequently a benumbed sensation about the latter. A sensation of heaviness and depression in the occipito-mastoid region was observed in one of Ménière's cases.

Tinnitus aurium of a very aggravated character is almost always complained of, and in some instances seems to precede the attack. Deafness may be absolute as a consequence of the first attack, including total abolition of bone conduction. In other instances, which are somewhat frequent, the hearing is lowered, still, however, retaining bone conduction, at least for tuning-forks of a medium pitch, but at each subsequent attack it becomes still further lowered, until ultimately it is likely to be nearly or quite destroyed, including bone conduction. Absence of the power to hear certain high or low tones has been observed frequently enough by a large number of surgeons, this being a well-known symptom of labyrinthine disease, where the cochlea has been invaded. The mind is always clear in the more typical forms of Ménière's disease, which serves to differentiate it from brain affections; moreover, there is no disease of cerebral or spinal nerves. Those cases complicated with brain or spinal disease, are scarcely to be received into the category of Ménière's disease. In falling, consequent on the attack, the patient sometimes fractures the skull at the base, or destroys the hearing by concussion, which places the subject rather in the category of traumatic diseases of the labyrinth, for an account of which, see Index. These cases, however, present many of the symptoms of Ménière's disease.

Cloudiness of vision, and limitation of the field from anæmia of the retina, has been noticed by several observers, as Ménière, Moos, Knapp, (l. c.) and others. In one case, Knapp noticed signs of atrophy of the optic nerve, and he raised the question as to whether the cause of the optic nerve trouble was not dependent on a cerebral condition, identical with that which might have developed the labyrinthine disease. Sometimes objects seem to waver before the vision (Ménière).

The vertigo in Ménière's disease is a most important symptom, being, of all the others, the most characteristic. The experiments of Flourens, Vulpian, Signol, Goltz, Brown-Séquard, and many others, have fairly settled the question of the agency of the semicircular canals in the production of the vertigo, at least of that form dependent on aural diseases. McBride, in the *Medical Times and Gazette*, January 22d and 29th, 1881, (*Arch. Otol.*, Vol. XI), claims that there is a cerebral centre, the excitation of which produces vertigo. "This centre can be excited by influences from the eye, ear, sensorial or visceral, as well as central alterations." Any source of irritation to the semi-circular canals, undoubtedly is likely to cause vertigo. It must not be forgotten that any influence which increases the intra-auricular pressure, is likely to thus irritate the semicircular canals. Nothing is better known than the agency of middle-ear disease in causing vertigo, through pressure on the labyrinth, and consequent irritation of the semicircular canals. Burnett, of Philadelphia, has written much on the vertigo, which has its origin in the condition of a morbid drum-cavity and membrana tympani. He relates a case quite at length, where a patient presented the symptoms of variable hearing and vertigo, dependent on pressure on the labyrinth, the result of collapsed membrana tympani, induced by a spasmodic condition of the tensor tympani. These middle-ear symptoms are often almost identical with those of Ménière's disease, but the consequences to the hearing are not as grave, or as sudden in their action, and the bone conduction is not as likely to be lowered. Physiologically speaking, however, the general disturbance to labyrinthine functions is very similar to that found in organic disease of the labyrinth. The nausea and vomiting sometimes found in middle-ear disease, may have a similar causation to that which occurs in labyrinthine disease; namely, irritation of the sympathetic. It seems hardly profitable in this place to discuss the question as to whether certain semicircular canals are diseased, as evinced by the peculiarity of the vertiginous movements; whether circular, the patient having a tendency to move about a circle, pointing to disturbance in the horizontal canal, or a tendency to move backwards and forwards, or actually to fall, in consequence of disease of the vertical or the oblique semicircular canal, or both of them.

This vertiginous movement is, next to the deafness, the most lasting symptom. According to Nave (Politzer's reference), the staggering gait has persisted in one instance for ten years.

It seems not to matter what the labyrinthine change may be, whether a hæmorrhage, an organized exudation, the consequences of traumatism, serous infiltration, hyperæmia, etc., as in any event, pressure and consequent irritation of the semicircular canals, with vertiginous movements, is likely to result. As is well known, the vertigo is aggravated by closing the eyes, attempting to walk, or to turn around, especially if the body has no support. Sometimes the symptom is so extreme that the patient cannot lie in bed without holding on to the sides of the bed, or supporting the head between the hands and looking at a fixed object. The vertiginous movement is generally in a direction away from the affected ear, although the opposite condition is sometimes seen. The tinnitus is very likely to last for the remainder of the patient's life, and is often a very distressing symptom. The deafness is sometimes recovered from in certain cases, notably the syphilitic, which indeed hardly belong to this category. In a typical case of Ménière's disease, that is, the primary labyrinthine affection, without involvement of other parts of the ear, or brain, where the hearing is once lost, it is rarely or never regained.

Diagnosis of Ménière's disease.—This is sometimes ex-

tremely difficult.   In the typical form of the disease, where
the patient has no dyscrasia of any kind, no affection of the
ears, but is suddenly attacked with some or all of the
Ménière symptoms, as vertigo, nausea, faintness, etc., with
considerable or complete deafness, accompanied by defective or abolished bone conduction, the diagnosis of
Ménière's disease is easily made.   In old cases of middle-
ear affection, in which the tubes are pervious and the middle-ear inflammation inactive, with moderate deafness, if
the patient be suddenly attacked with the Ménière symptoms, grow harder of hearing, with diminished or lost
bone conduction, it is reasonably certain that there has
been a sudden exudation into the labyrinth, without any
connection with the old middle-ear trouble.   The symptoms of vertigo, nausea, etc., being often indicative of
middle-ear trouble, the greatest care needs to be exercised
to eliminate this source of error in making a diagnosis.
Where the peculiar Ménière symptoms strongly resemble
those dependent on brain and spinal-cord disease, it will
be sufficient to investigate the latter organs, and evidence
of disease in them will probably show that the ear is unaffected.   In this instance we may, however, have a somewhat irregular form of Ménière's disease, secondary to
brain trouble, which may be difficult to diagnosticate.
Comparing the hearing after the attack with the amount
previously possessed, will give a hint as to the nature of
the trouble.

Some years since, I had a patient with chronic aural
catarrh, with considerable diminution of hearing.   After a
while she had symptoms of vertigo; if assisted to rise, she
would stand by leaning on an attendant, but if she were
unsupported, would instantly fall.   I diagnosticated Ménière's disease, as the tympanum and membrana were not
in a condition to occasion pressure on the labyrinth and
thus cause the symptoms.   At present I have a child
under my care who has had mastoid disease, chronic suppuration of the tympanum and polypi.   She is now nearly
convalescent, the mastoid having been operated upon, the
polypi in the tympanum removed, and the discharge has
nearly ceased.   I infer that there is nothing in the tympanum to press upon the labyrinth.   Recently she has
suddenly fallen several times from an attack of vertigo, and
I conclude that this is a Ménière symptom.   She is too
young to test the hearing critically.   The vertigo of an

epileptic seizure is preceded by an aura, and the hearing is unaffected, which is not the case in Ménière's disease.

Treatment.—In the outset, it may be indicated to make cold or cooling applications about the head and ears, with the view of diminishing any possible hyperæmia of the labyrinth. This may further be aided by warm baths, or even a warm foot-bath, purgatives, leeches about the ears, counter-irritation on the mastoid processes, or to the nape of the neck. Internally, bromide of potass. may diminish the tendency to cerebral hyperæmia, or ergot may be given for the same purpose. Later on, a thorough course of mercury and iodide of potass. may be given with a view of diminishing hyperæmia and absorbing any inflammatory products. The iodide of potass. may sometimes be administered in very large doses (a drachm or more three times a day) especially if there be any syphilitic taint about the patient. To still further aid absorption, Politzer (Text-book, English translation) recommends the daily subcutaneous injection of 4 to 10 drops of a 2 per cent solution of the muriate of pilocarpine, after the second or third week, when the more acute symptoms have subsided.

For the vertigo it is well to keep the patient in bed, his head somewhat elevated, and something near at hand for him to take hold of if necessary. Sometimes holding the head between the hands and fixing the eyes on a given object will serve to diminish the vertiginous sensation. Where there is very annoying tinnitus and vertigo, Charcot's method of administering quinine promises some hope of relief. It may be given in daily doses of from ten to sixteen grains for three or four weeks, and should at first apparently aggravate the tinnitus. At the end of this period it may be interrupted for fifteen days, then again commenced. This may be repeated a third time if necessary (*Arch. Otol.,* Vol. XI.). It is well not to push the excitant effects of the quinine too far. A drachm of hydrobromic acid in a tumbler of water, I have occasionally found serviceable in relieving the tinnitus. It may be repeated daily, or oftener if necessary. I do not know that its effect is more than temporary. I have nothing to say favorably, of electricity. By any method whatever, including Brenner's formula, it seems to have no influence in alleviating the symptoms.

In certain cases where there is a doubt about the diagnosis, means may be used to better the condition of the

membrana tympani and middle ear. By Siéglè's otoscope
the air may be rarified in the meatus, and the membrana
drawn outward, and possibly the tinnitus and giddiness
may be relieved, or division of the posterior fold of the
membrana may be done if there is any evidence of contraction of the membrane and consequent pressure on the oval
window.

To those who desire still further to study the subject of
Ménière's symptoms, I would state that Knapp, in the
Arch. Otol., Vol. II., has a very exhaustive article on the
subject with cases, including a transcription of Ménière's
original ones, and the literature of the subject fully re-
ported to date. J. Gottstein in the Arch. f. Ohrenheilk.
Bd. XVII., has an article on the neuropathic form of Mé-
nière's disease. There is also a thesis for the degree of
Doctor of Medicine, on this subject, by Palasne de Cham-
peaux, Paris, February 24, 1881. In the Am. Jour. Otol.,
July, 1881, is an article bearing somewhat on this subject by
H. N. Spencer, M.D. "Giddiness and Middle Ear Disease."
McBride, "Etiology of Vertigo," "Med. Times and Gazette,"
January, 1881. Bonnafont: "Some pathological conditions
of the tympanic cavity which give rise to nervous symptoms,
ascribed by Flourens and Goltz as belonging exclusively to
the semicircular canals." Comptes rendus de l'Acad. des
sciences, November 3, 1879, on Ménière's disease. A. Guye
of Amsterdam (translated by Spalding), in Arch. Otol. Vol.
IX., C. Fére et A. Dumais, "Note sur la Maladie de Ménière, et en particulier sur son traitment par la methode de
M. Charcot," Rev. de Méd., 1881. G. Brunner, on "Vertigo
in Affections of the Ear," Arch. Otol., Vol. II., No. 1, pp.
283-342, besides a number of other authors.

CERTAIN AFFECTIONS OF THE LABYRINTH AND TYMPANUM DEPENDENT ON TRAUMATISM.

The first and most important injury is that of fracture at
the base of the skull. From the nature of things, this involves some portion of the temporal bone, in which is
placed the organ of hearing. The most frequent direction
of the fracture is through the petrous portion of the temporal bone, where it is hollowed out to make room for the
auditory nerve at the inner meatus, and the cavities for
the reception of the vestibule, semicircular canals, cochlea,

and in adults, the tympanum. A single fracture, therefore, may involve the whole of the essential organ of hearing. It is hardly necessary to allude to the manner in which such fractures take place; direct violence upon any part of the skull by almost any conceivable method may be an adequate cause.

A fall, where the patient strikes on his feet, may, by a *contra-coup*, fracture the base of the skull in the region of the foramen magnum. A blow on one side of the head may produce a fracture of the opposite petrous portion by a *contra-coup*. It is scarcely necessary to go into detail in the explanation of the manner in which the labyrinth is injured. Hemorrhage into the labyrinth, inflammation, either purulent or proliferating, may ensue. Escape of cerebro-spinal fluid or labyrinth waters and rupture or complete disintegration of delicate labyrinthine structures are some of the more frequent consequences of this form of fracture.

There are few if any exceptions to the rule that total deafness sooner or later results from fracture of the petrous portion of the temporal bone.

The symptoms of fracture of the petrous portion of the temporal bone are as follows: if the patient has received a severe blow on the head from a fall or by any other means, he may or may not remain unconscious for a few hours or a day or two. There may be vomiting and involuntary evacuations; delirium and convulsions, accompanied by alteration in the size of the pupil, with or without intolerance of light, may be present. Other signs of inflammation of the brain may speedily follow, notably a febrile condition. Pain in the head is quite a constant symptom, and is often very severe, together with a peculiar sensation of feeling "wrong in the head." Numbness about the head and face is sometimes complained of. Often the patient has a feeling that the head is of great weight. Chills are sometimes complained of. Vertigo is a very constant symptom, and may continue long after the patient is otherwise convalescent. The latter symptom is often accompanied by so great a degree of incoördination as to make walking impossible. Tinnitus is almost always a symptom, and is very persistent. Profound deafness usually results, from which few or none recover. The patient is often ill in bed for a number of weeks, but occasionally he is able to keep about his work in a somewhat unsteady manner. The pain about the head and ear is sometimes aggravated by blowing the nose, swallowing, or by any movement of the jaw. If only one side be affected, the patient finds it more agreeable to lie on the opposite side. In a few cases there may be most of the signs of a purulent inflammation of the tympanum after a few days have elapsed.

Some of the *physical signs* are: bleeding and a watery discharge from the meatus, nose, or throat. The hemorrhage from the meatus, characteristic of fracture of the petrous portion, is considerable in quantity, and continues for some days. A fracture involving the region of the tympanum and osseous meatus does not admit of excessive or long-continued bleeding; moreover, the blood is likely to be arterial in quality, coming largely from the tympanic artery (Buck), while the hemorrhage from fracture of the petrous portion is likely to involve some of the venous sinuses. If the membrana is not involved in the fracture, the blood may be poured into the tympanum and discharged from the throat or nose. The serous-looking discharge is usually thought to be identical with the cerebro-spinal fluid, although, as there is a communication between the labyrinthine fluids and those of the subarachnoid space, this point is not certainly determined. Often the serous discharge may only be from the labyrinth. It must, however, be excessive in quantity and continued for some days to be an evidence of fracture of the petrous portion. When the tympanic walls and those of the osseous meatus only are fractured, there may be a discharge of bloody serum from the serous otitis media, which is often the result of the injury. In some instances this tympanal discharge may be excessive in quantity. The cerebro-spinal fluid may be distinguished from the serum, the result of a middle-ear inflammation, by the fact that the former contains double the quantity of chloride of sodium that is found in serum, and only a trace of albumen, so as not to be coagulable by heat and nitric acid. As is well known, serum contains a certain amount of albumen, which causes a perceptible coagulation on the application of heat and nitric acid.

On inspection of the auditory canal and meatus evidence usually exists, though not always, of the injury. The membrana tympani will frequently be found ruptured, more frequently in the upper anterior portion. As the fracture so frequently extends along the osseous meatus through the Glasserian fissure, a rupture of the soft parts of the meatus may be seen at that point, or there may only be a minute spot where blood is seen oozing. The meatus may be filled with blood-clots. The membrana, besides a rupture, may have blebs projecting from its surface, filled with blood (Buck). The serous discharge will generally be tinged with blood, whether it comes from the tympanum, labyrinth, or intra-cranial cavity. If there has been any considerable solution of continuity in the meatus or tympanum, granulations may be developed, requiring attention.

Fractures of the osseous meatus extending to the tympanum may not destroy the hearing, as several of Buck's cases (l. c.) show. A case also reported in the *Am. Jour. Otol.*, Vol. III., No. 2, points in the same direction. It occurred in the wards of George Buchanan, in the Glasgow Hospital. The patient was a man, who fell a distance of twenty feet, and an iron column upon which he was at work, fell upon him, striking him on the head. There was pain in the right temporal region; he was stupid, and bled from the right ear and posterior nares; a watery fluid was mixed with the flow of blood from the ear. There was facial paralysis of the right side and impairment of hearing of the same ear. The hæmorrhage from the ear continued for two days. The ear continued to discharge a transparent yellowish fluid for about six weeks, when this became purulent and gradually ceased under the use of astringents.

In this case the serous-looking discharge was from the middle ear, the result of inflammation, and the patient made a good recovery. The paralysis of the facialis was probably due to a fracture of the bony wall of the hiatus Fallopii. Another case, from Mr. Walmsley, in the London *Lancet*, Oct., 9th, 1880 (*Am. Jour. Otol.*, Vol. III., No. 2). A boy, æt. 6, fell from a swing, striking his head. He was brought to the hospital in an unconscious condition. There was copious bleeding from the ear, with facial paralysis of · the same side. On the following day consciousness returnd but he remained somewat stupid for several days longer. The profuse serous discharge from the ear ceased on the second day, and on the sixth the child was bright and sharp. Five weeks later he left the hospital perfectly well, excepting some facial paralysis. Hearing not affected by the injury.

Some of the reported cases of fracture about the base of

the skull show that the hearing was not impaired until some days after the injury, the deafness seeming to result from the inflammation following the injury. Again, falls on the head have produced sudden or subsequent deafness, without evidence of fracture. These may be explained by, first: Shock to the acoustic nerve from the concussion; and secondly, from the basilar inflammation resulting from the blow (*vide* Buck's cases, l. c.).

Any penetrating wound of the tympanum, when the labyrinth has been opened or encroached upon, may spoil the hearing, especially if the liquor Cotunnii is evacuated. Buck (l. c.) reports a case where a negro, with suicidal intent, fired a pistol, first into one ear, then into the other, without destroying life, but abolishing the hearing. Pieces of bullet were removed from the right ear. The left ear contained no shot, and the Doctor concluded that it had been previously removed. Pieces of dead bone were removed from the meatus, granulations and polypi resulted from the violent suppurative inflammation following the injury.

The following case, by Prof. Holmes, of Chicago, illustrates one of the phases of traumatism of the tympanum and labyrinth. It is from the *Tr. Internat. Otol. Soc.*, 1876. Mr. T., æt. 43, in 1871 accidentally thrust a penholder into his right ear. It caused great pain, which subsided after a few hours, but there was for ten days great dizziness, tinnitus and total deafness. Subsequently riding in the cars caused a return of the vertigo, which became so great that he could scarcely walk with safety. For five months this dizziness increased, so that he had to give up business. At this time the Doctor first saw him; he was in good general health, and there were no objective signs of inflammation about the ear. Just behind the centre of the membrana was what seemed a minute cicatrix. Ear totally deaf. Eustachian tubes normally pervious. The vertigo had increased to a wonderful degree; the slightest motion of the head produced a sensation of falling forward, as though he was making a somersault, and he had to grasp his chair to prevent falling. While lying on his back in bed, he had to grasp the side of the bed and keep his head absolutely motionless. The bed seemed to whirl rapidly forward. There was at times considerable nausea, and an unpleasant sensation about the head. By holding the head with both hands, it could be moved slowly. For some time the patient had taken pot. iodid., pot. bromid., hyd. bichlorid., diuretics and cathartics. Blisters and galvanism had been applied over the mastoid. The violent symptoms continued about three weeks, when they began slowly to disappear. Remedies did not seem to have contributed to his relief. After a few months he was able to take a voyage to Europe; the vertigo and tinnitus were slightly troublesome. Two and a half years after this there was very little tinnitus; the ear showed subjectively no signs of disease. On bending the head low and suddenly raising it, there was slight vertigo. Deafness of right ear complete.

A few indications for *treatment* present themselves. Most of these have been already hinted at in the previous recital of cases and symptoms. When there is considerable inflammation of the tympanum, leeches may be necessary, as in ordinary acute inflammation, with the warm applications, anodynes, and opiates appropriate to the management of that affection. Any foreign substance may be removed by means already suggested. (See Foreign Bodies in the Meatus). Granulations and polypi may be treated in the usual manner.

Any head symptoms may receive appropriate treatment, and the febrile symptoms be met in the usual manner. When there is great shock to the system, alcoholic stimulation or ammonia may be needed. At a later stage mercury may be given to dispose of any products which may be strangulating nerve filaments in the labyrinth. Sometimes large doses of iodide of potash may successfully appeal to any inflammatory symptoms at the base of the brain. I have said nothing about the prognosis as far as the life of the patient is concerned. Naturally many of the more serious fractures at the base of the brain destroy life by the purulent meningitis which so often results.

PRIMARY ACUTE INFLAMMATION OF THE MEMBRANOUS LABYRINTH.

Voltolini claims that such a disease as *Primary Acute Inflammation of the Membranous Labyrinth* exists (Monatsch f. Ohrenheilk, V. 9, *Blake's Reference*). The symptoms are as follows: the disease attacks only children who appear to be in perfect health, with no ear trouble; suddenly they are stricken down with a fever, chills, delirium, vomiting, vertigo, loss of consciousness, convulsions, face very red and inflamed, and pain in the head. Profound or absolute deafness comes on in one or two days from the commencement of the attack. The duration of these symptoms is only a few days, something less than a week, and in some instances a little longer. No lesion whatever seems left behind, except the hopeless deafness and the staggering gait; the latter may remain for some time. The intelligence is never affected in this disease, and no cranial nerves are involved. The disease much resembles a meningitis, but it is too brief in its duration, as the latter ordinarily continues for several weeks. Absence of lesion of any of the cranial nerves except that of the acousticus is relied upon by Voltolini to prove that there is no central disease. According to him this affection never destroys life; this point being an apparent refutation of the idea of its being associated with brain disease. It may strike the observer that this affection resembles in some respects the more typical forms of Ménière's disease—both attacking the patient suddenly, the intellect unaffected in each, and profound or total deafness the result in both. Moreover, in each disease *no cause* has been discovered for the sudden development of the symptoms. J. Gottstein, in the *Arch. Otol.*, Vol. X., p. 81 (Review), and *Arch. Otol.*, Vol. IX., p. 254 et seq. (translation by Spaulding), raises the question, with some apparent success, that Voltolini's cases are, many of them, only mild attacks of cerebro-spinal meningitis, which he denominates as one of the neuropathic forms of the Ménière group of symptoms. He illustrates this by a case, as follows: H. K., æt. 15, was suddenly attacked, Feb. 11th, 1879, with fever, violent headache, vomiting, and pain in his limbs. He lay in an apoplectic condition, and ate nothing. This state lasted one day. Late at night the child recovered consciousness and found that he was totally deaf.

On Feb. 21st sensorium clear, no fever, the patient being very weak. He only had moderate headache, and a dreadful roaring in both ears; on leaving his bed he staggered so much as to require to be led. No other signs of disease of the ears. He recovered very gradually, but his gait was uncertain in the dark for several months. Deafness seemed absolute. Buck (Diagnosis and Treatment of Ear Diseases), evidently follows in the same line, when, in speaking of one of the forms of Ménière's disease, he concludes that in young subjects the symptoms often depend on a mild attack of epidemic cerebro-spinal meningitis.

He cites a case bearing on this point, thus:

A girl about fourteen years of age was suddenly taken ill with dizziness, tinnitus, slight disturbance in the vision, photophobia, and finally vomiting. Two days after, she was very deaf, and during the same day this increased to apparently total deafness. The other symptoms soon disappeared, but the deafness and inability to maintain her equilibrium persisted for at least three months afterwards.. The middle ears were practically normal. As all these cases seem not amenable to treatment nothing is suggested.

DEAFNESS DEPENDENT UPON SCLEROSIS OF THE POSTERIOR COLUMNS OF THE SPINAL CORD.

This occurs with considerable frequency, accompanied by affections of the nerves in the vicinity of the acousticus, more especially the second, third, fifth and sixth nerves. The sight is consequently frequently affected. The impairment of hearing is usually on both sides, but one ear may be more affected than the other. It may go on to profound deafness, but it often stops short of it. There may be tinnitus, but it is frequently absent.

It is somewhat uncertain what the exact change in the acoustic nerve consists in; probably something similar to that which takes place in the posterior columns of the cord. Erb, Ziemssen's Handbuch, p. 142, states that in some cases of tabes he found definite atrophy of the auditory nerve to be the cause of progressive deafness, together with "disseminated sclerosis" (A. f. O. ii.). Could find no change in the auditory nerve in gray degeneration of the spinal cord (Politzer's Reference, p 725).

Thomas Buzzard, of London, in the *Lancet*, Sept. 24th, 1881, quotes Duchene, Bemak, Tobinard, and others, as having already pointed out that the acoustic nerve may also be affected in tabes dorsalis, but no one suspected the frequency of this complication until Pierrot called attention to it.

Since then McBride has found a greater or less degree of deafness, though sometimes only transitory in its appearance, in every case that came under his observation. Ormerod observed this symptom in five out of a total of thirteen cases.

McBride thinks it possible that the affection of the acousticus may in some cases be a leading symptom, just as in others the optic nerve may be affected, and he presumes that certain cases of "nervous deafness" may be accounted for by the presence of tabes dorsalis. (Archiv. Otol., XI., No. 2). J. Gottstein in the Arch. Otol., Vov. IX., No. 3, in speaking of Case V., just reported, thinks the affection in the ear should be connected with the symptoms of tabes which were present. He states that disturbances of hearing are rare, in cases of tabes, in comparison with those of sight. He quotes Eulenberg as having seen four cases of disturbance of hearing in sixty-four cases of tabes.

Gottstein does not believe the labyrinth to be the seat of the disease, but that the auditory nerve is "affected in the anatomical process of tabes," and that in Case No. V. the nerve was affected at its central origin, on account of the suddenness of the attack and the complete deafness resulting in both sides.

The following case is a good illustration of this affection. He came under my care at the hospital.

P. J. C., æt. 38; was formerly a policeman; has most of the symptoms of multiple sclerosis of the posterior columns of the spinal cord. Among them are diminished sight, contracted pupils, great deafness and extreme difficulty in walking, from the usual incoördination of locomotor ataxy. He is unable to walk with his eyes closed, without great danger of falling. His vision is $\frac{20}{40}$ in each eye; the fields are both considerably diminished concentrically (about 60° in the right eye, and a little more in the left); signs of atrophy of the nerve and choroid; pupils about the size of a pin hole.

The ear affection came on at about the time the sight was noticed to fail (about three months ago) and the hearing is growing worse. The hearing distance of the left ear is, clicking of finger nails at contact; shouting not heard. The tuning-fork heard close to the ear for eight seconds, but bone conduction is entirely absent. In the right ear the voice could be heard at one foot, if considerably elevated. Click of finger nails heard three fourths of an inch; slight improvement on inflation. Inflation does not improve the hearing of the left ear. The tuning-fork against the teeth was heard for 17 seconds in the right ear; by ærial conduction, eight seconds; on the mastoid of the same side it was heard twelve seconds. There were signs of old catarrhal otitis in both ears. The right ear seemed to be somewhat of a mixed case, but the left showed unmistakable evidences of labyrinthine trouble. Galvanism was tried without effect.

DEAFNESS FROM SUNSTROKE.

This is undoubtedly a rare affection. I find only occasional mention of it in literature. In Roosa's Treatise on the Diseases of the Ear, fourth edition, p. 536, is a case of deafness with effusion about the acoustic nerve. There were epileptic seizures. The patient recovered his hearing. Among other possible causes of the difficulty was mentioned sunstroke, but without enlarging on the subject. In the somewhat various cerebral conditions incident to sunstroke, it is not difficult to imagine changes occurring in the nerve or labyrinth. The following case seems to have at least been aggravated by the attack of sunstroke, but I am by no means certain that it was the principal item in the pathogenesis of the deafness. I am indebted to Dr. N. J. Hepburn, of the Manhattan Eye and Ear Hospital, for the case. It is as follows: T. F., æt. 37, laborer. Deafness in the right ear dating back four years, at which time he had a sunstroke. While working in the sun, he suddenly became dizzy and fell. He was taken to the police station, where he lay unconscious for two and half hours. He was not treated, and was able to return to his work at the expiration of one week. On reaching home on the day of the sunstroke he found that he could not hear with his right ear. There was also tinnitus; no history of syphilis. He came at the present time for treatment, as he is growing hard of hearing in his left ear. The Hearing Distance Right, is watch $\frac{0}{48}$; after inflation, $\frac{1}{48}$. H. D. Left, w. $\frac{45}{48}$; after inflation, w. $\frac{48}{48}$. With the tuning-fork the aërial conduction is better than bone conduction in each ear; when placed on the central incisors with both ears closed, it is heard longest in the left. With both ears open, is heard in the left by bone and aërial conduction; but is heard in the right by aërial conduction only; both ears show signs of a previous chronic otitis media, and there is some pharyngitis of the upper portion. I infer that there is middle-ear disease in both, and that there is labyrinthine or nervous trouble in the right, which seems to have dated from the time of the sunstroke.

DEAFNESS FROM MENTAL AND NERVOUS EXHAUSTION.

I have observed one case where deafness seemed to depend on *mental and nervous exhaustion*. A case may illustrate it. A. F., a lawyer, æt. 44, consulted me for some difficulty in hearing. No constitutional dyscrasia. There was some difficulty in hearing the replies of witnesses in the court room. A moderate amount of tinnitus aurium accompanied the other symptoms. Inflation slightly improved the hearing. There was some catarrh of the Eustachian tubes. Bone conduction was weakened in both ears, and aërial conduction corresponded well to his hearing for the voice. Not being able to otherwise account for his symptoms, I directed him to rest, improve his nutrition, exercise in the open air, take all the sleep possible, besides avoiding mental worry. In four weeks I was gratified to find that the hearing had returned, the tinnitus had ceased, and bone conduction was restored. I make no attempt at explanation, but record what evidently was a fact.

DEAFNESS FROM MUMPS.

It may be proper to devote more space to this subject than is usual in text-books on the ear, as at the present time (1883) considerable interest seems to be taken in the subject by otologists throughout the world. Whether the parotitis is on one or both sides, only one ear is usually affected. The deafnes is generally, though not always, complete, even to the abolition of bone conduction. This occurs usually during the early period of the parotitis—from the first to the fifth day. Very aggravating tinnitus aurium, vertigo, unsteadiness of gait, nausea, vomiting, etc., are accompanying symptoms. The vertigo frequently continues after most of the other symptoms have subsided, although it sometimes lasts but a few days, or it may be absent altogether. In a very large percentage of cases no signs of inflammation about the meatus or tympanum exist, although catarrh of the pharynx is a frequent accompaniment of the disease. The Eustachian tubes are occasionally obstructed, together with signs of sunken drum membrane. A few cases show suppuration of the tympanum. Pain is only occasionally experienced, most cases reported showing a singular absence of this symptom. A large percentage of the cases so far observed have been in adults, although children occasionally are rendered deaf from mumps.

Perhaps the best method for arriving at a correct estimate of the nature of deafness from mumps is to revert to the experience of some of those who have observed the affection. Mr. Dalby ("Diseases and Injuries of the Ear," p. 188) says: "On several occasions I have known partial deafness to take place during an attack of mumps, and in one instance total loss of hearing. The latter happened in a girl, æt. 7; could hear perfectly well before she had the mumps, but afterwards could not hear a sound. No history of hereditary syphilis." Toynbee, quoted by Turnbull, says: "The peculiar poison which causes the disease generally known by the name of mumps is very often the source of complete deafness, which, however, usually occurs in one ear only. In these cases the nervous apparatus is evidently affected, as the deafness comes on suddenly, is usually complete, and, as a general rule, no appearance of disease can be detected in the meatus, membrana tympani, or tympanic cavity." He admits that a slight degree of hearing may sometimes remain. George P. Field, "Diseases of the Ear," Lond. 1882, says, p. 248, "Where it (nervous deafness) originates in (from?) mumps, one side alone is usually implicated." He cites the case of a young man who, three months after a sharp attack of mumps, remained quite deaf, although no lesion of the membrana tympani could be detected. Dr. Knapp, in a discussion in the Trans. Internat. Med. Cong., 1881, Vol. III., p. 376, speaks of a case under his observation of a young lady who during an attack of mumps had become completely deaf in one ear without a trace of disease of the outer or middle ear. Bone conduction was also destroyed. These cases were very rare in his practice, but he had frequently seen suppuration of the middle ear as a complication of mumps, but which did not differ in nature from the same disease elsewhere. Knapp also reports a case of one-sided deafness from mumps in the Archiv. Otol., vol. xi. p. 232. It was a young lady æt. 15. She consulted him on May 25th, 1881. She had double parotitis. During this attack she noticed a diminution in the hearing in the right ear, accompanied by tinnitus. The membranes were normal. The hearing of the right ear was soon completely lost, together with the bone conduction. The fellow ear was unaffected. He also reports another case on p. 385. Theresa O., æt. 25, consulted him on Aug. 10th, 1882. The hearing was perfect previous to the attack of parotitis, and there had never been any disease of the ears. Six years since she had mumps on both sides. On the seventh day of the attack the hearing became lowered, and on the eighth day was completely lost. On the seventh, and on the following day there was excessive dizziness, which still exists to some degree. Bone and aërial conduction absent; middle ears healthy; patient does not remember whether tinnitus was present.

Hinton, " The Questions of Aural Surgery" (Lond., 1874), p. 220, says: "Next or perhaps equal in frequency to scarlatina in this respect (the spoiling of the hearing) stands mumps, which has an effect on the nervous apparatus of the ear which has as yet received no explanation, and affords no clue to the use of remedies, every part of the ear being normal, so far as examination can extend, but the functions are almost abolished." He states, however, that some cases show clear signs of tympanic disorder mixed with the nervous symptoms. "The similarity of the nerve affection that follows mumps to that which ensues upon parturition is very striking." I recently saw a case of total deafness at the Manhattan Eye and Ear Hospital, with the following history: Woman æt. 20. In early childhood had had suppurative otitis media with very little impairment of hearing. Seven years since she had an attack of mumps, during which she awoke from sleep one morning and found herself totally deaf. She has remained so ever since. Bone conduction entirely absent. The membranæ are cicatricial, and the left one has a slight perforation, and there is a little discharge. Her voice has the usual pitch of a person who has been deaf for a considerable period. I have seen two other cases of mumps where the deafness came on suddenly and without explanation from the condition of the middle ear. Roosa, "Treatise on the Ear," fourth edition, p. 539, says: " Parotitis sometimes . . . leaves the patient profoundly deaf," which " may be from a direct extension of the inflammatory process from the gland into the auditory canal, tympanic cavity, and nerve, or we may suppose that a metastatic inflammation of the membranous labyrinth has occurred. I have seen but one case where it was plain that nerve deafness resulted from parotitis. In this case the loss of hearing occurred without pain, and affected both ears." He quotes a case from Dr. H. D. Noyes, where an adult had loss of hearing, accompanied by inability to walk without staggering when his eyes were closed.

This occurred and persisted after an attack of mumps. There was also metastatic orchitis on the same side as the deafness. Dr. Buck reports two cases in the Am. Jour. Otol., July, 1881, and alludes to a third case he has seen, but the history of the last was imperfect. His first patient lost the hearing of the right ear completely on the fourth day of an attack of mumps. Bone conduction was also lost. In this case there were no symptoms of middle-ear trouble except an obstruction of the Eustachian tube and a slight sinking of the membrana tympani, some pharyngeal catarrh, but the hearing was not improved by the subsidence of these symptoms. The second patient lost his hearing in one ear on the third or fourth day of an attack of mumps. No middle-ear trouble or any signs of disease sufficient to account for the sudden loss of hearing; bone conduction abolished. These patients did not recover their hearing. He quotes Vogel from Ziemssen's Cyclopædia of the Practice of Medicine, vol. vi. p. 838: "Not at all infrequently the labyrinth and middle ear participate in the parotitis, in which case the pus probably passes directly by means of the vessels and nerves that go from the parotid gland to the ear. The ossicles of the ear thus become destroyed and, at the best, lifelong deafness results." Vogel further remarks that the facial nerve seems especially adapted to conduct the inflammation into the auditory apparatus. Against this theory is the fact that the facial nerve is not often paralyzed in mumps as suggested by Burnett.

In the Archiv. Otol., Apr., 1882, is reported a case of mumps, with total deafness, by Moos, and translated by Spalding, as follows: W. S., æt. 13. On the previous February had double parotitis. Hearing was destroyed on the fifth day of the disease. Right membrana somewhat more opaque than usual; left more concave than right. Bone conduction only in right ear by the tuning-fork C and C'; A' not heard. The author is inclined to favor the theory of metastasis. He concludes that the staggering gait and the profound deafness pointed to labyrinth disease. Brunner, a translated by Ayers, in the Arch. Otol., June, 1882, reports case of complete unilateral deafness from mumps as follows: A woman æt. 30 had mumps on both sides. Soon after she became deaf in the right ear.   This was accompanied by tinnitus and vertigo; bone conduction destroyed   No signs of disease in the Eustachian tube or membrana tympani.   Many discussions have been engaged in with a view of determining the mode in which the disease involves the ear.

Dr. Roosa seems to conclude that the meatus or tympanum, or both, become first inflamed, and that the internal ear suffers secondarily.   Most of the reported cases have no tympanal complications, and those which do have, are so slight as not to cause serious deafness in the first place, and in the second are not of a sufficiently serious nature to travel as far as the labyrinth.   The theory of passage by the facial nerve does not seem tenable, and we are left the old explanation of metastasis.   It is true that the throat is often involved, and evidently as a consequence of the parotitis, but it does not produce serious ear complications, except perhaps the somewhat unusual condition of suppurative inflammation of the tympanum.   Until a sufficient number of autopsies are made, the labyrinthine condition can only be inferred from its disturbed function.   It seems unnecessary to discuss the question as to whether the deafness depends on labyrinthine or nervous trouble.   Its completeness, suddenness, the absence of middle-ear disease sufficient to cause it, and usually the presence of vertigo, nausea, unsteadiness of gait, etc., with complete abolition of bone conduction, point to the inner ear as the part affected.

Treatment has had no influence, so far, in preventing the deafness or restoring the hearing after it has once been lost. In a few cases where there is pain from hyperæmia, leeches and other similar measures may be employed.   Iodide of potass. and mercury may be tried, but with very little hope of benefit.

As electricity accomplishes so little in nervous or-labyrinthine affections dependent on other causes, it hardly can be expected to do so in this connection.   It might, however, be tried.   Middle-ear symptoms may be met according to general indications.

DEAFNESS DEPENDENT ON GESTATION AND PARTURITION.

This, in a large number of instances, seems to be associated with a state of anæmia of the labyrinth, often accompanied by nervous exhaustion or shock. If the exciting cause is not of too powerful a nature, and is not often repeated, the hearing may completely recover. In numerous instances, however, the patient becomes harder of hearing at each accouchement, until she may be left in a condition of almost absolute deafness. The state of gestation may be accompanied by defective hearing from a similar cause. The patient may not take sufficient nourishment to sustain both the fœtus and the mother, hence the anæmia. I have often seen this aural condition immediately benefited by the ingestion of large quantities of nourishing food, and perhaps the addition of alcoholic stimulants. Great fatigue or worry of mind in the pregnant woman will aggravate any existing ear symptoms. Wilde (Text-book, American edition, p. 275) says "many females have *become* deaf immediately after parturition. In such cases I have generally observed a speckled opacity of the membrane." In the Trans. Am. Otol. Soc. for 1878, p. 180, Sexton states that Dr. Pierce, of Manchester, believes that·the deafness of pregnancy begins with that condition. Mr. Lenox Brown regards such cases as common, and believes they are not catarrhal, but a thickening of the mucous membrane. "Sometimes they are nervous in character." Toynbee (Text Book, p. 368) thinks that a woman may become totally deaf from the nervous exhaustion attendant on childbed; in some instances of the latter, the deafness has begun with the birth of the first child, and increased with each successive birth, until at last the nervous power was wholly lost. On p. 370 he reports a case as follows: "Mrs. B., æt. 40, pale, and of a nervous temperament, consulted me in 1850, on account of complete deafness in both ears. She stated that she had married in India ten years previously, and at the time of her marriage she could hear perfectly well. On the occasion of her first confinement, previous to which her hearing was still perfect, she suffered a good deal from exhaustion, and this was followed by a great degree of deafness, so that she could scarcely hear what was said to her, even when the voice was much raised. Upon getting up, and growing stronger, the deafness was so much relieved that she merely required to be spoken to a little louder than usual. During each successive confinement, in India, amounting in all to four, the deafness greatly increased, and after each recovery became more permanent, until on the last occasion she remained as deaf as at present, when she is obliged to have recourse to signs. Indeed, she has never heard the voice of her younger children, and can only, by the movements of their lips, understand their words."

A condition similar to this occurs with the vision; the optic nerve and retina sometimes undergoing complete atrophy, with resulting blindness. Little need be said about *treatment;* it will be readily suggested from the nature of things. Appropriate means may be used to maintain a proper nutritive condition and avoid, as far as may be, excessive loss of blood at the confinement, and hasten the patient's recovery.

DISEASES OF THE EAR DEPENDENT ON TYPHUS AND TYPHOID FEVERS.

These affections have usually been regarded as more or less labyrinthine in character, although, in the mention of the graver forms of otitis media in the exanthemata and the profound deafness which sometimes follows them, the ear diseases associated with typhus and typhoid fevers are included as belonging to a similar category. Most forms of otitis media and externa with pharyngeal and tubal catarrh and even mastoid complications, may exist as a consequence of these varieties of fever. Dr. Arthur Hartman, Arch. Otol., vol. ix., No. 1, p. 55, translation by Knapp, says: "In almost all cases of typhus exanthematicus the general hyperæmia of the head is accompanied by marked hyperæmia and swelling of the naso-pharyngeal mucous membrane. It seems that a congestion of the auditory organ forms an essential part of the general congestion of the head. If it persists after the convalescence of the general disease, exudation may occur in different forms, either most pronounced in the drum cavity, or the tubes, or the labyrinth." Suppurative otitis media is not as likely to occur as a consequence of these fevers as in those of the exanthemata. It is true that labyrinthine complications may occur as a consequence of the extension of a severe middle-ear inflammation to the internal ear, but evidence does not exist that the typhoidal patients are more liable to this accident than others. In some cases the patient becomes profoundly deaf without any middle-ear symptoms, as in a case reported by Toynbee, in "The Diseases of the Ear," English edition, p. 360, as follows: "Miss A. M. æt 16, saw me on March 1st, 1851. Eleven years previously she had an attack of typhus fever, and during the illness became so deaf as not to be able to hear the human voice. After the symptoms of fever had disappeared, the power of hearing slowly returned, until she was able to hear when loudly spoken to close to the head. There was no appearance of disease in either ear." Out of forty-two cases dependent on typhus, Hartman (l. c.), observed three, in which "the hardness of hearing and the diminution of the bone-conduction were so pronounced that a participation of the nervous apparatus in the disease had to be assumed. In one of these cases the objective examination showed no abnormity in the drum cavity; in the two others only slight alterations." He inferred that in the last-named cases there were inflammatory changes in the labyrinth— Burnette ("Treatise on the Ear") does not admit that there are labyrinthine complications in these affections, unless the tympanal symptons are neglected, when "a labyrinth disorder may be established." Wilde, on "Diseases of the Ear" holds the opinion that the symptoms are mostly those of the middle ear, but admits that there may be labyrinthine complications. What these are, seem, from lack of a sufficient number of autopsies, unknown. In none of the ear complications of typhus fever is it assumed that any possible inflammation of the brain influences the aural sympsems. Hartman (l. c.) observes that in three of his caseht rote was increase in the noises and hardness of hearing previously present, and that there was a recurrence of an otorrhœa in one case. In some instances an old otitis which was well-nigh cured had relapsed under the influence of the fever. The ear disease comes on during the febrile attack or soon after the patient has recovered.

The *prognosis* is favorable. Most of the patients make a good recovery. Although, as previously hinted at, an occasional deaf-mute may date his trouble to an otitis contracted during a typhoid or typhus fever. Little need be said concerning *treatment*. It may be done according to the exigencies of a given case; the fact of its febrile causation need not modify treatment.

INSTRUMENTS FOR AIDING THE HEARING.

By these, I refer to the different forms of hearing-trumpets, the audiphone, etc. The action of the hearing-trumpet is simply to collect and convey to the membrana tympani a larger number of sonorous undulations than would otherwise reach it. The undulations pass through the middle-ear mechanism precisely as in normal unaided hearing, with sometimes, as hereafter will appear, a certain amount of aid from bone conduction. In the construction of the trumpet there is no difficulty, at least at the near point, in collecting as many sound-waves as are desired; the problem is, rather, to collect undulations that are likely to reach the cochlea properly analyzed—that is, to be appreciated as articulate sounds, proper tones and pitch in music, etc. The difficulty with most ear-trumpets is that there is likely to be a constant roaring, like a sea-shell, when the trumpet is applied to the ear, and all loud sounds are so exaggerated as to be both indistinct and painful, especially as there is frequently a certain amount of hyperæsthesia of the auditory nerve in such cases. Tinnitus may also result, and in some cases, congestion of the tympanum from the concussion of the strong undulations.

To obviate this many expedients have been resorted to, such as avoiding the use of metal in the construction of the trumpet, and substituting some less resonant material; covering the mouth of the trumpet with a piece of metal in which are placed larger or smaller openings as suggested by Dr. C. J. B. Williams, even down to the size of those used in the wire sieve, of the North trumpet, and the use of reflectors, vibrators, etc., in the interior of the trumpet. Perhaps the most successful procedure of all these for diminishing this reverberation is the screen or sieve placed over the mouth of the trumpet, which is a most important feature of many instruments. Specific conditions of the ear have by many authors been regarded as a cause for selecting particular appliances for aiding the hearing. I am not certain that any person having a small amount of hearing remaining may not have this augmented by the use of an ear-trumpet, whether the defect is caused by labyrinthine trouble, thickened and adherent drum membrane, or perforation and even loss of one or more of the ossicula—it being simply a matter of collecting and conveying to the cochlea a larger number of undulations than would otherwise reach the nerve. The different forms of trumpet may be made of Japanned tin, horn, vulcanite, German silver, ebonite, etc. If the deafness is not profound, the ordinary conversation-

FIG. 89.—CONVERSATION TUBE.

tube, Fig. 89, may be used, as it is quite inconspicuous and easy to carry; moreover, by its flexible tube it enables the bowl to be placed near to the speaker's mouth, whereby much increase in the power and distinctness of the articu-

FIG. 90.—THE DIPPER TRUMPET.

lation is gained. Deaf patients are constantly disappointed at the very moderate gain to the hearing resulting from an attempt to hear a lecture or a concert.

The tip for the insertion of the trumpet into the meatus

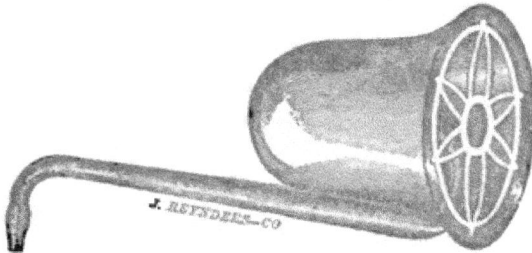

FIG. 91.—ORDINARY JAPANNED TIN TRUMPET.

should be gently pushed in, as in the event of its being crowded in too far it interferes with the best hearing; besides it irritates the canal and may produce unpleasant congestion of the tympanum. For listening to any distant sound a larger expansion of the trumpet-shaped extremity needs to be made like the one in Fig. 90 or Fig. 91, which are very similar to the famous Martineau trumpet. None of the smaller, so-called invisible, aids to hearing are of any use whatever. I have latterly come to the conclusion that the North earphone, Fig. 92, is the most useful of any hearing-trumpet I am acquainted with. It is made of four sizes, the illustration being somewhat less than half the diameter of the largest size. It may be used as in No. 1, with the short tip inserted into the meatus. It will be seen that a rubber tube is drawn over it, so as to cause it to adjust better to the ear. When in posi-

FIG. 92.—NORTH'S EARPHONE.

tion the bowl rests against the temple, thus aiding the hearing by bone conduction. In other instances it will be more convenient to use No. 2 with the flexible tube attached, although in this instance no aid is obtained from bone conduction. On the top of No. 2 may be seen one of the coarse wire meshes which the inventor calls a "refiner." Three of these are used, each of a different-sized mesh; the smallest is extremely fine. The object of these is to soften or tone down the overpowering reverberations, rendering articulate sounds much more distinct. It is somewhat a matter of experiment what exact combination of these refiners is to be made; in some cases, where the hearing is not too much lowered and there is great sensitiveness to strong vibrations, all three of the refiners may be used. In other instances only the coarsest-meshed refiner is used. The same observation may be made, to a certain degree, in reference to the selection of the particular-sized instrument to be used. It will be seen that the patient may "try" each size with all the combinations of refiners, until one is found best adapted to the given case. The instrument admits of quite a large number of combinations. I have been unable to study the interior of the *earphone*, but it is described by the inventor as having a series of tubings, refiners, drumheads, and detonating wires: the latter being forked and running from the bottom where it is attached to the reflector, seen on the depressed portion, through the full depth of the instrument. I am indebted to D. M. Coe, 295 4th Ave., the agent, for the description and woodcuts. I came to a knowledge of its value from the experience of some of my patients, who had used it with greater satisfaction than other instruments which they had tried.

RHODES' AUDIPHONE OR OSTEOPHONE (THOMAS).

This is illustrated in Fig. 93, as it is applied by its extremity to the incisor teeth of the upper jaw. In Fig. 94 is seen the mode in which the proper tension is accomplished by means of cords. It is composed of vulcanite, and is twenty-five to thirty cm. long, and twenty to twenty-five cm. broad. The convex surface catches the vibrations, which are carried through the teeth and bones of the face and cranium, to the acoustic nerve, acting purely by bone conduction. The inventor of this instrument at one time visited the Manhattan Eye and Ear Hospital to exhibit it. Various experiments were made, which were altogether negative in result. I remember that I applied it to my upper teeth with both ears tightly closed. I could hear conversation at about three feet without any aid, and it was impossible to increase this distance by the use of the audiphone; neither did sounds appear louder or more distinct, nor could any increase of sonorous undulations be perceived. My opinion of its utility was unfavorable. Since that time I have tried it in many cases, and have not been able to produce a sufficient amount of improvement to the hearing to justify a continuance of its use. Since that time I have heard that Dr. Agnew found a patient who could hear preaching at about twenty feet with the audiphone when not a word could be distinguished without it.

Fig. 3. The Audiphone properly adjusted to the upper teeth; ready for use. (Side view.)

FIG. 93

Dr. Carmalt of New Haven, Conn., reports a somewhat similar case. Dr. Knapp, Arch. Otol., Vol. IX., No. 1, p.

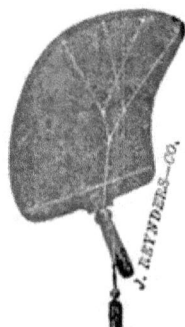

Fig. 2. The Audiphone in tension; the proper position for hearing.

FIG. 94.

89, sums up the results of the use of the audiphone in fourteen cases, comparing the instrument with the most approved form of ear trumpet. With the audiphone there was no increase in the hearing in twenty per cent of the cases, a slight increase in twenty per cent, and a moderate increase in sixty per cent; whereas the ear trumpet increased the acuteness of hearing in all cases, slight in eight per cent, moderate in thirty-five per cent, and great in fifty-eight per cent, so it is readily inferred that the ear trumpet is greatly superior to the audiphone as an aid to hearing.   The Japanese fan, made of quite stiff material, is sometimes superior to the Rhodes audiphone.   It requires no adjustment, the patient is not suspected of carrying any aid to hearing, and it costs only a few cents, whereas the audiphone costs from ten to twenty dollars, or more.

As hopelessly deaf people are anxious to try anything that affords a chance of relief, and as the fan and audiphone do *sometimes* benefit, they may be recommended.

INDEX.